STATISTICAL PROCESS CONTROL
The Deming Paradigm and Beyond

SECOND EDITION

James R. Thompson
Jacek Koronacki

CHAPMAN & HALL/CRC

A CRC Press Company
Boca Raton   London   New York   Washington, D.C.

Library of Congress Cataloging-in-Publication Data

Thompson, James R., 1938-
    Statistical process control: the Deming paradigm and beyond / James R. Thompson, Jacek Koronacki.—2nd ed.
        p. cm.
    Includes bibliographical references and index.
    ISBN 1-58488-242-5 (alk. paper)
        1. Process control—Statistical methods. 2. Production management—Quality control. I. Koronacki, Jacek. II. Title.

TS156.8 .T55 2001
658.5′62′015195—dc21

2001043990

Visit the CRC Press Web site at www.crcpress.com

© 2002 by Chapman & Hall/CRC

No claim to original U.S. Government works
International Standard Book Number 1-58488-242-5
Library of Congress Card Number 2001043990
Printed in the United States of America   2 3 4 5 6 7 8 9 0
Printed on acid-free paper

To my wife, Ewa Majewska Thompson
James R. Thompson

To my wife, Renata Houwald Koronacka and to my children,
Urszula and Krzysztof
Jacek Koronacki

Contents

Preface to the First Edition

"May you live in interesting times" can be a curse if one lives in a society perceived to be so perfect that improvement can only be marginal and not worth the trauma of change. If, on the other hand, one lives in a society that is based on constant improvement, living in "interesting times" presents opportunity. For good or ill, it is clear that people these days live in very interesting times indeed. The struggle between the West and the Communist World is over. Yet, with the triumph of the Western system, there comes the challenge of seeing what happens now that the political and military conflict is ended. Enormous residues of intellectual energy are now freed to be focused on peacetime pursuits.

It is interesting to note that one of the basic optimization postulates of *statistical process control* (SPC) was developed by Vilfredo Pareto (1848-1923), who was trained as an engineer but is best known for his economic and sociological works. According to Pareto's Maxim, the many failures in a system are the result of a small number of causes. General malaise is seldom the root problem. It follows that in order to improve a system, skilled investigators are required to find and correct the causes of the "Pareto glitches."

We are, in this book, concerned about the orderly process of optimization which is the nuts and bolts of statistical process control. But a few words about the social theory of Pareto are in order, since this theory gives insight to the important part SPC is likely to play in the post Cold War world.

Pareto perceived the inevitability of elites in control of society. Extrapolating from Pareto's works, particularly the massive *Mind and Society*, the American political scientist, James Burnham, writing in the late 1930s and early 1940s, observed the presence of such elites in Fascist, Communist and Bourgeois Capitalist societies. These elites were based on the expertise to seize and maintain power, rather than on excellence in the arts, sciences and technology. Burnham pointed out that Pareto had started out with views similar to those of his father, who had resisted Bourbon control of Italy in favor of a Jeffersonian republic. In early middle age, Pareto's point of view had changed fundamentally from a kind of Scottish Enlightenment optimism to one of cynicism when he noted the great mistake of Aristotle. Aristotle assumed that once human beings had understood the Aristotelean logic system, everybody would embrace it eagerly. Yet, Pareto observed that human beings tend to make their most important decisions based on gut instincts, passions, and narrow self and group interests. It is not simply that people are inclined to use personal utility functions as opposed to those of the greater society. Frequently, a boss will attempt to assimilate all the information relevant to making a decision, and then, at the end of the day, make a decision which appears whimsical, without particular relevance to any perceivable utility function. Pareto devoted most of his mature years to trying to understand how bosses arrive at decisions. Today we note reconfirmation of Pareto's views when we see how seemingly irrational managerial decisions have caused so much havoc in the otherwise enviable industrial model of Japan.

In *The Managerial Revolution*, Burnham observed a movement of the members of nomenklaturas back and forth between top posts in seemingly unrelated areas. So, for example, a university president might be named president of an industrial corporation. A key Party official might be given first a post in the Ministry of Justice and then move on to become a general. There was no "bottom line" in terms of effectiveness within the current post that led to lateral or vertical promotion within the society. Essentially, a new feudalism was developing, with all the stagnation that entails. Upon reflections on Burnham's book, his fellow former Trotskyist colleague, George Orwell, wrote the profoundly pessimistic *1984* about a society in which all notions of human progress had been sacrificed for the purpose of control by the Inner Party.

Skipping forward a few decades, it is interesting to see how accurate Pareto's perceptions had been. We note, for example, in the 1980s, the promotion to Chief of State of the Soviet Minister of Agriculture, a post in which he had produced no growth in productivity. We note in America how the CEO of a large soft drink company became the CEO of a high tech computer company, even though he was unable to write a computer program. Until fairly recently, Burnham's predictions appeared to be woefully justified.

But a sea change has taken place. The abysmal record of the Soviet system to produce even the logistics necessary for maintenance of its military power has driven that former Minister of Agriculture, Mikhail Gorbachev, from his office as emperor of the Soviet empire, and has, in fact, led to the overthrow of that empire. There is no evidence that Gorbachev's meekly accepting his loss of power was due to any moral superiority over his old mentor, Yuri Andropov, who had crushed the Hungarian workers in 1956. Ultimately, it seems, there was a bottom line for the productivity of the Soviet nomenklatura. Their mismanagement of the various productivity areas of society was no longer acceptable in the high tech world of the 1980s.

During this period, the workers in the Soviet empire, from the shipwrights in Gdansk to the coal miners in the Kuzbas, realized that they were the productive members of their societies and that they could present their feudal masters with a choice between mass murder and abdication of their power. They organized strikes. Such strikes had been unsuccessful in the past, but the nomenklatura realized that this time destruction of their skilled workers would leave their already disastrous economies completely unfeasible. Technology had advanced too far to replace skilled workers readily. The new technology has given skilled labor a power most do not yet perceive.

In *1984*, a despairing Winston Smith writes in his diary

> *If there is hope, it lies with the proles.*

In the case of the Soviet Union and its empire, Orwell's hero appears to have been prophetic.

Yet, in the United States, the nomenklaturas, composed largely of armies of attorneys and MBAs, continue to dominate much of society. They bankrupt one company, merge another, and slither confidently on. Henry Ford hired his employees for life. Few employees of contemporary American companies can assume their current company will employ them until retirement, or even that their company will survive until they retire. MBAs are trained in marketing and finance, almost never in quality and production. American attorneys, who once upon a time proudly bore the title of "counsellor," now have a low prestige consistent with their current contributions to society. Workers are enjoined to improve their performance "or else," but they are not given any inkling of how this improvement is to be achieved. And little wonder, for seldom does American management know enough about their production processes to tell their employees how to improve them.

Increasingly free markets appear to be presenting American managers with a bottom line that they can no longer avoid. The slogan of "buy American" begins to ring hollow indeed. A worker engaged in a well-managed company has difficulty in seeing why he should use his hard earned money to buy an inferior domestic product at a higher price rather than a superior foreign import. American consumers may be on the verge of creating a situation where the nomenklatura in this country will begin to be replaced with managers, who are life members of their companies, experienced in what they do, willing to learn and explain to their colleagues how to improve the process of production. It is unfortunate for all concerned, however, that the directors of most failing American companies seem to spurn the rather well defined statistical process control protocols which could, very frequently, save their businesses. Rather, we seem to have a situation of natural selection. Companies with managers who have adopted the philosophy of statistical process control tend to survive. Those with managers who do not are likely to fail. But, like dinosaurs in a Darwinian jungle, most managers refuse to adapt to the new world of high quality and the paradigms required to participate in it. There may be better ways to improve the general quality of American production than simply watching those firms imprudently managed perish. But at least the discipline of a nearly free market does keep the quality of surviving firms on a path of improvement. A pity that so many American managers are unwilling to learn statistical process control, for, if they did, they might very well deliver themselves, their workers and their stockholders from the inevitability of disaster.

Unfortunately, the organic force for the maintenance of control by nomenklaturas continues. In Poland, most of the factory directors have been pensioned off (handsomely, by Polish standards). But the new managers (following advice from experts from the United Nations, the World Bank, the United States, etc.) have not been recruited from the ranks of the foremen, blue collar supervisors generally very familiar with the running of the plant. Rather, to a significant degree, the new directors have been recruited from the junior league of the old nomenklatura. These are individuals, generally in their 30s and 40s, who have supplemented their training with accelerated American-style MBA programs. Thus, big state run or newly privatized companies are being controlled, in large measure, by persons who are attempting to make the transition from one nomenklatura to another nomenklatura. The thousands of new small companies in Poland, freed from stifling governmental controls and unimpeded by nomenklatura personnel, are thriving. But the big companies are not generally improving; they are, in fact, going bankrupt with astounding regularity. One of the tragedies of the newly free countries of Central and Eastern Europe appears to be that, having been subjected for decades to the depredations of a Soviet-style nomenklatura, they now must suffer the inefficiencies of an American-style one.

It is the hope of revolutionaries to leapfrog the failed paradigms of the past. Our hope for the newly free countries of the former Soviet empire is that they will adopt the new style of management, advocated by Professor W. Edwards Deming, sooner rather than later.

It is interesting to note a practical consequence (observed by Shewhart and Deming) of Pareto's Maxim: the failures in systems can be viewed, mathematically, as a problem in contaminated distributions. This fact provides us with a tool that can lead to the replacement of the quasi-feudal managerial systems Pareto predicted by that nurturing system of continual improvement, that is the hallmark of statistical process control.

Statistical process control has nothing to do with attitude adjustment, slogans or boosterism of any sort. It is based on concepts which, though not as theoretically well understood, are as substantial as Newtonian physics. The basic notion of a few rather than many causes of failure in a system was perceived at least as long ago as Pareto. The profound observation that machines operate in fundamentally different ways than do people goes back at least to Henry Ford.

Walter Shewhart perceived in the early 1930s that Pareto's qualitative observation about causes of failure could be quantitized as a model of mixtures of distributions. In the real world, there appear to be switches in time, which periodically transfer the generating process into a distribution not typical to the distribution when the dominant distribution is the driving force behind the process. These epochs in time exhibit product with differences in average measurement and/or variability of measurement from the product produced during "in control" epochs. The control charts developed by Shewhart enabled him to identify these "out of control" epochs. Then, by backtracking, he was frequently able to discover the systemic cause of a "Pareto glitch" and remove it, thus fixing the system. Based on the observation of Ford that a fixed machine tends to stay fixed, Shewhart was able to build the basic paradigm of SPC, which is essentially a kind of step-wise optimization of a system.

Taken at first glance, there is no particular reason to be excited about Shewhart's paradigm. SPC sounds, at first blush, about as likely as Pyramid Power and Transcendental Meditation to improve a system of production. Pareto's Maxim is not intuitively obvious to most. But then, Galileo's observation that objects fall to Earth at a velocity independent of their mass does not sound, at first hearing, obvious. Of course, Galileo's conjecture admits of relatively simple experimental validation. The verification of the utility of SPC tends to require implementation in a rather large and costly system of production.

It would be interesting to undertake a careful historical investigation to determine just how much of the Shewhart algorithm was foreshadowed by the production techniques of Henry Ford and by the German manufacturer, Ferdinand Porsche, who was strongly influenced by Ford's work. Although World War II did see some implementation of Shewhart's paradigm in American war production, there is also significant evidence that the influence of the Soviet style of optimization by slogan and psychology grew in the United States very rapidly during this period. On balance, World War II may very well have left America farther away from the Shewhart paradigm than it had been previously.

At any rate, it is clear that Deming's massive implementation of SPC in Japan after World War II brought Japan quickly into a position first of challenging, then surpassing, American automobile and audio/video production. And the Japanese competitive success vis-à-vis the Germans has also been clear in these areas. One recent study indicates that the total time spent in producing a Lexus LS 400 is less than the average rework time spent on a competitive German product. Whatever portion of the Shewhart paradigm was presaged by Ford and Porsche, it is hard to avoid the conclusion that it was much less than its potential, and that it was Deming who carried out the equivalent of Galileo's gravity experiment on a massive scale. Perhaps there is a valid comparison between Shewhart and Adam Smith, who had perceived the power of the free market. But there appears to be no single implementer of the free market who was as important in validating *The Wealth of Nations* as Deming has been in validating the paradigm of Shewhart. There has never been, in world history, so large scale an experiment to validate a scientific hypothesis as Deming's Japanese validation and extension of the statistical process control paradigm of Shewhart.

Revisionists in quality control abound. If Deming has his Fourteen Points, other quality control "experts" also have theirs. If Deming specifically warns, in his Fourteen Points, against sloganeering and posters, others specifically advocate such devices in theirs. If Deming argues for dumping the productivity-destroying paradigm of "quality assurance" and going to optimization of the production process via SPC, others argue that SPC and QA are just two different tools in the quality arsenal and promise a smooth, painless transition from QA to SPC. Some argue that they have gone far beyond Deming by the implementation of continuous feedback concepts according to the paradigms of classical control theory, thereby demonstrating that they really haven't much of a clue what Pareto, Deming and Shewhart had discovered.

Multibillion dollar corporations in the United States are as likely to consult revisionist gurus as they are to consult those who implement the mixture paradigm of Shewhart. Somehow, American CEOs seem to think that "all these guys are implementing the quality control system of the Japanese. We need to pick an expert who has a presentation which is well packaged and management friendly." Such CEOs would probably do their employees and stockholders a considerable service if they would at least do a bit of research to see how revisionist gurus are regarded by the Japanese.

Returning to the analogy with Adam Smith, it is certainly possible to argue against the free market for many reasons. Lack of effectiveness is not one of them. We may quarrel with the simple SPC paradigm of Deming, but not on the basis of its record of performance. There is more to the evolution of technology than SPC, just as there is more to the improvement in the living standard of the population than is presented by the free market. But those managers who neglect the system optimization paradigm hinted at by Pareto, postulated by Shewhart, and implemented and validated by Deming, do so at the hazard of their futures and those of their companies.

In 1989, James Thompson and Jacek Koronacki began investigations as to how statisticians might assist in the economic development of post-Marxist Poland. Koronacki had translated Thompson's short course notes (used in a number of industrial settings in Texas) into Polish, and it was decided to use these notes first to train instructors, and then as a basis of within factory teaching and consultation. In June of 1991, Kevin McDonald, president of the International Team for Company Assistance, headquartered in Warsaw, provided United Nations funding to hire one dozen Polish Ph.D. statisticians for a period of 6 months, during which period they would go into ten Polish companies recently privatized or on the verge of privatization and introduce the Deming approach. The resulting consortium, entitled the Quality Control Task Force, led by Koronacki, has been functioning since that time. Two colleagues from the Department of Statistics at Rice University, Marek Kimmel and Martin Lawera, together with Thompson, have provided on-site and remote consultation to the group since the summer of 1991.

The current book is an evolutionary development, starting with the short course notes developed over a decade of consulting and lecturing on quality improvement, including recent experiences in Poland, and adding a mathematical modeling background not generally employed in industrial courses. The book is organized so that the earlier part of the book can be utilized by persons only interested in the practical implementation of the paradigm of statistical process control. Some of the material in the later part of the book deals with topics that are of ongoing interest in our research and that of our students. By including mathematical and statistical appendices, we have attempted to write a book which, say, a foreman might utilize over time as he wishes to develop both his practical and theoretical insights into statistical process control. Problems are given at the end of each chapter. For university instruction, the book is appropriate both for advanced undergraduate and graduate level courses.

Chapter 1 represents an overview of the practical implementation of statistical process control. The intuitive contaminated distribution approach taken in this chapter is appropriate for use in industrial short courses and has been so employed in both Poland and Texas.

We consider in Chapter 2, as a beginning, the data available in the present state of quality activity in most American firms, namely that of quality assurance. Though such data are much less desirable than modularized measurement data, it is a natural starting point for quality investigators dealing with the world as they find it, as opposed to how they might wish it to be.

Chapter 3 considers in some detail the theory of contaminated distributions which forms the model basis of most statistical process control. The performance of Shewhart Control Charts is considered in a variety of practical situations. Various procedures for robustification of the parameters characterizing the uncontaminated process distribution are considered.

In Chapter 4 we examine a variety of sequential procedures favored by some as alternatives to the Shewhart Control Charts. These include CUSUMs, Shewhart CUSUMs, Acceptance-Rejection CUSUMs, Page CUSUMs and Exponentially Weighted Moving Averages.

Chapter 5 presents a number of exploratory and graphical techniques frequently useful for troubleshooting in those situations which do not readily lend themselves to standard SPC paradigms.

In Chapter 6 we present a number of optimization techniques useful for designing experiments and modifying process conditions for enhanced production performance. Included among these are the simplex algorithm of Nelder and Mead and the rotatable designs of Box, Hunter and Draper.

Chapter 7 deals with the subject of examining time indexed multivariate data for clues to quality improvement. A compound test is suggested as an alternative to the generally standard approach of testing one dimension at a time. A robust procedure for estimating the mean of the dominant distribution in a data contamination situation is proposed. A nonparametric test for shift of location is considered.

Appendices A and B give an overview of the linear algebra and mathematical statistics used in the rest of the book. The inclusion of these appendices is an attempt to make the book as self-contained as possible.

For reasons of practicality, we have attempted to create a book which does not need an accompanying software diskette. The standard statistical procedures in SPC are not particularly involved. The standard SPC charts can be handled very nicely with an inexpensive hand-held calculator, such as the TI-36X solar, which we tend to introduce into the firms with whom we consult.

There are a number of excellent spreadsheet-based statistical packages which would be of use in handling, say, the linear models sections in Chapter 6 as well as dealing more quickly with the problems in the first five chapters. These include the various versions of SYSTAT and Statview. Indeed, simple spreadsheet packages such as Lotus 1-2-3 and Excel can easily be adapted to assist with many of the problems. For the simulations in Chapter 7, we used programs written in C (by Martin Lawera). Lawera's work with Thompson in the development of the "king of the mountain" algorithm for finding the location of the multivariate "in control" distribution is duly noted.

The book was typeset using a Macintosh IIfx, utilizing the *Textures* LaTex processing program of Blue Sky Research, using graphics from SYSTAT 5.2, MacDraw Professional and MacPaint II.

The support of the Army Research Office (Durham) under DAAL-03-91-G-0210 is gratefully acknowledged. The support given to Polish members of the Quality Control Task Force by the International Team for Company Assistance has facilitated the introduction of SPC into Polish production.

We particularly wish to thank our Rice colleagues Marek Kimmel and Martin Lawera (both originally from Poland) for their valuable work with the QCTF in Poland. The graphs in Chapter 7 and the trimmed mean algorithm therein are due to Martin Lawera, as are the reference tables at the end of the book.

We also extend our thanks to Gerald Andersen, Wojciech Bijak, Barry Bodt, Diane Brown, Barbara Burgower, Jagdish Chandra, Gabrielle Cosgriff, Dennis Cox, Miroslaw Dabrowski, Piotr Dembinski, Katherine Ensor, James Gentle, Stuart Hunter, Renata Koronacka, Robert Launer, Kevin McDonald, Charles McGilchrist, Jan Mielniczuk, Marek Musiela, Michael Pearlman, Joseph Petrosky, Paul Pfeiffer, Zdzislaw Piasta, Rick Russell, Michael Sawyers, David Scott, Beatrice Shube, Andrzej Sierocinski, Malcolm Taylor, Ewa Thompson, John Tukey, Matt Wand, Geoffrey Watson, Jacek Wesolowski, Edward Williams and Waldemar Wizner.

James R. Thompson and Jacek Koronacki
Houston and Warsaw, September 1992

Preface to the Second Edition

The first edition of this book, *Statistical Process Control for Quality Improvement*, was published nearly ten years ago. Nevertheless, we find our earlier work a fair attempt (and, as near as we have been able to find, the only book length attempt) at mathematically modeling the Deming paradigm (SPC) for continuous quality improvement.

Deming was an optimizer, not a policeman. Philosophically, the Quality Assurance paradigm (QA), which has dominated American manufacturing since World War II, has about as much in common with SPC as a horse and buggy has to a Mercedes. Both of these have wheels, brakes, and an energy source for locomotion, but it would be less than useful to think of them as "essentially the same," as so many managers still seem to consider QA and SPC to be the same. Both QA and SPC use control charts, but to very different purposes. QA wants to assure that bad units are not shipped. SPC wants to assure that bad units are not created in the first place, and that the units are being produced by a system in a continuing state of improvement. If this point is not understood by the reader of this book, it is not from any lack of trying on the part of the authors. Anyone who has walked into both QA establishments and SPC establishments knows the extreme difference in the sociology of the two types. In the QA establishment, the worker is being watched for poor performance. In the SPC establishment, the worker is a manager, calmly focused on the improvement of the process, and with constant recognition of his contributions.

Again, the Deming paradigm must not be confused with the touchy-feely boosterism associated with the "Quality Is Free" movement. Although SPC is one of the best things ever to happen to making the workplace a friendly and fulfilling environment, its goal is to improve the quality of the goods delivered to the customer by a paradigm as process oriented as physics. It was the insight of Deming that has led to the realization that one can use the fact that a lot exhibits a mean well away from the overall mean to indicate that something specific is wrong with the system. Using this technique as a marker, the team members can backtrack in time to see what caused the atypicality and fix the problem. As time progresses, relatively minor problems can be uncovered, once the major causes of jumps in variability have been found and removed.

For reasons of user friendliness, Deming advocated the already venerable run charts and control charts as a means of seeking out atypicality. Thus, Deming's methodology, on the face of it, does appear to be very much the same thing as that advocated by the Quality Assurance folks or the New Age "Quality Is Free" school. Perhaps Deming himself was partly to blame for this, for he never wrote a model-based explanation for his paradigm. Moreover, Deming was advocating the use of old tests familiar to industrial engineers to achieve quite a different purpose than those of the older QA school: namely, Deming used testing to achieve the piecewise optimization of an ill-posed control problem. And, as anyone who has used SPC on real problems will verify, SPC works.

This book is an updated and extended version of the first edition, with an increased length of roughly 25%. Criticisms by our colleagues and students of that earlier endeavor have been taken into account when preparing this book, as have been our own new experiences in the fascinating area of quality improvement in the manufacturing, processing and service industries. As in the first edition, we have tried to give examples of real case studies flowing from work we have ourselves undertaken. Consequently, as beyond the "production line" examples, we include an example of problems encountered when a new surgical team was brought into the mix of teams dealing with hip replacement. There is an example showing problems experienced by a company involved in the production of ecologically stable landfills. A look is given at a possible start-up paradigm for dealing with continuous improvement of the International Space Station. Thus, one aim of this book is to convince the reader that CEOs and service industries need SPC at least as much as it is needed on production lines. Deming viewed SPC as a managerial tool for looking at real world systems across a broad spectrum. So do we.

Revisions of the former book include discussions, examples and techniques of particular interest for managers. In addition, the new edition includes a new section recapitulating in Chapter 1 how properly to understand and react to variability within a company; new section on process capability in Chapter 3; on the Pareto and cause-and-effect diagrams, as well as on Bayesian techniques, on bootstrapping and on the seven managerial and planning tools (also known as the Japanese seven new tools) in Chapter 5; and on multivariate SPC by principal components in Chapter 7.

Usually, Professor Deming discussed methodologies for use with systems rather mature in the application of SPC. In the United States (and more generally) most systems in production, health care, management, etc., are untouched by the SPC paradigm. Consequently, we find it useful to look at real world examples where SPC is being used on a system for the first time. It does the potential user of SPC no good service to give the impression that he or she will be dealing with "in control" systems. Rather, our experience is quite the contrary. Startup problems are the rule rather than the exception.

Exploratory Data Analysis and other minimal assumption methodologies are, accordingly, in order. In this new edition, we introduce Bayesian techniques for the early stages of operation of a complex system. This is done in the context of a real world problem where one of us (Thompson) was asked by NASA to come up with a speculative quality control paradigm for the operation of the International Space Station. NASA, which uses very sophisticated reliability modeling at the design stage, generally does not use SPC in the operation of systems. We show how even a very complex system, untouched by SPC, can be moved toward the Deming Paradigm in its operation by the use of Bayesian techniques.

The design and operational problems of optimization are quite different. American companies frequently have excellent design capabilities, but forget that a system, once built, needs continually to be improved. On the other hand, the SPC professional should realize that a horse buggy is not likely to "evolve" into a Lexus. Design and continuous operational optimization, over the long haul, must both be in the arsenal of the successful health care administrator, industrial engineer, and manager. Deming knew how to combine design and operational optimization into one methodologically consistent whole. In this edition, we discuss his unifying approach by referring to the so-called Shewhart-Deming Plan-Do-Study-Act cycle and, based on it, spiral of continual improvement. We also elaborate on means to help design an innovation, namely on the so-called seven managerial and planning tools.

Because measurement statistics in quality control activities are generally based on averaging, there is a (frequently justifiable) tendency to assume the statistics of reference can be based on normal theory. Rapid computing enables us to use the nonparametric bootstrap technique as a means of putting aside the assumption of normal theory when experience hints that deviation from normality may be serious. Furthermore, rapid computing enables us to deal with multivariable measurement SPC. It is true that most companies would greatly improve their operations if they used even one dimensional testing. Nevertheless, experience shows that multivariate procedures may provide insights difficult to be gleaned by a battery of one dimensional tests. Most SPC today is still being done away from a computer workstation. That is changing.

The support of the Army Research Office (Durham) under DAAL-03-91-G-0210, DAAH04-95-1-0665 and DAAD19-99-1-0150 is gratefully acknowledged. We wish to thank Andrzej Blikle, Barry Bodt, Sidney Burrus, Roxy Cramer, Kathy Ensor, Sarah Gonzales, Jørgen Granfeldt, Chris Harris, Marsha Hecht, Richard Heydorn, Olgierd Hryniewicz, Stu Hunter, Renata Koronacka, Marek Kimmel, Vadim Lapidus, Robert Launer, Martin Lawera, Andrea Long, Brian Macpherson, Jan Mielniczuk, Jim Murray, Ken Nishina, Philippe Perier, Rick Russell, Janet Scappini, Bob Stern, Ewa Thompson, Ed Williams and Waldemar Wizner.

James R. Thompson and Jacek Koronacki
Houston and Warsaw, Christmas 2001

Chapter 1

Statistical Process Control: A Brief Overview

1.1 Introduction

The common conception about quality control is that it is achieved by diligence, a good attitude and hard work. Yet there are many companies where the employees display all these attributes and the quality of the product is poor. An example of this is the construction of the famous Liberty ships of World War II. These were ships hastily constructed to transport supplies to some of America's allies. Largely due to the fact that everyone — designers, welders, shipwrights, painters, engineers, etc.— had a keen sense that they were engaged in an activity which was essential for the survival of the United States, there was strong motivation. Unfortunately, keenness was not enough, and these ships were prone to sinking, sometimes immediately after being launched. Naturally, in the case of a wartime emergency, it could be argued that it is quite reasonable to sacrifice quality in order to increase production. If America had insisted that Liberty ships be perfect, then the war might have been lost. There is some merit in this argument, and to one extent or another the argument can be used in any production setting. There are orders to be filled by a specific date. If a company cannot make the deadline, then it is quite possible that the order will be given to another company. Short range concerns may, in some cases, overwhelm the long range goals of delivering a product of the best possible quality to a customer.

In general, however, the intelligent application of the philosophy of

1

statistical process control will enable us to seek steady improvement in the quality of a product even while dealing with the day-to-day crises which are to one extent or another an unavoidable part of "staying alive" in the highly competitive world of a high tech society.

At present, there appears to be a kind of revisionist theory to the effect that there is a smooth transition from the policing-based "quality assurance," which makes up the bulk of "quality control" as practiced in the United States and Europe, and the "statistical process control" paradigm generally associated with the name of W. Edwards Deming. One reads a great deal about "quality assurance" leading naturally to "quality improvement" (under which category statistical process control is supposed to be only one of many techniques). Some of the more successful companies in the lucrative business of teaching "quality control" to the unwary certainly take such a point of view.

In reality, however, the statistical process control paradigm is quite different from the end product inspection strategies associated with "quality assurance." Almost anyone who does quality control consulting in real-world settings finds that the QA people are already firmly installed as the resident experts in QC, and that it is their ineffectiveness to achieve improvement which has given "quality control" a generally bad reputation among management and staff alike. Even worse, we frequently find that the "quality control" group in a factory has already co-opted the language of statistical process control to mean something entirely different from its classical definition, to mean, as often as not, just the same old end inspection stuff, with some "touchy-feely" industrial psychology thrown in for good measure. So, when one talks about using control charts, the QA staff trots out ream upon ream of their own charts dating back, in some cases, decades. That these charts are generally based on arbitrary tolerance levels, that they are not recorded at sensing stations throughout the production process, that they are used to decide who is OK and who is not rather than for improvement appears to matter little to them. They are doing everything right, marching on as quickly as possible to the nirvana of "zero defects."

Statistical process control, as we understand it, has as its irreducible core three steps:

(1) Flowcharting of the production process.

(2) Random sampling and measurement at regular temporal intervals at numerous stages of the production process.

(3) The use of "Pareto glitches" discovered in this sampling to backtrack in time to discover their causes so that they can be removed.

Statistical process control is a paradigm for stepwise optimization of the production process. That it works so well is not self-evident, but is empirically true. During the course of this book, we shall attempt to give some insights as to why the SPC paradigm works.

The given wording of the core three steps of SPC and of the SPC paradigm in general is for production processes. Let us emphasize most emphatically, however, that it could equally well be for service processes and also for management processes. An example of a process in a service setting, health care, is discussed in Section 2.4. Examples of management processes are alluded to in later sections of this chapter.

1.2 Quality Control: Origins, Misperceptions

There are a number of paradoxes in quality control. A "paradoxical" subject, of course, ceases to be a paradox once it is correctly perceived. It is not so much our purpose in this chapter to go into great detail concerning control charts and other technical details of the subject as it is to uncover the essence of the evolutionary optimization philosophy, which is always the basis for an effective implementation of quality control. Most of us, including many quality control professionals, regard quality control as a kind of policing function. The quality control professional is perceived as a kind of damage control officer; he tries to keep the quality of a product controlled in the range of market acceptability. The evolutionary optimization function of quality control is frequently overlooked. Yet it is the most important component.

The major aspect of much quality control philosophy in most American companies is based on worker motivation and attitudinal adjustment. Workers, it is deemed, are too lazy, too careless, too insensitive to the mission of the enterprise. The mission of the quality control professional is based on the notion of getting workers to be regular, sober, and keen.

Such a notion has always been paramount in the Soviet Union. The example of the Hero of Socialist Labor Stakhanov is remembered. Stakhanov was extremely adept at getting coal picked out of a coal seam. Some Soviet bureaucrat seized upon the brilliant idea of setting Stakhanov up as an example to his fellow workers. The idea was that although it might not be reasonable to expect every worker to perform at 100% of Stakhanov's efficiency, there was some fraction of 100% which could be set as a lower bound for satisfactory performance. Of course, some other bureaucrat decided that it might be a good idea to give Stakhanov a bit of assistance as he was setting the official standard of good performance. Thus Stakhanov was presented with a rich, easily accessible coal seam, and he had two "helpers" as he was picking away. So Stakhanov set a really fine standard, and workers who did not perform at an arbitrary fraction of the standard were docked, and if they fell below an arbitrary floor, they could be sent to Siberia to work without salary in the Gulags. The Stakhanovite program was one of the principal causes for worker unrest in the Soviet Block, and was the subject of one of the strongest anti-Communist films (produced in Peoples Poland during a brief thaw), *Man of Marble.*

Unfortunately, even in contemporary America, it is apparent that variants of the Stahkanovite program are still employed. Time and motion studies, piecework, quotas, etc., are an integral part of all too many businesses. However, in the best companies, the goal has been to work smart rather than hard. By cooperation with management and technological innovation, workers in these firms have brought their per person production well above that of Stakhanov. Statistical process control is a paradigm for achieving quality by which we can continuously learn from our experiences to work smarter rather than harder.

To find the exact time where the notion of quality control began is not easy. We might recall the craft guilds of medieval Europe. The function of such guilds was oriented toward the training of apprentices and journeymen so that they might become competent masters of the craft. In order to become a master of the craft of coppersmithing would typically require ten years or more of learning by doing.

Yet, in the sense in which we generally understand quality control, its true beginnings were in 1798 when Eli Whitney developed his musket factory. This was the first time in which a nontrivial production based on the notion of interchangeable parts was used. It was noted, with concern by some members of Congress, that the fledgling Republic might be drawn into the Napoleonic Wars. At that time, America had few producers of muskets, and a revolution in production was required if the musket shortage was to be solved. This may sound strange, since our textbooks are full of stories about Daniel Boone and other stalwart Indian fighters. But the reality of the situation is that few Americans even in the early days of the Republic were frontiersmen. Most spent their lives peacefully engaged in agriculture and other pursuits. The sizes of the armies engaged in the American Revolution, moreover, had been small when compared with the massive formations engaged in the European conflict. The musket craftsmen of the day turned out weapons one at a time. The process was lengthy and risky. The failure to achieve appropriate fits of the various components caused many a musket to explode during the test firing.

Whitney conceived the idea of making a musket out of approximately 50 key parts. Each of the 50 parts was milled on a machine in such a way that each copy was nearly identical to any other. At the test assembly, Whitney invited members of Congress to pick parts at random from numbered containers. He then assembled the musket, loaded it, and test fired it, while the Congressmen withdrew to a discreet distance. After the process had been carried out several times, it dawned on the Congressional delegation that something rather remarkable had been achieved. Just what would have happened to the United States during the War of 1812 absent Whitney's development of the first modern assembly line process is a matter of conjecture.

A century after Whitney's musket factory, Henry Ford developed the modern automobile. The key to the success of the Ford was its reliance on the internal combustion engine, with its high power to weight ratio. The Model T had approximately 5,000 parts.

It is interesting to note that Ford's Model T contained approximately 100 times as many parts as Whitney's musket. This is a true quantum leap in complexity. Many scoffed at the idea that such a complex system could work. Indeed, using anthropomorphic analogies, there is little chance that 5,000 human beings could be relied upon to perform precise tasks at precise times, over and over again millions of times. Ford demonstrated that machines do not perform like people. This fact continues to elude the attention of many, even those who are chief executive officers of large industrial corporations.

The importation of Ford technology into Germany created the high technological society for which that nation is rightly known. The Volkswagen created by Ferdinand Porsche was a downsized version of the Ford vehicle, and its name was taken from Ford's name for the Model T, "the People's Car." Unfortunately and to the distress of pacifist Ford, the Germans learned Ford's methods all too well, applying his techniques to the production of a highly efficient war industry brutally used against mankind. During World War II, the Germans generally maintained higher standards of quality control than did the Americans, who had developed the notion of the assembly line in the first place. It has been argued [6] that in a real sense World War II America moved perceptibly toward Soviet-style production, characterized by grandiose schemes, poor planning, and quality achieved by empty slogans and pressure on the workers.

As we shall see, it is a serious mistake to use workers as well disciplined automatons. In a high technology society, most work is done by machines. The function of a worker is not regularity, anymore than the function of a scientist is regularity. A worker, like a scientist, should be an innovator. To attempt to manage an industrial system by stressing the workers is generally counterproductive. The work is done by machines. Machines, unlike human beings, are incredibly regular and, with some very high tech exceptions, completely unable to innovate. The function of "quality control" is to provide the maximum amount of creativity at all levels to achieve a constantly improving standard of performance. Quality control, correctly perceived, is an orderly system for monitoring how well we are doing so that we can do better.

These observations apply to quality control philosophy in general, whether in a production setting or in service and management settings. Only the role of machines in the production setting is played in the two latter settings by procedures to be followed by employees when they process their tasks related to an external customer or report results to management.

1.3   A Case Study in Statistical Process Control

Let us consider a favorite example of the effect of intelligent quality control presented in *Quality, Productivity and Competitive Position* [1] by W. Edwards Deming. Nashua Corporation, a manufacturer of carbonless paper, would appear to be in an excellent competitive position to market its product in Europe. Located in New Hampshire, Nashua had a ready source of wood pulp for paper; the expensive coating material for the paper was made completely from materials in the United States. Nashua had the very latest in equipment. The manufacture of carbonless paper is rather high tech. Paper is coated with photosensitive material on a web 6 to 8 feet wide traveling at a speed of 1,100 linear feet/minute. A key part of the process is to coat the paper with the minimal amount of photosensitive material consistent with good quality of performance. Parts of the paper with too little material will not reproduce properly. In late 1979, Nashua was using 3.6 pounds per 3,000 square feet.

It would appear that Nashua was in a strong competitive position indeed. There was, however, a small problem. The Japanese, without trees for paper and without the minerals for the paper coating, and far from Europe, were able to sell carbonless paper to the Europeans for less cost than Nashua. Even worse, the quality of the Japanese paper was better (i.e., more uniform and, hence, less likely to jam and better for reproduction) than that of the Nashua product. What was to be done? One immediate answer preferred by some American industrialists and union leaders goes as follows. "It is unfair to ask American companies to compete with low Japanese wage structures. Put a tariff on the Japanese product." There are problems with this argument and solution.

First of all, labor cost differentials are not generally dominant in a highly automated industry. The differential might make up something like 5% of the competitive advantage of the Japanese, an amount clearly not equal to the Nashua advantages of ready raw materials and relative closeness to markets. Furthermore, suppose the Japanese decided to retaliate against an American tariff on carbonless paper by putting a tariff on lumber from Washington and Oregon, purchasing their material rather from British Columbia. Americans who had to buy a more expensive American product would be giving a subsidy to Nashua, raising the cost of production in their own businesses and, hence, marginally making their own products less competitive on a world market. And since a major portion of Nashua's target market was in Europe rather than in the United States, an American tariff would be an impossibility there.

Demanding a tariff on carbonless paper simply was not a feasible option for Nashua. The most likely option in late 1979 appeared to be to go into bankruptcy. Another was to buy a new coating head, at a cost of $700,000 plus loss of production during installation. The solution used by Nashua was to try to find out what was going on in an orderly fashion. It is always a good idea, when a competitor producing the same product with equipment and logistical support inferior to our own, is able to best us both in terms of quality and price, to learn what he is doing better than we are.

A modular investigation involves measuring the input and output at each module of the process. In the Nashua case, it turned out that the major problem and its solution were incredibly simple. The output record of the coating machine had been used to achieve instantaneous feedback control. Thus, if the paper was being coated too thickly, an immediate adjustment was made in the coating machine. A first step in quality control involves a modularization of the system of the process. Such a modularization is shown in Figure 1.1.

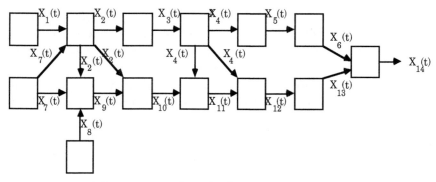

Figure 1.1. Typical System Flowchart.

It turned out that the coating machine reached equilibrium only after some time. Instantaneous control, then, was a terrible idea, leading to an institutionalized instability causing high variability in the product and causing a waste of substandard paper and a loss of expensive coating material. The solution to Nashua's problem was to cease "overcontrolling" the coating process. Letting the coating machine do its job produced immediate dividends. First of all, it was unnecessary to purchase a new coating head; hence, a saving of $700,000 was made possible. More importantly, very quickly Nashua was able to produce a uniform product using a coating rate of 3.6 pounds per 3,000 square feet. This made the Nashua product competitive. But this was not the end of the story. Quality control is a continuous evolutionary optimization process. Noting that the coating costs $100,000 per tenth of pound per year, Nashua began to take advantage of its high tech coating machine. In less than a year, it was able to reduce the coating rate to 2.8 pounds per 3,000 square feet, producing an annual savings of $800,000. By 1985, the coating rate had been reduced to 1.0 pound per 3,000 square feet [4].

The consequences of the above example are clear. An unprofitable American company escaped ruin and became a savvy enterprise which can stand up to competition from any quarter. No recrimination was taken against those who had overcontrolled the system. A superior, more profitable product benefited both the company and its customers. The price paid was not increased stress for the workers. Indeed, the improvements in production eliminated a good deal of stress on the part of everyone involved. The company had simply engaged in that which Americans have typically done well: finding out problems, describing them and solving them.

The Nashua example is fairly typical, in that the problem involved was not a general malaise across the production line, but a failure at one particular point of the process. It is not that everything else except overadjustment of the coating machine was perfect. But the other failings were relatively insignificant when compared to that of overadjustment.

1.4 If Humans Behaved Like Machines

Imagine a room filled with blindfolded people which we would wish to be quiet but is not because of the presence of a number of noise sources. Most of the people in the room are sitting quietly, and contribute only the sounds of their breathing to the noisiness of the room. One individual, however, is firing a machine gun filled with blanks, another is playing a portable radio at full blast, still another is shouting across the room, and, finally, one individual is whispering to the person next to him. Assume that the "director of noise diminution" is blindfolded also. Any attempt to arrange for a quiet room by asking everyone in the room to cut down his noise level 20% would, of course, be ridiculous. The vast majority of the people in the room, who are not engaged in any of the four noise making activities listed, will be annoyed to hear that their breathing noises must be cut 20%. They rightly and intuitively perceive that such a step is unlikely to do any measurable good. Each of the noise sources listed is so much louder than the next down the list that we could not hope to hear, for example, the person shouting until the firing of blanks had stopped and the radio had been turned off.

The prudent noise diminution course is to attack the problems sequentially. We first get the person firing the blanks to cease. Then, we will be able to hear the loud radio, which we arrange to have cut off. Then we can hear the shouter and request that he be quiet. Finally, we can hear the whisperer and request that he also stop making noise. If we further have some extraordinary demands for silence, we could begin to seek the breather with the most clogged nasal passages, and so on. But generally speaking, we would arrive, sooner or later, at some level of silence which would be acceptable for our purposes. This intuitively obvious analogy is a simple example of the key notion of quality control. By standards of human psychology, the example is also rather bizarre. Of the noise making individuals, at least two would be deemed sociopathic. We are familiar with the fact that in most gatherings, there will be a kind of uniform buzz. If there is a desire of a master of ceremonies to quieten the audience, it is perfectly reasonable for him to ask everyone please to be quiet. The fact is that machines and other systems tend to function like the (by human standards) bizarre example and seldom behave like a crowd of civilized human beings. It is our tendency to anthropomorphize systems that makes the effectiveness of statistical process control appear so magical.

1.5 Pareto's Maxim

A cornerstone of SPC is an empirical observation of the Italian sociologist Vilfredo Pareto: *in a system, a relatively few failure reasons are responsible for the catastrophically many failures.* Let us return again to the situation described in Figure 1.1. Each one of the boxes represents some modular task in a production process. It is in the nature of the manufacturing process that it be desirable that the end product output $X_{14}(t)$ be maintained as constant as possible.

For example, a company which is making a particular type of machine bolt will want to have them all the same, for the potential purchasers of the bolt are counting on a particular diameter, length, etc. It is not unheard of for people to pay thousands of dollars for having a portrait painted from a simple photograph. The artist's ability to capture and embellish some aspect he perceives in the photograph is (quite rightly) highly prized. A second artist would produce, from the same photograph, quite a different portrait. No one would like to see such subjective expression in the production of bolts. If we allowed for such variability, there would be no automobiles, no lathing machines, and no computers. This fact does not negate aesthetic values. Many of the great industrial innovators have also been major patrons of the arts. Modularity demands uniformity. This fact does not diminish the creative force of those who work with modular processes any more than a net interferes with the brilliance of a tennis professional. And few workers in quality control would wish to have poems written by a CRAY computer. The measurement of departures from uniformity, in a setting where uniformity is desired, furnishes a natural means of evolutionary improvement in the process.

Now by Pareto's Maxim, if we see variability in $X_{14}(t)$, then we should not expect to find the source of that variability distributed uniformly in the modules of the process upstream. Generally, we will find that one of the modules is offering a variability analogous to that of the machine gun firing blanks in an auditorium. The fact that this variability is very likely intermittent in time offers the quality control investigator a ready clue as to where the problem is. Naturally, there may be significant time lags between $X_3(t)$, say, and $X_4(t)$. The sizes of the lags are usually well known; e.g., we usually know how far back the engine block is installed before final inspection takes place. Thus, if a particular anomaly in the final inspection is observed during a certain time epoch, the prudent quality control worker simply tracks back the process in time and notes variabilities in the modules which could impact on the anomaly in the proper time frame. Although they are useful for this purpose, statistically derived control charts are not absolutely essential. Most of the glitches of the sort demonstrated in Figure 1.2 are readily seen with the naked eye. The Model T Ford, which had roughly 5,000 parts, was successfully monitored without sophisticated statistical charts. We show, then, the primitive *run chart* in Figure 1.2. Once we have found the difficulty which caused the rather substantial glitch between hours 4 and 8, we will have significantly improved the product, but we need not rest on our laurels. As time proceeds, we continue to observe the run charts.

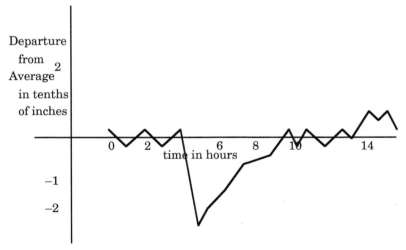

Figure 1.2. Run Chart.

We observe a similar kind of profile in Figure 1.3 to that in Figure 1.2. However, note that the deviational scale has been refined from tenths to hundredths of inches. Having solved the problem of the machine gun firing blanks, we can now approach that of the loud radio. The detective process goes forward smoothly (albeit slowly) in time. Ultimately, we can produce items which conform to a very high standard of tolerance.

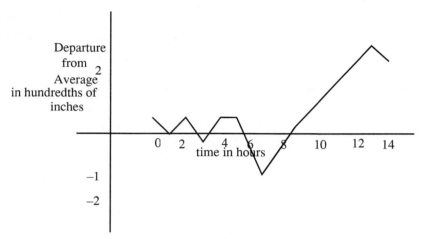

Figure 1.3. Run Chart of Improved Process.

The run charts showed in Figures 1.2 and 1.3 allude again to an industrial setting. It is easy, however, to construct such charts in other settings too. For example, for accounts payable, one can depict percent of unpaid invoices month by month. Equally easily, one can monitor monthly sales, seasonally adjusted if needed. One can gather numbers of customers serviced in consecutive periods, percents of items returned by customers, numbers of incoming or outgoing telephone calls, energy consumption, etc., etc. It is in fact amazing how often and how many numerical data on our daily activities in a business can be recorded to be later used to advantage. A friend of one of the authors, a British SPC consultant and assistant to Deming in his last years of work, made patients of a cardiologist record their blood pressure data to help the doctor better see, diagnose and react to the glitches thus revealed.

Another Difference between Machines and People. A second paradox in quality control, again due to our tendency to treat machines and other systems as though they were human beings, has to do with the false perception that a source of variability, once eliminated, is very likely to occur again as soon as we turn our attention to something else. In many human activities, there is something like an analogy to a man juggling balls. As he hurls a red ball upward, he notices that he must immediately turn his attention to the yellow ball which is falling. When he hurls the yellow ball upward, he notes that the green ball is falling. By the time he throws the green ball upward, he must deal with the red ball again. And so on. If a human decides to give up smoking, he is likely to note an increase in his weight. Once he worries about weight control, he may start smoking again.

Happily, machines suffer from no such difficulty. One of the authors drives an 11-year-old Volvo. The engine has experienced many millions of revolutions and still functions well. No one is surprised at such reliability with mechanical devices. The ability to execute identical operations many times without fail is in the very nature of machines. Interestingly, there are many systems in which similar reliability is to be expected. In highly automated industrial systems, once a source of variability has been eliminated, it is unlikely to be a problem again unless we are very careless indeed. The in-line, four-cylinder Volvo engine, developed over 40 years ago, still functions very well. A newer, more sophisticated engine may be much less reliable unless and until we have carefully gone through the quality control optimization process for it as well. Innovation is usually matched by new problems of variability not previously experienced in older systems. This does not speak against innovation; rather it warns us that we simply do not reasonably expect to have instantaneous quality control with a new system. Companies and other organizations that expect to go immediately from a good idea to a successful product are doomed to be disappointed.

The Basic Paradigm for Quality. The basic paradigms of both Whitney and Ford were essentially the same:

(1) Eliminate most potential problems at the design stage.

(2) Use extensive pilot study testing to eliminate remaining undiscovered problems.

(3) Use testing at the production stage to eliminate remaining glitches as a means of perfecting the product, always remembering that such glitches are generally due to defects in a few submodules of the production process rather than general malaise.

Ford's Four Principles of Manufacturing. Henry Ford codified his ideas in his four principles of manufacturing [3]:

(1) An absence of fear of the future or veneration of the past. One who fears failure limits his activities. Failure is only the opportunity to begin again. There is no disgrace in honest failure; there is disgrace in fearing to fail. What is past is useful only as it suggests ways and means for progress.

(2) A disregard of competition. Whoever does a thing best ought to be the one to do it. It is criminal to try to get business away from another man – criminal because one is then trying to lower for personal gain the

condition of one's fellowmen – to rule by force instead of by intelligence.
(3) The putting of service before profit. Without a profit, business cannot expand. There is nothing inherently wrong about making a profit. Well-conducted business enterprise cannot fail to return a profit, but profit must and inevitably will come as a reward for good service – it must be the result of service.

(4) Manufacturing is not buying low and selling high. It is the process of buying materials fairly and, with the smallest possible addition of cost, transforming those materials into a consumable product and giving it to the consumer. Gambling, speculating and sharp dealing tend only to clog this progression.

Ford's four principles no longer sound very modern. In the context of what we are accustomed to hear from America's contemporary captains of industry, Ford's four principles sound not only square but rather bizarre. That is unfortunate, for they would not sound bizarre to a contemporary Japanese, Taiwanese, Korean or German industrialist.

1.6   Deming's Fourteen Points

The modern paradigm of quality control is perhaps best summarized in the now famous fourteen points of W.E. Deming [1], who is generally regarded as the American apostle of quality control to Japan:

(1) Create constancy of purpose toward improvement of product and service, with a plan to become competitive and to stay in business. Decide to whom top management is responsible.

(2) Adopt the new philosophy. We are in a new economic age. We can no longer live with commonly accepted levels of delays, mistakes, defective materials and defective workmanship.

(3) Cease dependence on mass inspection. Require instead statistical evidence that quality is built in, to eliminate need for inspection on a mass basis. Purchasing managers have a new job and they must learn it.

(4) End the practice of awarding business on the basis of price tag. Instead, depend on meaningful measures of quality, along with price. Eliminate suppliers that can not qualify with statistical evidence of good quality.

(5) Find problems. It is management's job to work continually on the system (design, incoming materials, composition of material, maintenance, improvement of machine, training, supervision, and retraining).

(6) Institute modern methods of on the job training.

(7) Institute modern methods of supervision of production workers. The responsibility of foremen must be changed from sheer numbers to quality. Improvement of quality will automatically improve productivity. Management must prepare to take immediate action on reports from foremen concerning barriers such as inherited defects, machines not maintained, poor tools, fuzzy operational definitions.

(8) Drive out fear, so that everyone may work effectively for the company.

(9) Break down barriers between departments. People in research, design, sales, and production must work as a team, to foresee problems of production that may be encountered with various materials and specifications.

(10) Eliminate numerical goals, posters and slogans for the work force, asking for new levels of productivity without providing methods for improvement.

(11) Eliminate work standards that prescribe numerical quotas.

(12) Remove barriers that stand between the hourly worker and his right to pride of workmanship.

(13) Institute a vigorous program of education and training.

(14) Create a structure in top management which will push every day on the above 13 points.

The paradigm of Deming is sufficiently general that it might be interpreted as not being applicable to a specific industry at a particular time. We are all familiar with vacuous statements to the effect that we should somehow do a good job. But we should examine the Deming Fourteen Points carefully before dismissing them. Below we consider the points one at a time.

(1) Management is urged to plan for the long haul. We cannot simply lurch from crisis to crisis. We need to consider what we believe the company really is about. What do we plan to be doing next quarter, next year, five years, ten years from now. It goes without saying that our plans will require constant modification. But we need to ask what our direction is. What new technologies will likely affect us. We need to plan a strategy that will make the firm unique. We need to be pointed on a course which will make us the best in the world in what we do. If we have no such course plotted, it is unlikely that we will simply stumble upon it.

(2) A common assumption in some American industries is that a significant proportion of the things they do will simply be defective. Once

again, we return to the misconception that machines function as people do. They do not. If we move methodically toward improved production, we need not fear that our improved excellence will be fleeting. A human juggling balls is very different from a production schedule optimally designed. Our goal must be systematic improvement. Once we achieve a certain level of excellence, it is locked in forever. If, on the other hand, we assume that prospective clients will always be satisfied to return defective items or have us repair them, we are due, down the road, to an unpleasant surprise. Sooner or later, a competitor will discover how to produce a product like ours but without the defects.

(3) Once we have achieved the middle level of excellence, we will not require extensive inspection. Again, this fact points to the fact that work done by machines will generally be satisfactory once we have determined how to make them operate properly. One of the major benefits of a well designed system of statistical process control is that it will rather quickly liberate us from carrying out exhaustive (or nearly exhaustive inspection). Similarly, we should be able to demand from those companies who supply us a steady stream of products of excellent quality.

(4) Companies which constantly change their suppliers on the basis of marginally lower prices are playing a lottery with quality. We have the right to expect that those companies to whom we sell will think long and hard before replacing us with a competitor. This is not being polite but simply rational. "If it ain't broke, don't fix it" is a good rule. If our supplier is delivering high quality products to us at a reasonable price, then, on the offer by another supplier to deliver us the same products for less, we would probably be wise, as a first step, to meet with our original supplier to see whether there might be some adjustment downward in the price of his product. Even if no such adjustment is offered, we probably will be well advised to stick with the original supplier, particularly if the saving involved in the switch is modest and/or if we have no firm evidence that the product being offered is of quality comparable to that from the current supplier.

(5) Once a production process has been brought to the middle level of excellence, where day-to-day crisis management is not the norm, we will have arrived at the region where we can increase our understanding of the process and attempt to improve it. Many American firms always operate in crisis mode. In such a mode, it is not easy to effect even the simplest of improvements, such as rearranging the production line to minimize the necessity of transporting intermediate products from building to building.

(6) In a modern industrial setting, a worker is a manager, perhaps of other workers, perhaps of machines. The worth of a worker to the company is based on the cumulative knowledge and experience of that worker. If the company is beyond the crisis mode of operation, then there is time to enhance worker knowledge by a well thought-out system of on the job training. There are few more profitable investments which the company might make than those spent in expanding the ability of the worker to deal effectively with the challenges of his current job and of those further down his career path.

(7) A foreman should not be a taskmaster, looking to improve production by stressing the workers. This is simply not the way to achieve improved performance, as the failure of economic systems based on such a paradigm indicates. A foreman is a manager of other managers, and his job is to coordinate the goals of the section and to entertain suggestions as to how these might best be achieved.

(8) So far from increasing productivity, arguments from management along the lines, "You have to increase your output by 10%, or you are likely to lose your job," are very counterproductive. It is not hard to figure out why. In addition to the natural frustrations associated with any job, management has just added another frustration: a demand that the worker increase his output without being shown how to do so. An upbeat environment in which management provides concrete help for the increase of productivity, without threats, has proved the best way to achieve maximum quality and productivity.

(9) Every organization requires some form of structure. Over time, however, changes in products and technology tend to render the boundaries between departments less and less relevant. One solution is to have reorganization of the firm every six months or so. Another, much more appropriate one, is to permit free flow of information and cooperation between departments. Once again, experience shows that such an approach does not impair organizational discipline and does enhance the total quality and productivity of the company.

(10) No society has been bigger on goals than that of the Soviet Bloc. No society has been bigger on posters, slogans and the like. The result has been disastrous. A principal duty of management is to show how productivity can been improved. It is unnecessary to emphasize the fact that productivity enhancements are crucial to the life of a firm and to those of the workers. Everyone knows that.

(11) This point is a reemphasis of the previous one. It is so important that it deserves special emphasis. Workers need to be shown how to

perform better. Simple demands that they do so are generally counter-productive.

(12) No human being wishes to be a cog in a machine. Every worker in an effectively managed firm is important. This is reality, not just a slogan. A major source of improvement in the production process is discoveries by workers as to how things might be done better. Such discoveries will be encouraged if those who make the discoveries are given full credit.

(13) Beyond on the job training, a firm is well advised to enhance the ability of workers to improve their skills by in house courses.

(14) A quality control program which is treated by top management as a pro forma activity is not likely to be effective. Top management needs to be convinced as to the importance of the program and needs to take as much a lead in it as they do in marketing and administration.

1.7   QC Misconceptions, East and West

Problems in production are seldom solved by general broad spectrum exhortations for the workers to do better. An intelligent manager is more in the vein of Sherlock Holmes than that of Norman Vincent Peale. This rather basic fact has escaped the attention of most members of the industrial community, including practitioners of quality control. To give an example of this fact, we note the following quotation from the Soviet quality control expert Ya. Sorin [5]:

> *Socialist competition is a powerful means of improving the quality of production. Labour unions should see to it that among the tasks voluntarily assumed by brigades and factories ... are included those having to do with improving quality control .... In the course of socialist competition new forms of collective fight for quality control are invented. One such form is a "complex" brigade created in the Gorky car factory and having as its task the removal of shortcomings in the design and production of cars. Such "complex brigades" consist of professionals and qualified workers dealing with various stages of production. The workers who are members of such brigades are not released from their basic quotas. All problems are discussed after work. Often members of such*

brigades stay in factories after work hours to decide collec-
tively about pressing problems.

In the above, we note the typical sloganeering and boosterism which is
the hallmark of a bad quality control philosophy. The emphasis, usually,
of bad quality control is to increase production by stressing the workers.
The workers are to solve all problems somehow by appropriate attitudinal
adjustment. Quality control so perceived is that of a man juggling balls,
and is doomed to failure.

Of course, Soviet means of production are proverbial for being bad.
Surely, such misconceptions are not a part of American quality control?
Unfortunately, they are. Let us consider an excerpt from a somewhat
definitive publication of the American Management Association's *Zero
Defects: Doing It Right the First Time* ([7], pp 3-9):

> *North American's PRIDE Program is a positive approach to*
> *quality that carries beyond the usual system of inspection. It*
> *places the responsibility for perfection on the employee con-*
> *cerned with the correctness of his work. Its success depends*
> *on the attitude of the individual and his ability to change*
> *it. This program calls for a positive approach to employee*
> *thinking .... A good promotional campaign will help stimu-*
> *late interest in the program and will maintain interest as the*
> *program progresses. North American began with a "teaser"*
> *campaign which was thoughtfully conceived and carried out*
> *to whet the interest of employees prior to the kick-off day ....*
> *Various things will be needed for that day – banners, posters,*
> *speeches, and so forth.... Many of the 1,000 companies having*
> *Zero Defects programs give employees small badges or pins*
> *for signing the pledge cards and returning them to the pro-*
> *gram administrator. Usually these badges are well received*
> *and worn with pride by employees at all times on the job*
> *.... In addition, it is good to plan a series of reminders and*
> *interest-joggers and to use promotional techniques such as is-*
> *suing gummed stickers for tool boxes, packages and briefcases.*
> *Coffee cup coasters with printed slogans or slogans printed on*
> *vending machine cups are good. Some companies have their*
> *letterhead inscribed with slogans or logos which can also be*
> *printed on packing cases.*

There is little doubt that an intelligent worker confronted with the
Zero Defects program could think of additional places where manage-

ment could stick their badges, pins and posters. A worker, particularly an American worker, does not need to be tricked into doing his best. In a decently managed company, employee motivation can be taken as a given. But a human being is not a flywheel. The strong point of humans is their intelligence, not their regularity. Machines, on the other hand, can generally be counted on for their regularity. The correct management position in quality control is to treat each human worker as an intelligent, albeit erratic, manager of one or more regular, albeit nonreasoning, subordinates (machines). Thus, human workers should be perceived of as managers, problem finders and solvers. It is a misunderstanding of the proper position of human workers in our high-tech society which puts many industries so much at the hazard.

1.8 White Balls, Black Balls

We begin with a simple example. In Figure 1.4, we show ten lots of four balls each. Each ball in lots 1 through 10 is white. There is no evidence to suggest that the process which produced the balls in any of the lots is different from any of the others.

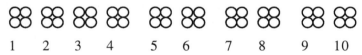

| | 1 | | 2 | | 3 | | 4 | | 5 | | 6 | | 7 | | 8 | | 9 | | 10 |

Figure 1.4. Perfect White Ball Production.

In Figure 1.5, we show ten lots of four balls each. We note that in nine of the lots, the balls are all white. In the tenth lot, all the balls are black. There is no doubt that the balls come from lots which are significantly different in terms of color. The production process for lot 10 appears significantly different from that which produced the balls in lots 1 through 9.

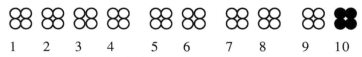

| | 1 | | 2 | | 3 | | 4 | | 5 | | 6 | | 7 | | 8 | | 9 | | 10 |

Figure 1.5. White Ball Production With One Bad Lot.

Next, in Figure 1.6, we show ten lots of four balls. In the tenth lot, one of the balls is black, and it would appear likely that the manufacturing process used to produce the balls is different in lot 10 than in the other lots.

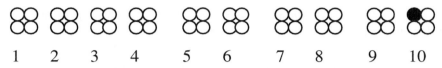

Figure 1.6. White Ball Production: Trouble In Tenth Lot.

In Figure 1.7, the situation becomes more ambiguous. Black balls appear in each of the lots except the fourth. White balls appear in each of the lots except for the eighth.

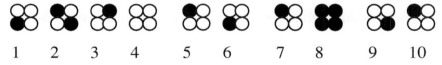

Figure 1.7. White Ball Production: Troubles Throughout.

We could, of course, say that the manufacturing process behaved quite differently in lot 4 than it did in lot 8. But if we are attempting to characterize changes across the manufacturing process, without knowing any more than the data before us, we would probably note that out of a total of 40 balls, 13 are black and 27 are white, and that there seems to be a random process at work by which black and white balls are produced in random fashion from lot to lot, with around 13/40 of these black and 27/40 white. In other words, there appears to be a difference in the conclusions we draw in Figures 1.4, 1.5, 1.6 and 1.7. In Figure 1.4, where the balls are all of the same color, we have no reason to doubt that the same production process was used for each of the lots. In Figure 1.5, there appears to be a big difference in the production process for lot 10, where the balls are all black. In Figure 1.6, we note one black ball in the tenth lot, and state, without a great deal of confidence, that we believe the production process was different for lot 10 than for lots 1 through 9. In Figure 1.7 we see such a variability in the process, that we are inclined to guess that each of the ten lots was produced by the same variable production process.

In the consideration of Figures 1.4 - 1.7, we see an apparent paradox. In Figure 1.4, which produced only white balls, we believe that each of the ten lots was produced by the same process. In Figure 1.7, which exhibited a widely varying number of white and black balls, there appeared no reason to question that the ten lots were each produced by a common process. In Figure 1.5, where one of the lots consisted exclusively of black balls, it appeared that lot 10 was probably produced by a different process than that of the lots 1-9. Figure 1.6 is, perhaps, the most ambiguous. The one black ball in lot 10 may be indicative of a basic change in the production process, or it may simply be the result of a small level of variability across the production process in all the lots.

It is interesting to note that the most uniform figure, Figure 1.4, and the most variable, Figure 1.7, seem to indicate that in the two figures, the production process does not change from lot to lot. Figure 1.6, which produced all white balls except for one black ball, is, nonetheless, more likely to have exhibited a change in the production process than the highly variable Figure 1.7.

The above example indicates a basic concept of statistical process control. In SPC, we examine the process as it is, rather than as we wish it to be. We might desire to produce only white balls. However, we must deal with the production record before us. In Figure 1.7, the performance is disappointing if our goal is the production of exclusively white balls. But there appears little information in the data as to where the production problem occurs in time. The process is "in control" in the sense that it appears that the process, though variable, is not particularly more variable in one lot than in any other. Let us consider the situation in Figure 1.8. Here, there are four black balls out of 40. There appears no guide, just from the data before us, as to what we can do to improve the situation. In other words, there appears to be no "Pareto glitch" which we can use as a guide to improving the production system.

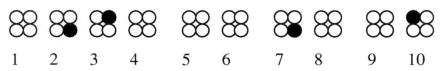

1   2   3   4   5   6   7   8   9   10

Figure 1.8. White Ball Production: Troubles Throughout.

Recall, however, the situation in Figure 1.5. Here, we also saw only four black balls out of forty. However, the fact that all these balls occurred in lot 10 may well give us some guide to improving the production process. Figure 1.5 exhibits a "Pareto glitch." Both Figures 1.5 and 1.8 exhibit the same proportion of black balls (10%). However, Figure 1.5 exhibits a situation where we can ask meaningful questions, such as, "Why is it that each of the nine lots was exclusively white except for lot 10, which exhibited only black balls?" Fortunately, the situation in Figure 1.5 is very common in industrial production. When we seek out the cause for the "Pareto glitch" we will probably be able to fix it.

Unfortunately, black balls are frequently produced as well, and each black ball is equally bad. In the real world, things are seldom black and white. Good items are not color coded as white, bad ones not coded as black. Generally, we have to rely on measurements to determine how we are doing, and measurements do not so easily fall into "good" and "bad" categories as did the white and black balls recently considered.

The previous example concerning the production of white balls is a study in black and white. We are trying to produce white balls, and any white ball is equally perfect. The real world situation of quality control frequently starts out as a "white ball, black ball" dichotomy. In a medical setting, a patient who survives a heart bypass operation and returns home is a "white ball." One who dies before getting back home is a "black ball." But as the process is refined, we will tend to use more sophisticated measurements than life and death. For example, we might consider also heart function one month after surgery as measured against some standard of performance. Length of postoperative stay in hospital is another measure. The refinement of measurement beyond life and death is almost always desired as we go down the road of optimizing treatment for bypass patients. But it brings us into the real world of shades of gray.

In an industrial setting, statistical process control will generally have to start with an in place "quality assurance" protocol. Thus, in the manufacturing of bolts, there will be two templates representing the upper and lower limits of "acceptable" diameter. A bolt is satisfactory if it fits into the upper limit template and fails to fit into the lower limit template. As time progresses, however, quality control will move to measuring diameters precisely, so as to provide greater possibility for feedback to improve the production.

Table 1.1			
Lot	Bolt 1 smallest	Bolt 2	Bolt 3	Bolt 4 largest
1	9.93	10.04	10.05	10.09
2	10.00	10.03	10.05	10.12
3	9.94	10.06	10.09	10.10
4	9.90	9.95	10.01	10.02
5	9.89	9.93	10.03	10.06
6	9.91	10.01	10.02	10.09
7	9.89	10.01	10.04	10.09
8	9.96	9.97	10.00	10.03
9	9.98	9.99	10.05	10.11
10	9.93	10.02	10.10	10.11

In Table 1.1, we show measurements of thicknesses in centimeters of 40 bolts in ten lots of four each. Our goal is to produce bolts of thickness 10 centimeters. We have arranged the bolts in each of the ten lots from the smallest to the largest. In Figure 1.9, we display the data graphically. There does not appear to be any lot which is obviously worse than the rest.

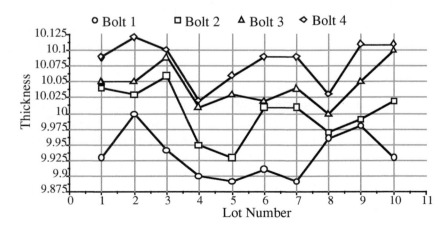

Figure 1.9. Thicknesses In Ten Lots Of Bolts.

Table 1.2			
Lot	Bolt 1 smallest	Bolt 2	Bolt 3	Bolt 4 largest
1	9.93	10.04	10.05	10.09
2	10.00	10.03	10.05	10.12
3	9.94	10.06	10.09	10.10
4	9.90	9.95	10.01	10.02
5	9.89	9.93	10.03	10.06
6	9.91	10.01	10.02	10.09
7	9.89	10.01	10.04	10.09
8	9.96	9.97	10.00	10.03
9	9.98	9.99	10.05	10.11
10	10.43	10.52	10.60	10.61

On the other hand, in Table 1.2 and in Figure 1.10, we note the situation when .500 centimeters are added to each of the four observed measurements in lot 10. Simply by looking at Figure 1.10, most people would say that something about the process must have changed during the production of the tenth lot. This is an analog of the four black balls produced in the tenth lot in Figure 1.5.

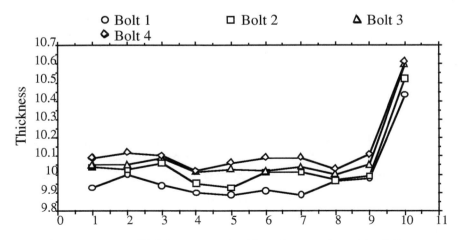

Figure 1.10. Thicknesses In Ten Lots: One Bad Lot.

Table 1.3				
Lot	Bolt 1	Bolt 2	Bolt 3	Bolt 4
	smallest			largest
1	9.93	10.04	10.05	10.09
2	10.00	10.03	10.05	10.12
3	9.94	10.06	10.09	10.10
4	9.90	9.95	10.01	10.02
5	9.89	9.93	10.03	10.06
6	9.91	10.01	10.02	10.09
7	9.89	10.01	10.04	10.09
8	9.96	9.97	10.00	10.03
9	9.98	9.99	10.05	10.11
10	9.93	10.02	10.10	10.61

In Table 1.3 and the corresponding Figure 1.11, we note the situation in which .500 centimeters have been added to only one of the measurements in lot 10. The situation here is not so clear as that in Figure 1.10. The one bad bolt might simply be occurring generally across the entire production process. One cannot be sure whether the process was degraded in lot 10 or not. We recall the situation in Figure 1.6, where there was only one black ball produced.

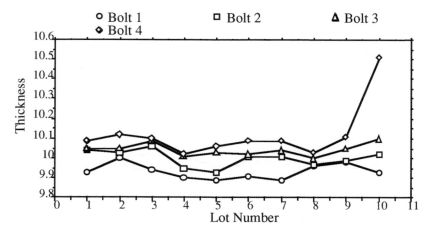

Figure 1.11. Thicknesses In Ten Lots: One Bad Bolt.

Table 1.4			
Lot	Bolt 1 smallest	Bolt 2	Bolt 3	Bolt 4 largest
1	9.93	10.04	10.05	10.59
2	10.00	10.03	10.55	10.62
3	9.94	10.06	10.09	10.60
4	9.90	9.95	10.01	10.02
5	9.89	9.93	10.03	10.56
6	9.91	10.01	10.02	10.59
7	9.89	10.01	10.04	10.59
8	10.46	10.47	10.50	10.53
9	9.98	9.99	10.05	10.61
10	9.93	10.02	10.10	10.61

Next, in Table 1.4 and the corresponding Figure 1.12, we observe a situation where .500 centimeters have been added to one bolt each in lots 1,3,5,6,7,9, and 10 and to two of the bolts in lot 2 and four of the bolts in lot 8. Note the similarity to Figure 1.7, where there were so many black balls distributed across the lots that it seemed unlikely that we could be confident in saying the process was performing better one place than another: it was generally bad but since equally bad everywhere, could not be said to be "out of control."

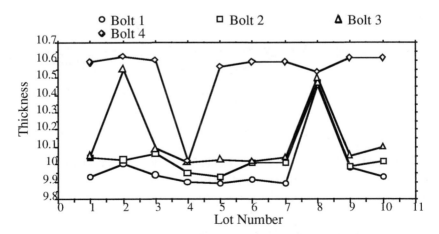

Figure 1.12. Thicknesses: System Generally Variable.

Table 1.5			
Lot	Bolt 1 smallest	Bolt 2	Bolt 3	Bolt 4 largest
1	9.93	10.04	10.05	10.09
2	10.00	10.03	10.05	10.62
3	9.94	10.06	10.09	10.60
4	9.90	9.95	10.01	10.02
5	9.89	9.93	10.03	10.06
6	9.91	10.01	10.02	10.09
7	9.89	10.01	10.04	10.59
8	9.96	9.97	10.00	10.03
9	9.98	9.99	10.05	10.11
10	9.93	10.02	10.10	10.61

In Table 1.5 and Figure 1.13, we show the situation where .500 centimeters have been added to one bolt each in lots 2,3,7, and 10. We note that unlike the situation in Figure 1.10, where we also had four bad bolts, but all in one lot, there seems to be little guide in Figure 1.13 as to any "Pareto glitch" in the production process.

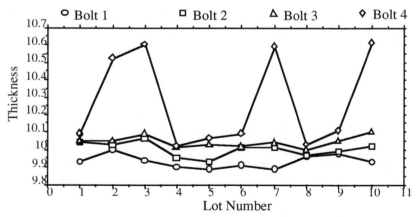

Figure 1.13. Thicknesses: Several Wild Bolts.

The Run Chart. Figures 1.9 - 1.13 are rather crowded. We have elected to show all of the forty thickness measurements. If there were 10 thicknesses measured for each of the ten lots, then the figures would be very crowded indeed. We need to come up with one figure for each lot which gives us a measure of the average thickness for the lot. The most common measure is the sample mean. Returning to Table 1.1, let us compute the sample mean for each lot. To obtain the average for lot 1, for example, we compute

$$\bar{x} = \frac{9.93 + 10.04 + 10.05 + 10.09}{4} = 10.028 \qquad (1.1)$$

Table 1.6					
Lot	Bolt 1 smallest	Bolt 2	Bolt 3	Bolt 4 largest	$\bar{x}$
1	9.93	10.04	10.05	10.09	10.028
2	10.00	10.03	10.05	10.12	10.050
3	9.94	10.06	10.09	10.10	10.048
4	9.90	9.95	10.01	10.02	9.970
5	9.89	9.93	10.03	10.06	9.978
6	9.91	10.01	10.02	10.09	10.008
7	9.89	10.01	10.04	10.09	10.008
8	9.96	9.97	10.00	10.03	9.990
9	9.98	9.99	10.05	10.11	10.032
10	9.93	10.02	10.10	10.11	10.040

We show the run chart of means of the ten lots of bolts from Table 1.1 (Figure 1.9) in Figure 1.14.

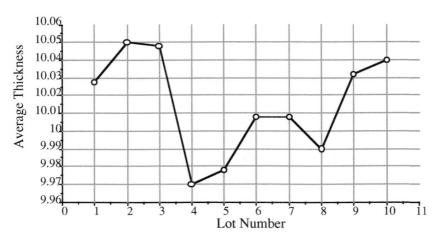

Figure 1.14. Run Chart of Means.

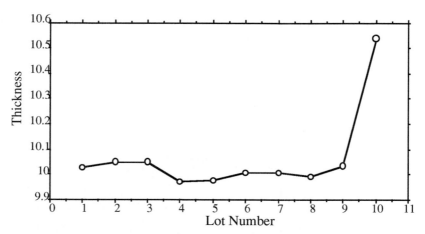

Figure 1.15. Thicknesses In Ten Lots Of Bolts.

We now compare the situation in Figure 1.14 with the run chart in Figure 1.15, based on the data from Table 1.2 graphically displayed in the corresponding Figure 1.10. We note the fact that simply using the run chart of means, it is clear that the production process changed dramatically in run 10 of Figure 1.15. In contrast, there does not appear to be any indication of such a dramatic change during any of the runs in Figure 1.14. The concept of the run chart is quite intuitive and predates the modern science of statistical process control by some decades. In many situations, a run chart of lot means will point out the "Pareto glitch" as surely as formal SPC charts. This simple chart should be viewed as the prototype of the more sophisticated techniques of statistical process control. If a firm does not employ anything more complex than run charting, it will be far better off than a firm that does no statistical process control at all.

Still there are situations where simply tracking the sample means from lot to lot is insufficient to tell us whether a "Pareto glitch" has occurred. For example, let us reconsider the data in Figure 1.10. If we subtract 2 centimeters from the lowest measurement in each lot, 1 centimeter from the next lowest measurement in each lot, then add 2 centimeters to the highest measurement in each lot, and 1 centimeter to the second highest measurement in each lot, then the lot averages do not change from those in Figure 1.15. On the other hand, if we plot this data, as we do in Figure 1.16, it is no longer so clear that the production process has changed in the tenth lot.

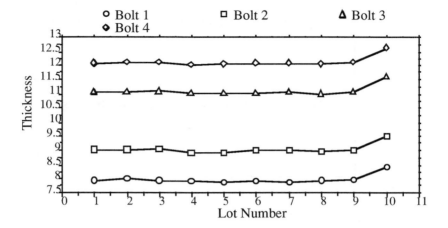

Figure 1.16. Thicknesses In Ten Lots Of Bolts.

So the two data sets, that in Figure 1.10 and that in Figure 1.16, have the same running mean charts, but the data in Figure 1.10 showed a production change for run 10, whereas the data in Figure 1.16 did not. The mean chart does not always tell the whole story.

Lot	Bolt 1 smallest	Bolt 2	Bolt 3	Bolt 4 largest	$\bar{x}$
1	9.93	10.04	10.05	10.09	10.028
2	10.00	10.03	10.05	10.12	10.050
3	9.94	10.06	10.09	10.10	10.048
4	9.90	9.95	10.01	10.02	9.970
5	9.89	9.93	10.03	10.06	9.978
6	9.91	10.01	10.02	10.09	10.008
7	9.89	10.01	10.04	10.09	10.008
8	9.96	9.97	10.00	10.03	9.990
9	9.98	9.99	10.05	10.11	10.032
10	7.93	9.02	11.10	12.11	10.040

Table 1.7

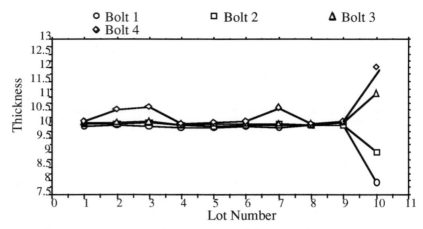

Figure 1.17. Ten Lots: Variable Thicknesses.

To give another example where the chart of running means fails to tell us everything, let us return to the data set of Table 1.1. We modify, in Table 1.7, the data for lot 10 by subtracting 2 centimeters from the smallest measurement, 1 centimeter from the second smallest, and then adding 2 centimeters to the largest measurement and 1 centimeter to the second largest.

We note that the mean for lot 10 has not changed. However as we observe the graph from Table 1.7 in Figure 1.17, we would be very much inclined to say the variability of the measurements in lot 10 is such that we would be inclined to conjecture a change in the production process has taken place.

Once again, in our example, we have referred to an industrial context. It is clear, however, that the same charts could be obtained (except for the units used) if lots comprised of 4 similar stores in similar locations and weekly sale volumes were measured for each store, or 4 similar offices of a travel agency were recorded as to their monthly output of a type, etc.

1.9 The Basic Paradigm of Statistical Process Control

We see that in addition to the average thickness in each lot of bolts, we need also to pay attention to the variability of the production process. A major task in statistical process control is to seek lots which exhibit behavior which is significantly different from those of the other lots. Once we find such lots, we will then be able to investigate what caused them to deviate from the norm. And it must be pointed out that our notion of "norm" in process control is not some predetermined standard, but rather the measurements associated with the great bulk of the items being produced.

Sometimes we will not be trying to "fix" an imperfection of the production process at a given time. For example, we may find that a lot exhibits tensile strength 20% above that of the other lots. In such an instance, we will be trying to find out why the nonstandard lot was so much better than the norm.

This is the basic paradigm by which we proceed in SPC:
(1) Find a Pareto glitch (a nonstandard lot);
(2) Discover the causes of the glitch;
(3) Use this information to improve the production process.

This basic paradigm is as effective as it is simple. Most people who hear it explained to them for the first time feel that it is simply too good to be true. The paradigm of SPC works and the fact that it works makes for one of the most cost effective pathways to excellence in modern industrial production. We note again how little statistical process control really has in common with the end inspection schemes of quality assurance. Both paradigms use similar statistical tools. However, the goal of quality assurance is to remove bad items at the end of production, before they get to the customer. In order to be really effective, QA requires 100% inspection. And the effectiveness simply consists in removing the results of flawed procedures, not in correcting them. It is this fundamental difference in the philosophies of quality assurance and quality improvement (i.e., statistical process control) that needs to be grasped if one is to institute modern quality control. The quality control teams in all too many enterprises have the false notion that they are on the cutting edge of QC technique, when, in actuality, they are little more advanced than the industrial commissars of Stakhanov's day.

1.10 Basic Statistical Procedures in Statistical Process Control

We have seen that the notion of variability is key in our search for Pareto glitches. We have yet to give concrete procedures for dealing with variability. We do so now. There is an obvious rule for deciding which lots to investigate. We could simply investigate them all. Naturally, this is neither practical nor desirable. Recall furthermore that we are not dealing with issues which are as clear as finding black balls when we are attempting to produce white ones. We will decide about whether a lot is unusual or not based on measurements. This means that sometimes we will incorrectly pick a lot as indicating a divergence from the usual production process when, in fact, it really represents no such divergence. In other words, we must face the prospect of "false alarms."

One standard is to pick a rule for declaring a lot to be out of the norm so that the chance of a "false alarm" is roughly one in five hundred. This generally keeps the number of lots whose production is to be investigated within manageable bounds and still gives us the kind of stepwise progression to excellence, which is the pay-off of an effective SPC strategy.

Postponing going into mathematical details to Chapter 3, we develop below the essential rules for carrying out statistical process control. We return to the data in Table 1.2 (and Figure 1.10).

Table 1.8						
Lot	Bolt 1 smallest	Bolt 2	Bolt 3	Bolt 4 largest	$\bar{x}$	s
1	9.93	10.04	10.05	10.09	10.028	.068
2	10.00	10.03	10.05	10.12	10.050	.051
3	9.94	10.06	10.09	10.10	10.048	.074
4	9.9	9.95	10.01	10.02	9.970	.056
5	9.89	9.93	10.03	10.06	9.978	.081
6	9.91	10.01	10.02	10.09	10.008	.074
7	9.89	10.01	10.04	10.09	10.008	.085
8	9.96	9.97	10.00	10.03	9.990	.032
9	9.98	9.99	10.05	10.11	10.032	.060
10	10.43	10.52	10.60	10.61	10.540	.084

We have already explained how the sample mean $\bar{x}$ is computed. For lot 10, it is

$$\bar{x} = \frac{10.43 + 10.52 + 10.60 + 10.61}{4}. \tag{1.2}$$

Now in order to find an unusual lot, we must find the average for all lots via

$$\bar{\bar{x}} = \frac{10.028 + 10.05 + \ldots + 10.54}{10}. \tag{1.3}$$

The standard deviation, s, is a bit more complicated to obtain. It is a measure of the spread of the observations in a lot about the sample mean. To compute it, we first obtain its square, the sample variance

$$s^2 = \frac{(10.43 - 10.54)^2 + \ldots + (10.61 - 10.54)^2}{3} = .007056. \qquad (1.4)$$

(We observe that, whenever we are computing the sample variance, we divide by one less than the number of observations in the lot.) The sample standard deviation is then easily obtained via

$$s = \sqrt{s^2} = .084. \qquad (1.5)$$

We need to obtain a value of s which is the approximate average over all the sampled lot. We obtain this value via

$$\bar{s} = \frac{.068 + .051 + \ldots + .084}{10} = .0665. \qquad (1.6)$$

Now let us recall that our goal is to find a rule for declaring sample values to be untypical if we incorrectly declare a typical value to be untypical one time in 500. For the data in Table 1.2, we will generally achieve roughly this goal if we decide to accept all values between $10.0652 + 1.628(.0665) = 10.173$ and $10.0652 - 1.628(.0665) = 9.957$.

In Figure 1.18, we show the lot means together with the upper and lower control limits. It is seen how lot 10 fails to fall within the limits and hence we say that the production system for this lot is out of control. Thus we should go back to the production time record and see whether we can find out the reason for lot 10 to have gone nonstandard. We note from Table 1.9 below the multiplication factors used in the above computation of the limits. We see that the factor 1.628 comes from the A_3 column for a lot size of 4. More generally, the acceptance interval on the mean is given by

$$\bar{\bar{x}} \pm A_3(n)\bar{s}. \qquad (1.7)$$

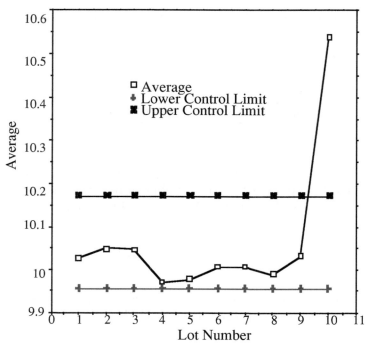

Figure 1.18. Mean Control Chart.

This is generally the most important of the quality control charts.

n	$B_3(n)$	$B_4(n)$	$A_3(n)$
2	0.000	3.267	2.659
3	0.000	2.568	1.954
4	0.000	2.266	1.628
5	0.000	2.089	1.427
6	.030	1.970	1.287
7	.118	1.882	1.182
8	.185	1.815	1.099
9	.239	1.761	1.032
10	.284	1.716	.975
15	.428	1.572	.789
20	.510	1.490	.680
25	.565	1.435	.606

Table 1.9

It is not always the case that the mean control chart will detect a system out of control. Let us consider, for example, the data in Table 1.7.

Table 1.7 (enhanced)					
Lot	Bolt 1 smallest	Bolt 2	Bolt 3	Bolt 4 largest	$\bar{x}$	s
1	9.93	10.04	10.05	10.09	10.028	.068
2	10.00	10.03	10.05	10.52	10.050	.051
3	9.94	10.06	10.09	10.60	10.048	.074
4	9.90	9.95	10.01	10.02	9.970	.056
5	9.89	9.93	10.03	10.06	9.978	.081
6	9.91	10.01	10.02	10.09	10.008	.074
7	9.89	10.01	10.04	10.59	10.008	.085
8	9.96	9.97	10.00	10.03	9.990	.032
9	9.98	9.99	10.05	10.11	10.032	.060
10	7.93	9.02	11.10	12.11	10.040	1.906

We compute the average mean and standard deviation across the table via

$$\bar{\bar{x}} = \frac{10.028 + 10.05 + \ldots + 10.04}{10} = 10.015 \qquad (1.8)$$

and

$$\bar{s} = \frac{.068 + .051 + \ldots + 1.906}{10} = .249 \qquad (1.9)$$

respectively. The control limits on the mean are given by LCL = 10.015 - 1.628(.249) =9.610, and UCL = 10.015 + 1.628(.249) = 10.420. Plotting the means and the mean control limits in Figure 1.19, we note that no mean appears to be out of control.

We recall from Figure 1.17 that there is a clear glitch in lot 10, which does not show up in the mean control chart in Figure 1.19. We need to develop a control chart which is sensitive to variability in the standard deviation. The control limits on the standard deviation are given by LCL = 0 (.249) = 0.0 and UCL = 2.266 (.249) = .564.

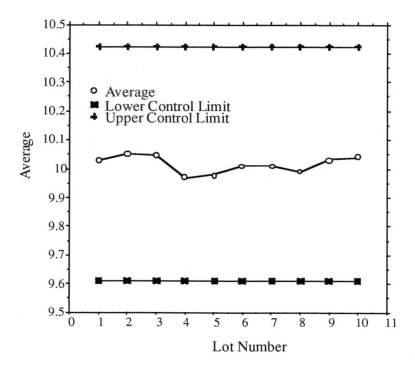

Figure 1.19. Mean Control Chart.

As we note in Figure 1.20, the standard deviation in lot 10 is clearly out of control. More generally, the lower control limit for the standard deviation is given by

$$\text{Lower Control Limit} = B_3(n)\bar{s} \qquad (1.10)$$

and the upper control limit is given by

$$\text{Upper Control Limit} = B_4(n)\bar{s}. \qquad (1.11)$$

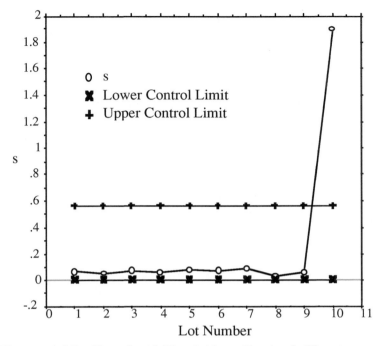

Figure 1.20. Standard Deviation Control Chart.

1.11   Acceptance Sampling

Although it is generally a bad idea to base a quality control program on an acceptance-rejection criterion, in the most primitive stages of quality control such a strategy may be required. Unfortunately, starting with World War II, a preponderance of quality control in the United States has been based on acceptance-rejection.

The idea in applying statistical process control to an acceptance-rejection scenario is that, instead of looking at measurements, we simply look at the variability in the proportion of items defective. Let us return to the data in Table 1.2. Suppose that an item is considered defective if it is smaller than 9.92 or greater than 10.08. In order to apply a statistical process control procedure to data which is simply characterized as defective or not, we note the proportion of defectives in each lot. The overall proportion of defectives is our key statistic. For example in Table 1.2, the overall proportion is given simply by

$$\hat{p} = \frac{.25 + .25 + .50 + \ldots + 1.00}{10} = .375. \qquad (1.12)$$

Table 1.2 (enhanced)				
Lot	Bolt 1 smallest	Bolt 2	Bolt 3	Bolt 4 largest	p
1	9.93	10.04	10.05	10.09	.25
2	10.00	10.03	10.05	10.12	.25
3	9.94	10.06	10.09	10.10	.50
4	9.90	9.95	10.01	10.02	.25
5	9.89	9.93	10.03	10.06	.25
6	9.91	10.01	10.02	10.09	.50
7	9.89	10.01	10.04	10.09	.50
8	9.96	9.97	10.00	10.03	0.00
9	9.98	9.99	10.05	10.11	.25
10	10.43	10.52	10.60	10.61	1.00

To obtain the control limits, we use

$$\text{LCL} = \hat{p} - 3\sqrt{\frac{\hat{p}(1 - \hat{p})}{n}} \tag{1.13}$$

and

$$\text{UCL} = \hat{p} + 3\sqrt{\frac{\hat{p}(1 - \hat{p})}{n}}, \tag{1.14}$$

respectively. It is crucial to remember that n *is the size of the lot being tested for typicality.* That means that here $n = 4$. Naturally, the approximation to normality for such a small sample size is very unrealiable. Nevertheless, if we perform the standard normal theory computation, we find that following are given for the two control limits.

$$\text{UCL} = .375 + 3\sqrt{\frac{.375(1 - .375)}{4}} = 1.101, \tag{1.15}$$

and

$$\text{LCL} = .375 - 3\sqrt{\frac{.375(1 - .375)}{4}} = -.3511. \tag{1.16}$$

Thus there simply is no way to reject a lot as untypical, since the proportion of failures must alway be between 0 and 1. But what if we use the precise value obtained from the binomial distribution itself? We recall that, if X is the number of failed items in a lot,

$$P(X) = \binom{n}{X} p^X (1 - p)^{n-X}. \tag{1.17}$$

In the current example, the probability that all four items fail if $p = .375$ is almost .02. It might be appropriate under the circumstances to accept an increased possibility of false alarms (declaring the lot is untypical when it is, in fact, typical) of .02. Such compromises are frequently necessary when using rejection data. In such a case, we might wish to consider the tenth lot as being "out of control." The point to be made here is that the use of failure data is a very blunt instrument when compared to use measurement data. This is particularly the case when, as frequently happens, the lot sizes are very small.

1.12   The Case for Understanding Variation

Let us summarize briefly what has been said so far. First of all, one has to agree that managing by focusing on end results or on end inspection has nothing to do with quality improvement. As one outstanding SPC consultant has succinctly put it: *such a management is like driving along the road by watching the white line in the rear-view mirror.* Instead, one has to look upstream, that is to shift the focus to the source of quality, this source being the design of product and processes which lead to the product. Indeed, there is no other way to influence quality. So understood, quality improvement has to be arranged in an orderly and economically sound way. Not incidentally, the very first (and up-to-date although published in 1931) guide on the subject, written by Walter Shewhart, bears the title *Economic Control of Quality of Manufactured Product.* But then, this amounts to just one more reason that it must be the responsibility of organization's management, from its CEO's to foremen on the shop floor, to implement both overall strategy and everyday activity for quality improvement. And more than that, it is in particular necessary for the CEO's to better their understanding of what process control is and to improve the ways they manage the organization. We shall dwell more on these last issues in the sequel.

Having thus agreed that the strategy for continuous improvement should encompass all of the organization and that it has to form an integrated and selfconsistent system, one has to decide what is the real core of the strategy to be implemented. We have argued in the preceding Sections that it is a wisely arranged reduction of variability or variation within processes and within the organization as a whole.

Variation within any process, let alone a system of processes, is inevitable. It can be small, seemingly negligible or large, it can be easy to measure or not, but it always **is**. It adds to complexity and inefficiency of a system if it is large. Indeed, the larger the variability of subprocesses, the more formidable task it becomes to combine their outputs into one product (or satisfactory service). Under such circumstances, it is particularly transparent that managing by results is a hindrance on the way to improvement, as Deming's Fourteen Points amply demonstrate. On the other hand, reduction of variability within subprocesses, i.e., looking upstream and working on improvement of the sources of quality, does not loose the aim of the whole system from sight. To the contrary, it helps achieve the overall aim and leads to reduction of system's complexity.

In turn, one has to address the issue of how to arrange for wise reduction in variation of a process. Following Walter Shewhart, we have argued that there are two qualitatively different sources of variability: *common cause variation*, and *special cause variation*. As regards the latter, it is this cause of variation which leads to Pareto glitches. We have already discussed at some length how to detect them and how to react to them, in particular in Sections 1.9 and 1.10. Pareto glitches, also called *signals*, make the process unpredictable. We often say that the process is then *out of statistical control* due to *assignable* causes. What we in fact mean by the latter term is intentional – we want these causes to be assignable indeed, for we want to find and eliminate them in order to improve the process (fortunately, they most often can be found and hence removed).

As we already know, a process which has been brought to stability or, in other words, to a state of *statistical control*, is not subject to special cause variation. It is subject only to common cause or inherent variation, which is always present and cannot be reduced unless the process itself is changed in some way. Such a process is predictable, although it does not have to perform satisfactorily. For instance, both the white ball production as depicted in Figure 1.7 and the bolts production as observed in Table 1.4 and Figure 1.12 are in control processes. Both processes are next to disastrous, but they are such consistently and stably. In each of these two situations, we are faced with a highly variable but consistent production process with no sudden change in its performance. It is possible, e.g., that the bolts are turned on an old and faulty lathe. In service industry or, e.g., in processing invoices in a company, high common cause variation is often due to inadequacy of adopted procedures. In any case, if no signal due to a special cause is revealed when running a process, it is wrong to treat a particular measurement or lot as a signal and waste time and effort on finding its alleged cause. In the situations mentioned, proper action would be to change the process itself, i.e., the lathe or, respectively, the adopted procedures.

Another example of an in control process, which exhibits only natural or inherent variability, seems to be provided by data from Table 1.1, shown in Figure 1.9 and summarized into the corresponding run chart of means in Figure 1.14 (we urge the Reader to calculate control limits for both the mean and standard deviation control charts for these data to see if the process is indeed stable). This time, the process variation is much smaller than in the case of data from Table 1.4, which is not to say that further improvement of the process would not be welcome.

Assuming that the upper control limit for lot means for data from Table 1.1 lies above the value of 10.05 (as it most likely does), it would be wrong to treat means larger than, say, 10.04 as atypical signals. However mistaken, a practice of this sort is surprisingly popular among managements. One can hardly deny that the executives like to set goals as "the lot mean thickness of bolts should not be larger than 10.04" (or, more seriously, that drop in monthly sales should not exceed $a\%$, amount of inventory should not exceed b units, monthly late payments should amount to less than $c\%$, off budget expenditures be less than $d\%$, etc.) with no regard to common cause variation of the process in question whatsoever. Worst of all is the fact that lots 2 and 3 in Figure 1.14, if flagged as atypical signals, will not only lead to waste of energy on finding an alleged special cause, but that such a cause will sooner or later be declared found. As a rule, this misguided policy brings an opposite result to that intended — after removal of the alleged special cause (which often amounts to policing the workforce), common cause variation becomes larger.

Reacting to common causes as if they were special ones is known as *tampering* or the *error of the first kind*, in contrast to the *error of the second kind*, which consists in disregarding signals and their special causes, and treating the latter as if they were common causes.

All in all, control charts have been found an excellent and in fact irreplaceable means to find whether a process is in control and to guide the action of bringing a process to statistical control when needed. If stable, the process reveals only common cause variation, measured by *process capability* (see Section 3.8), whose reduction requires improvement of the process itself. As is now clearly seen, the two actions, one of bringing a process into stability and another of improving a stable process, are qualitatively very different.

When summarizing in Section 1.5 the basic paradigm for quality, as perceived already by Whitney and Ford, we have combined the two actions mentioned into one whole. Upon closer inspection, one can note that the first two points given there refer to a process at the design stage and the third to that process at the production stage. Actually, while the third point refers to improving an out of control process, the first two can be claimed to pertain to a stable one which is to be redesigned to bring further improvement. The so-called *Plan – Do – Study – Act (PDSA) cycle*, developed by Shewhart and later refined by Deming, provides a particularly elegant prescription for improvement of a stable process (see Figure 1.21). In order to make a change, one has first to *Plan* it. It is recommended that such a plan be based, if possible, on a mathematical model of the process under scrutiny. Some optimization techniques, in particular for regression models, as well as issues of design of experiments are treated in Chapter 6. After the change has been planned, *Do* it, if possible on a small scale. The next step is to *Study* the effects of change. Here one has to remember that conclusions concerning the effects can be drawn only after the process in question has again reached a stable state. Finally, one can, and should, *Act* accordingly: adopt the change if it has proved successful or re-run the pilot study under different conditions or abandon the change and try some other. While it sounds simple, the methodology presented is not only most powerful but the only reasonable one. In fact, it is a basis for the *spiral of continuous improvement*. At each level of an organization, it is to be used repeatedly, as one one of the two core steps in an iterative process of quality improvement. Each new process has to be brought to statistical control, and this is one of the two core steps, to be followed by the implementation of the PDSA cycle, which is again to be followed by bringing the improved process to stability. It is often said that the spiral described is obtained by successive turns of the quality wheel. (The Shewhart–Deming cycle is sometimes abbreviated as PDCA, with *Check* replacing *Study* term.)

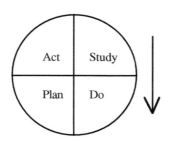

Figure 1.21.Shewhart–Deming PDSA Cycle.

Let us conclude this Section by emphasizing once again that without proper understanding of variability one's efforts to improve quality are doomed to failure (even incredible luck can help one once or twice, but this luck cannot change the overall outcome for a company). Generally, without such understanding, one is doomed either to tampering or to committing errors of the second kind. We are all sinful and all like reports on sales, inventory, productivity, stocks, etc., if not in the form of monthly reports for company's CEOs, then in the press. Such reports give numbers for the current month and the same month one year ago, and, if we are lucky, the previous month. They are next to useless, harmless if we read them for the sake of curiosity only, but counterproductive if they are a basis for business decisions. The fact of the matter is that numbers, seen apart from their context, are meaningless. In the situations hinted to, such context is provided by the time series of data from consecutive months. It is only because the CEO keeps past data, more or less accurately, in memory that his or her judgments if a particular number reflects common or special cause variation are not patently wrong. And let us add one more remark: it is not only that the data should be seen in their context as a time series, but also that the data investigated should mostly refer to processes, not to end results.

1.13   Statistical Coda

Our basic strategy in the detection of Pareto glitches is to attempt to answer the following basic question: *Do the items in a given lot come from the same statistical distribution as the items in a previous string of "typical lots"?* In order to answer this question, it is possible to use a variety of very powerful mathematical tools, including nonparametric density estimation. However, in SPC, we must always be aware of the fact that it is not essential that we detect every "nontypical" (and hence "out of control") lot. Consequently, we will frequently find it satisfactory to answer the much more simple question: *Is it plausible to believe that the sample mean of the items in a lot is consistent with the average of sample means in the previous string of lots?* Now, we recall from the Central Limit Theorem (see Appendix B) that for n large

$$\frac{\bar{x} - \mu}{\sigma/\sqrt{n}} \approx Z \qquad (1.18)$$

where Z is a normal random variable with mean 0 and variance 1. Z will lie between 3 and -3 roughly 99.8% of the time. Generally speaking, if the number of previous lots is lengthy, say 25 or more, the average of the averages of those sample lots, $\bar{\bar{x}}$, will give us an excellent estimate of μ. The average $\bar{s}$ of the sample standard deviations from the previous lots is a good estimate of σ when multiplied by a unbiasing factor $a(n)$ (see (3.36)). So, a natural "in control" interval for $\bar{x}$ is given by

$$LCL(\bar{x}) = \bar{\bar{x}} - 3a(n)\frac{\bar{s}}{\sqrt{n}} \le \bar{\bar{x}} + 3a(n)\frac{\bar{s}}{\sqrt{n}} = UCL(\bar{x}). \qquad (1.19)$$

The multiplication of the unbiasing term $a(n)$ by 3 and the division by $\sqrt{n}$ gives as the acceptable interval (see Table 1.9)

$$\bar{x} \pm A_3(n). \qquad (1.20)$$

Similarly, a sample standard deviation also has the property that, if n is large

$$\frac{s - E(s)}{\text{standard deviation of } s} \approx Z \qquad (1.21)$$

where Z is a normal random variable with mean 0 and variance 1. In (3.58), we show that this enables us to find an interval in which s should fall roughly 99.8% of the time if the items in the lot have the same underlying variance as those in the string of lots from which we estimate $\bar{s}$. Referring to Table 1.9, this yields us as the "in control" interval for a new lot s where lot size is n:

$$LCL(s) = B_3(n)\bar{s} \le s \le B_4(n)\bar{s} = UCL(s). \qquad (1.22)$$

Finally, for acceptance-rejection data, the Central Limit Theorem also gives us that for large lot size n the proportion of defectives in a lot $\hat{p}$

$$\frac{\hat{p} - p}{\sqrt{\frac{p(1-p)}{n}}} = Z \qquad (1.23)$$

where Z is a normal random variable with mean 0 and variance 1. If the number of lots in a string of prior lots in which the proportion of defectives has been estimated, then their average $\hat{\bar{p}}$ is a good estimate for p. So we have

$$LCL(\hat{p}) = \hat{\bar{p}} - 3\sqrt{\frac{\hat{\bar{p}}(1 - \hat{\bar{p}})}{n}} \le \hat{p} \le \hat{\bar{p}} - 3\sqrt{\frac{\hat{\bar{p}}(1 - \hat{\bar{p}})}{n}} = UCL(\hat{p}). \qquad (1.24)$$

We should never lose sight of the fact that it is the large number of lots in the string prior to the present one which enables us to claim that

$\bar{\bar{x}}$ is a good estimate of μ, $a(n)\bar{s}$ is a good estimate of σ, and $\hat{p}$ is a good estimate of p. But the n in the formulae for the control limits is always the size of an individual lot. Consequently, the normal approximation may be far from accurate in determining the control limits to give a one chance in 500 of rejecting an "in control" lot. When we use the control limits indicated above, we expect to reject an in control item at a proportion different from the nominal 0.2% probability which would apply if the CLT was in full force. This is particularly the case in the acceptance-rejection situation. Nevertheless, the plus or minus 3 sigma rule generally serves us very well. Our task is to find lots which are untypical, so that we can backtrack for possible flaws in the system and correct them. To achieve this task, for one dimensional testing, the old plus or minus 3 sigma rule generally is extremely effective.

References

[1] Deming, W. E. (1982). *Quality, Productivity and Competitive Position.* Center for Advanced Engineering Studies, pp. 16-17.

[2] Falcon, W. D. (1965). *Zero Defects: Doing It Right the First Time,* American Management Association, Inc.

[3] Ford, H. (1926). *My Life and Work.* Sydney: Cornstalk Press, p. 273.

[4] Mann, N. R. (1985). *The Keys to Excellence: the Story of the Deming Philosophy,* Los Angeles: Prestwick Books.

[5] Sorin, Ya. (1963). *On Quality and Reliability: Notes for Labor Union Activists.* Moscow: Profizdat, p. 120.

[6] Thompson, J. R. (1985). "American quality control: what went wrong? What can we do to fix it?" *Proceedings of the 1985 Conference on Applied Analysis in Aerospace, Industry and Medical Sciences,* Chhikara, Raj, ed., Houston: University of Houston, pp. 247-255.

[7] Todt, H. C. "Employee motivation: fact or fiction," in *Zero Defects: Doing It Right the First Time,* Falcon, William, ed., New York: American Management Association, Inc., pp. 3-9.

Problems

Problem 1.1. The following are the data from 24 lots of size 5.

Lot	measurements	Lot	measurements
1	995, 997, 1002, 995, 1000	13	1002, 1004, 999, 996, 1000
2	990, 1002, 997, 1003, 1005	14	1003, 1000, 996, 1000, 1005
3	1003, 1005, 998, 1004, 995	15	996, 1001, 1006, 1001, 1007
4	1002, 999, 1003, 995, 1001	16	995, 1003, 1004, 1006, 1008
5	1001, 996, 999, 1006, 1001	17	1006, 1005, 1006, 1009, 1008
6	1004, 1001, 998, 1004, 997	18	996, 999, 1001, 1003, 996
7	1003, 1002, 999, 1003, 1004	19	1001, 1004, 995, 1001, 1003
8	1001, 1007, 1006, 999, 998	20	1003, 996, 1002, 991, 996
9	999, 995, 994, 991, 996	21	1004, 991, 993, 997, 1003
10	994, 993, 991, 993, 996	22	1003, 997, 998, 1000, 1001
11	994, 996, 995, 994, 991	23	1006, 1001, 999, 996, 997
12	994, 996, 998, 999, 1001	24	1005, 1000, 1001, 998, 1001

Determine whether the system is in control.

Problem 1.2. The following are the data from 19 lots of size 5.

Lot	measurements	Lot	measurements
1	831, 839, 831, 833, 820	11	832, 836, 825, 828, 832
2	829, 836, 826, 840, 831	12	834, 836, 833, 813, 819
3	838, 833, 831, 831, 831	13	825, 850, 831, 831, 832
4	844, 827, 831, 838, 826	14	819, 819, 844, 830, 832
5	826, 834, 831, 831, 831	15	842, 835, 830, 825, 839
6	841, 831, 832, 831, 833	16	832, 831, 834, 831, 833
7	816, 836, 826, 822, 831	17	827, 831, 832, 828, 826
8	841, 832, 829, 828, 828	18	838, 830, 822, 835, 830
9	831, 833, 833, 831, 835	19	841, 832, 829, 828, 828
10	830, 838, 835, 830, 834		

Determine whether the system is in control.

Problem 1.3. The following are the sample means and sample standard deviations of 24 lots of size 5.

Lot	$\bar{X}$	s	Lot	$\bar{X}$	s
1	90.008	0.040	13	90.007	0.032
2	90.031	0.062	14	89.996	0.021
3	89.971	0.060	15	89.546	0.066
4	90.002	0.047	16	89.627	0.056
5	89.982	0.038	17	89.875	0.042
6	89.992	0.031	18	89.800	0.053
7	89.968	0.026	19	89.925	0.054
8	90.004	0.057	20	90.060	0.047
9	90.032	0.019	21	89.999	0.058
10	90.057	0.022	22	90.068	0.032
11	90.030	0.058	23	90.042	0.041
12	90.062	0.061	24	90.073	0.026

Determine whether the system is in control.

Problem 1.4. The following are the sample means and sample standard deviations of 20 lots of size 5.

Lot	$\bar{X}$	s	Lot	$\bar{X}$	s
1	146.21	0.12	11	146.08	0.11
2	146.18	0.09	12	146.12	0.12
3	146.22	0.13	13	146.26	0.21
4	146.31	0.10	14	146.32	0.18
5	146.20	0.08	15	146.00	0.32
6	146.15	0.11	16	145.83	0.19
7	145.93	0.18	17	145.76	0.12
8	145.96	0.18	18	145.90	0.17
9	145.88	0.16	19	145.94	0.10
10	145.98	0.21	20	145.97	0.09

Determine whether the system is in control.

Remark: In problems 1.5-1.10, we report on initial stages of implementing quality control of the production process of the piston of a fuel pump for Diesel engine. The pumps are manufactured by a small subcontractor of a major U.S. company. It has been decided that, initially, four operations, all performed on a digitally controlled multiple-spindle automatic lathe, should be examined. The operations of interest, numbered from 1 to 4, are indicated in Figure 1.22. The examination was started by selecting 28 lots of pistons. The lots, each consisting of five pistons, were taken at each full hour during four consecutive working days.

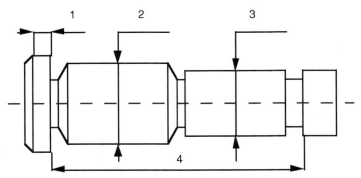

Figure 1.22. Piston Of A Fuel Pump.

Problem 1.5. The table below shows the sample means and sample standard deviations of lengths of the front part of pistons (length 1 in Figure 1.22) for 28 lots of size five. (Measurements are in millimeters.)

Lot	$\bar{x}$	s	Lot	$\bar{x}$	s
1	2.960	.018	15	3.034	.047
2	3.054	.042	16	2.994	.044
3	3.010	.072	17	2.996	.071
4	3.032	.087	18	3.046	.023
5	2.996	.018	19	3.006	.035
6	2.982	.013	20	3.024	.051
7	2.912	.051	21	3.012	.059
8	3.000	.039	22	3.066	.021
9	3.036	.029	23	3.032	.011
10	2.954	.022	24	3.024	.080
11	2.976	.036	25	3.020	.058
12	2.998	.019	26	3.030	.045
13	2.982	.046	27	2.984	.059
14	2.956	.045	28	3.054	.046

Comment on whether the system is in control.

Problem 1.6. The table below shows the sample means and sample standard deviations of diameter 2 of pistons (see Figure 1.22) for 28 lots of size five. (Measurements are in millimeters.)

Lot	$\bar{x}$	s	Lot	$\bar{x}$	s
1	7.996	.017	15	7.964	.039
2	8.028	.039	16	7.964	.045
3	8.044	.075	17	7.944	.060
4	8.036	.075	18	7.956	.035
5	8.008	.044	19	7.992	.019
6	8.018	.032	20	8.032	.044
7	7.962	.064	21	7.974	.080
8	8.022	.059	22	7.946	.029
9	8.020	.025	23	7.978	.038
10	7.992	.030	24	7.986	.058
11	8.018	.052	25	8.016	.059
12	8.000	.034	26	7.988	.074
13	7.980	.023	27	7.962	.041
14	7.952	.073	28	8.002	.024

Comment on whether the system is in control.

Problem 1.7. The table below shows the sample means and sample standard deviations of diameter 3 of pistons (see Figure 1.22) for 28 lots of size five. (Measurements are in millimeters.)

Lot	$\bar{x}$	s	Lot	$\bar{x}$	s
1	6.794	.011	15	6.809	.036
2	6.785	.017	16	6.773	.016
3	6.730	.051	17	6.762	.047
4	6.750	.034	18	6.776	.032
5	6.746	.021	19	6.782	.019
6	6.758	.015	20	6.758	.032
7	6.683	.053	21	6.770	.038
8	6.746	.056	22	6.788	.031
9	6.780	.053	23	6.744	.023
10	6.758	.013	24	6.756	.029
11	6.752	.047	25	6.736	.051
12	6.758	.057	26	6.750	.034
13	6.760	.031	27	6.746	.037
14	6.766	.037	28	6.736	.018

Comment on whether the system is in control.

Problem 1.8. The table below shows the sample means and sample standard deviations of length 4 of pistons (see Figure 1.22) for 28 lots of size five. (Measurements are in millimeters.)

Lot	$\bar{x}$	s	Lot	$\bar{x}$	s
1	59.250	.018	15	59.210	.010
2	59.240	.014	16	59.214	.018
3	59.246	.013	17	59.200	.007
4	59.278	.043	18	59.206	.011
5	59.272	.008	19	59.202	.008
6	59.222	.022	20	59.210	.023
7	59.170	.019	21	59.208	.019
8	59.220	.027	22	59.190	.010
9	59.210	.009	23	59.200	.007
10	59.200	.011	24	59.208	.019
11	59.220	.011	25	59.206	.022
12	59.210	.010	26	59.196	.011
13	59.212	.013	27	59.200	.007
14	59.214	.015	28	59.212	.011

Comment on whether the system is in control.

Problem 1.9. After the control charts from problems 1.5-1.8 were constructed and analyzed, some corrections of the production process have been introduced. The next 21 lots of size 5 of pistons were selected. The measurements of diameter 3 and length 4 (see Figure 1.2) are summarized in the following two tables.

		Diameter 3			
Lot	$\bar{x}$	s	Lot	$\bar{x}$	s
1	6.756	.027	12	6.734	.021
2	6.744	.011	13	6.732	.008
3	6.760	.021	14	6.756	.017
4	6.726	.010	15	6.764	.021
5	6.728	.008	16	6.756	.020
6	6.748	.016	17	6.798	.024
7	6.742	.022	18	6.776	.020
8	6.758	.022	19	6.764	.020
9	6.758	.030	20	6.778	.024
10	6.726	.016	21	6.774	.020
11	6.754	.020			

		Length 4			
Lot	$\bar{x}$	s	Lot	$\bar{x}$	s
1	59.200	.012	12	59.208	.013
2	59.214	.015	13	59.200	.007
3	59.210	.014	14	59.200	.008
4	59.198	.008	15	59.204	.011
5	59.206	.005	16	59.202	.011
6	59.220	.010	17	59.212	.013
7	59.200	.008	18	59.208	.015
8	59.196	.011	19	59.210	.016
9	59.200	.007	20	59.194	.005
10	59.204	.009	21	59.200	.010
11	59.204	.009			

Determine whether the two data sets are in control. Compare the control charts obtained with those from problems 1.7 and 1.8, respectively. In particular, are the intervals between the lower and upper control limits narrower than they were before the corrections of the production process?

Problem 1.10. In the table below, the measurements of diameter 2 (see Figure 1.22) for lots 50 to 77 are summarized. Before the measurements were taken, some corrections, based on the analysis of earlier data, had been made. The lots' size is 5. Determine whether the system is in control. Compare the control charts obtained with those from Problem 1.6.

Lot	$\bar{x}$	s	Lot	$\bar{x}$	s
50	8.038	.044	64	7.984	.021
51	7.962	.039	65	8.006	.038
52	8.002	.008	66	8.034	.031
53	8.010	.019	67	8.044	.023
54	8.024	.034	68	8.005	.022
55	8.014	.023	69	7.994	.031
56	7.974	.038	70	7.958	.018
57	7.986	.028	71	8.006	.011
58	8.040	.029	72	8.022	.019
59	7.986	.028	73	8.004	.018
60	7.974	.028	74	7.990	.029
61	7.994	.011	75	8.000	.019
62	7.974	.023	76	7.994	.027
63	7.962	.030	77	7.998	.031

Problem 1.11. In a plant which is a major supplier of shock absorbers for Polish railway cars, the system of statistical process control includes measurements of the inner diameter of a cylinder turned in a metal casting. The diameter is specified as $91.4 - .2$ mm. Lots of size 5 are taken. The head of the plant's testing laboratory verified that the process is in control and that the last control limits are 91.361 and 91.257 for

the lot means and .076 for the lot standard deviations. He decided that the line workers and foreman responsible for the operation monitor the process using the given limits as the control limits for subsequent lots. Any Pareto glitch observed should call for an immediate action by the line. The results for new lots are as follows.

Lot	$\bar{x}$	s	Lot	$\bar{x}$	s
1	91.305	.076	14	91.310	.064
2	91.259	.038	15	91.337	.052
3	91.308	.048	16	91.336	.047
4	91.296	.046	17	91.328	.029
5	91.290	.066	18	91.320	.036
6	91.291	.039	19	91.350	.029
7	91.307	.065	20	91.368	.011
8	91.350	.043	21	91.308	.078
9	91.262	.060	22	91.330	.048
10	91.255	.020	23	91.331	.050
11	91.259	.031	24	91.348	.043
12	91.309	.030	25	91.246	.051
13	91.311	.033			

No cause of the glitches on lots 10 and 20 (on the $\bar{X}$ chart) and on lots 1 and 21 (on the s chart) was found. After detecting still another glitch on lot 25, the foreman asked the plant's foundry to examine casting moulds. A failure of one of the moulds in use was detected and corrected. The following are the summary statistics of subsequent lots:

Lot	$\bar{x}$	s	Lot	$\bar{x}$	s
26	91.290	.040	39	91.319	.034
27	91.301	.009	40	91.290	.038
28	91.271	.028	41	91.320	.018
29	91.300	.064	42	91.318	.051
30	91.331	.035	43	91.291	.031
31	91.333	.046	44	91.290	.010
32	91.308	.047	45	91.309	.046
33	91.298	.020	46	91.287	.034
34	91.337	.026	47	91.309	.020
35	91.326	.052	48	91.311	.055
36	91.297	.031	49	91.328	.052
37	91.320	.036	50	91.275	.028
38	91.290	.043			

a. Since the cause of the Pareto glitches was found, delete lots 1, 10, 20, 21 and 25, and use the other of the first 25 lots to calculate control limits. Are the first 25 lots (excluding lots 1, 10, 20, 21 and 25) in control, relative to the limits obtained?

b. Calculate the limits for lots 26 to 50 and verify whether these lots are in control.

Problem 1.12. Comment on whether the following system is in control.

Lot	Lot Size	Number Defectives	Lot	Lot Size	Number Defectives
1	100	5	10	100	3
2	100	7	11	100	9
3	100	3	13	100	14
4	100	0	14	100	2
5	100	1	15	100	10
6	100	13	16	100	2
7	100	5	17	100	1
8	100	5	18	100	3
9	100	2	18	100	5

Chapter 2

Acceptance-Rejection SPC

2.1   Introduction

The paradigm of quality control is largely oriented toward optimization of a process by the use of monitoring charts to discover "Pareto glitches" which can then be backtracked until assignable causes for the glitches can be found and rectified. The goal is simply real and measurable improvement. "Zero defects" is seldom, if ever, a realistic goal of the quality improvement specialist.

For example, if we apply the techniques of statistical process control in the management of a cancer care facility, we can certainly expect measurable improvements in the mortality rate of the patients. However, a goal of zero mortality is not a realistic one for achievement by the normal techniques of statistical process control. Such a goal is achievable only by a major medical breakthrough.

A manufacturer of micro-chips may realistically expect failure rates measured in failures per million once SPC paradigms have been in place for some time. A manufacturer of automobiles may expect failures measured in failures per thousand. The director of a general health care facility may expect patient death rates measured in deaths per hundred admissions. Each type of process has its natural limit of improvement of the failure rate, absent scientific breakthrough. The use of the paradigm of statistical process control does, very frequently, produce improvements so substantial as to appear miraculous. The current state of Japanese automobiles, and that of electronics compared to what they were 40 years ago, is such an example. But such improvements are obtained by steady monitoring and searching for causes of variation, not by the setting of arbitrary utopian goals.

Few things can foil quality improvement more surely than the announcement of unrealistic QC goals without any reasonable road map as to how they are to be achieved. When it becomes obvious that the "pie in the sky" is not forthcoming, workers in the system rightly "turn off" the quality enhancement program as just so much management huffing and puffing.

In some failed quality control cultures, there is a tendency to replace unattainable goals on relevant variables by attainable goals on irrelevant variables.

Some years ago, a new, highly motivated manager of a newly built wood products factory in the Soviet Union set out to have the best run wood products factory in the empire. He spurred the workers on with promises of bonuses for their enthusiastic cooperation. His plans for efficiency went so far as using the sawdust produced as fuel to power the factory.

At the end of his first year, fully expecting to find the government evaluation of the plant to be very high, he was crushed to find his factory ranked dead last among wood products plants in the Soviet Union.

The next year, he redoubled his efforts. He apologized to the workers, explaining that while their plant was moving forward, no doubt others were as well. He made new plans for accelerated improvement. At the end of the second year, justifiably pleased with the factory's progress, he was informed by Moscow that his plant was still last in the Soviet Union and that his productivity had fallen behind that of the first year. No bonuses for the workers and the danger of demotion of the plant manager.

Distraught to desperation, the young manager boarded the next train to Moscow. Perhaps his plant was at the bottom of the list, but there was no way the productivity of his plant the second year was less than that of the first.

Upon meeting with his bosses in Moscow, he presented graphs and tables to demonstrate how productivity had improved from the first to the second year. Unimpressed, his superiors noted they had also double-checked their computations and productivity had indeed declined.

Desperately, the manager asked what criteria were being used for productivity measurement. He discovered that the criterion was a rather simple one, namely the pounds of waste sawdust per worker taken from the plant each year.

The happy ending to the story is that the third year, the bright, but now somewhat cynical, manager achieved his goal of seeing his wood products factory ranked among the highest in the Soviet Union.

Almost equally bizarre situations regularly occur in the United States. For example, as part of the "zero defects" program, employees may be asked to sign cards pledging themselves to the goal of zero defects. When a certain proportion of workers have signed the cards, a day of celebration is decreed to mark this "important event." Note that it is not the achievement of zero defects which is being celebrated, but rather the somewhat less significant pledging of a commitment to zero defects (whatever that means).

2.2   The Basic Test

As we have pointed out earlier, statistical process control using failure data is a rather blunt instrument. However, at the start of almost any statistical process control program, the data available will frequently be end product failure rate. Whether or not a production item is satisfactory or not will frequently be readily apparent, if by no other feedback than by the judgment of the end user of the product.

Consequently, if we have past records to give us the average proportion of failed products, $\hat{p}$, we can use the information to see whether a new lot of size n and proportion of failures $\hat{p}$ appear to exhibit an atypical proportion of failed items. If the difference is truly significant, then we can use the information to suggest the presence of a Pareto glitch. Looking at the various factors which could have led to the glitch may then provide us the opportunity to find an assignable cause for the glitch. If the assignable cause has led to a significantly higher fraction of failed items (as is usually the case), then we can act to remove it. If the assignable cause has led to a significantly lower fraction of failures, then we may consider whether the cause should be made a part of our standard production protocols.

An investigator who wanted never to miss an assignable cause would have to declare a lot to be out of control whatever the measurement results might be. Such an approach could be described as management by constant crisis. Consistent with the old story of the shepherd constantly shouting "Wolf!, " a quality control investigator who is always declaring the system to be out of control will generally be unable to detect an out of control system when it occurs.

Contrariwise, an investigator who only wanted to be sure he never turned in a false alarm would pass every lot as being in control. Such an investigator could (and should) be replaced by a recording saying, "Yea, verily, yea," over and over.

Clearly, we must steer a course between these two extremes. As a matter of fact, experience has shown that we should steer rather closer to the investigator who accepts everything as being in control rather than the one who rejects every lot. We recall that the main function of statistical process control is not as a police action to remove defective items. To continue the analogy, a good detective would be much more interested in finding and destroying the source of a drug ring than simply arresting addicts on the streets. He will be relatively unimpressed with the discovery of casualties of the drug ring except insofar as such discoveries enable him to get to the source of the distribution system.

Just so, the statistical process control investigator is looking for a lot which is almost surely not typical of the process as a whole. To err on the side of not crying wolf unless we are nearly certain a lot is out of control (i.e., not typical of the process as a whole), it is customary to set the alarm level so that a lot which really is typical (and thus "in control") is declared to be out of control roughly only once in 500 times. The exact level is not usually very important. A test may have the false alarm rate as high as one chance in 50 or as low as one chance in 5,000 and still be quite acceptable for our purpose.

Let us suppose that past experience indicates that 5% of the items produced are not acceptable. We have a lot of size 100 and find that 7 are not acceptable. 7% is certainly higher than 5%. But, if the lot is typical, we would expect a higher value than 5% half the time. What is the probability that 7 or more defective items will be found in a lot of size 100 drawn from a population where the probability of any item being bad is .05? In accordance with the formula for a binomial distribution (see Appendix B), we simply compute

$$
\begin{aligned}
P(X \geq 7) &= 1 - P(X \leq 6) \qquad\qquad (2.1) \\
&= 1 - \sum_{j=0}^{6} \binom{100}{j} (.05)^j (.95)^{100-j} \\
&= .234.
\end{aligned}
$$

If we choose to intervene at such a level, we shall be intervening constantly, so much so that we will waste our efforts. We need a higher level of failure rate before taking notice. Suppose we decide rather to wait until the number of defectives is 10 or more. Here the probability of calling special attention to a lot which actually has an underlying failure rate of 5% is given by

$$
\begin{aligned}
P(X \geq 10) &= 1 - P(X \leq 9) \\
&= 1 - \sum_{j=0}^{9} \left(\begin{array}{c} 100 \\ j \end{array} \right) (.05)^j (.95)^{100-j} \\
&= .02819.
\end{aligned}
\tag{2.2}
$$

By our predetermined convention of 1 in 500 (or .002), this is still too high, though probably we would not be seriously overchecking at such a rate. The issue in statistical process control is not to have the alarm ring so frequently that we cannot examine the system for assignable causes of apparent Pareto glitches. When we get to 12 defectives, the probability goes to .0043. Then, we arrive at 13, the probability drops just below the .002 level, namely to .0015. And at 14, the probability is only .00046. It is quite reasonable to set the alarm to ring for any number of defectives past 10 to 14. The "one in 500" rule is purely a matter of convenience. We should note, moreover, that it might be appropriate for the alarm to ring when there were too few failures. When we reach a level of 0 defectives, for example, if the underlying probability of a failure is truly .05, then we would expect to see such a low number of defectives with probability only $.95^{100} = .0059$. It might well be the case that a statistical process control professional would want to see if perchance some really positive innovation had occurred during this period, so that standard operating procedure could be modified to take advantage of a better way of doing things. In such a case, where we wish to use both the low failure rate and the high failure rate alarms, we could set the low alarm at 0, say, and the high at 12, for a pooled chance of a "false alarm" equal to .0059 + .0043, that is, for a pooled "false alarm" rate of .0102 or roughly one in a hundred. Looking at the size of our quality control staff, we might make the judgment that we simply did not have enough people to check at a level where the process was carefully examined for major defects 1% of time when good lots were being produced. Then we might properly decide to check the process only when 13 or more of the items in a lot were faulty. Essentially, there is no "right" answer as to where we should set the alarm level.

2.3   Basic Test with Equal Lot Size

Let us consider the sort of situation encountered in the production of bolts of nominal diameter of 10 millimeters. Our customer has agreed with the factory management that he will accept no bolts with diameter below 9.8 millimeters, none above 10.1 millimeters. The factory has been using the old "quality assurance" paradigm, i.e., simply use 100% end inspection, where a worker has two templates of 9.799 mm and 10.101 mm. Any bolt which fits into the 9.799 template is rejected, as is any which will not fit into the 10.101 template.

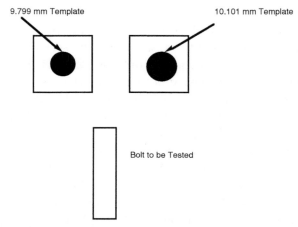

Figure 2.1. Templates for Finding Defective Bolts.

This is the kind of paradigm which is essentially avoided by factories which have been using the SPC (statistical process control) approach for some time. But it is the kind of paradigm still used in most factories in the industrial world. When one is starting up an SPC operation, this is the kind of data with which we have to work. And such data generally contains valuable first step SPC information.

Let us consider a consecutive sample of 40 lots of 100 bolts each inspected in a classical quality assurance system. Typically, these observations will represent 100% inspection. The lots are indexed by time. In a "quality assurance" program, feedback from the proportion of defectives in the output stream is generally casual. The pattern in Table 2.1 pictured in Figure 2.2 is not untypical. The proportion of defectives hovers around 5% . Starting with lot 11, the proportion drifts upward to 15%. Then, the proportion drifts downward, perhaps because of intervention, perhaps not. In general, the graph in Figure 2.2 is of marginal utility if it is used only retrospectively after all 40 lots have been examined.

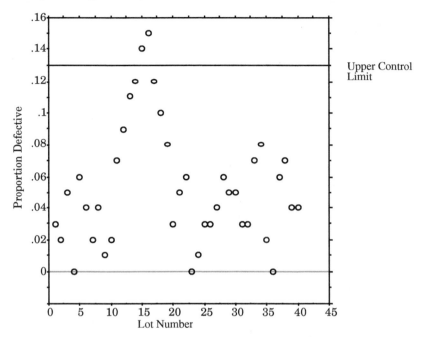

Figure 2.2. Acceptance-Rejection Control Chart.

Let us note how such information might be used to improve the production process. We elect to use as our upper control limit a proportion of 13%. This boundary is crossed on lot 15. Suppose that as a result of a point above the upper control limit, we engage in a thorough examination of the production process. We discover that a lubrication fault in the milling machine in the form of intermittent overheating is present. We correct the lubrication problem, and the resulting proportion of defectives is demonstrated in Figure 2.3.

The resulting proportion of defectives drops to around 1%. Such an excellent outcome is, interestingly enough, not as untypical as one might suppose. The gains available by time based control charting are far out of proportion to the labor required for their creation. Note that looking at the full 40 points after the entire lot has been collected is unlikely to discover the overheating problem. It is generally difficult to find "assignable causes" from a cold data set. Such a problem will simply generally get worse and worse as time progresses until the bearings burn out and the machine is replaced.

Table 2.1		
Lot	Number Defectives	Proportion Defectives
1	3	.03
2	2	.02
3	5	.05
4	0	.00
5	6	.06
6	4	.04
7	2	.02
8	4	.04
9	1	.01
10	2	.02
11	7	.07
12	9	.09
13	11	.11
14	12	.12
15	14	.14
16	15	.15
17	12	.12
18	10	.10
19	8	.08
20	3	.03
21	5	.05
22	6	.06
23	0	.00
24	1	.01
25	3	.03
26	3	.03
27	4	.04
28	6	.06
29	5	.05
30	5	.05
31	3	.03
32	3	.03
33	7	.07
34	8	.08
35	2	.02
36	0	.00
37	6	.06
38	7	.07
39	4	.04
40	4	.04

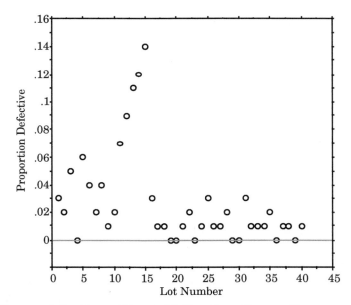

Figure 2.3. Run Chart Following Correction.

As we have noted, obtaining an exact significance level is not generally important in SPC. The usual rule of thumb is that the control limits are given by the mean plus and minus 3 times the standard deviation. In the case of a Gaussian variate X with mean μ and standard deviation σ we have

$$P(|\frac{X-\mu}{\sigma}| \geq 3) = .0027. \tag{2.3}$$

As we note in the Appendix, for the binomial variate X from a distribution with probability p and sample size n, for a sufficiently large sample, the normalized variate Z defined below is approximately Gaussian with mean 0 and standard deviation 1.

$$Z = \frac{X - np}{\sqrt{np(1-p)}} = \frac{\frac{X}{n} - p}{\sqrt{\frac{p(1-p)}{n}}}. \tag{2.4}$$

To obtain an upper control limit, we can use

$$\frac{UCL - np}{\sqrt{np(1-p)}} = 3, \tag{2.5}$$

giving,

$$UCL = np + 3\sqrt{np(1-p)}. \tag{2.6}$$

In the example from Table 2.1, with bolts of lot size 100 and $p = .05$, we have

$$UCL = 100(.05) + 3\sqrt{100(.05)(.95)} = 11.538. \tag{2.7}$$

So, we would reject a lot with 12 or more defectives as untypical of a population with a proportion of defectives equal to .05. Next, suppose that the lot sizes are much larger, say $n = 1000$. In that case the upper control limit is given by

$$UCL = 1000(.05) + 3\sqrt{1000(.05)(.95)} = 70.676. \tag{2.8}$$

We note that for the sample size of 100, we have an upper control limit of 12% defectives. For the sample size of 1,000, the upper control limit is 7.1% defectives. We recall that in both cases, we seek a probability of passing the UCL when the proportion of defectives is 5% of roughly .002. Accordingly, as the sample size increases, an ever smaller departure from 5% will cause us to investigate the process for possible problems.

For Acceptance-Rejection SPC, the lower control limit is frequently not used at all. An abnormally low proportion of defectives is less likely to indicate a need for process modification than is a high one. In the case of the lower control limit, for $n = 100$,

$$LCL = 100(.05) - 3\sqrt{100(.05)(.95)} = -1.538, \tag{2.9}$$

which we naturally truncate to 0. In this case, there is no lower control limit. But for $n = 1000$,

$$LCL = 1000(.05) - 3\sqrt{1000(.05)(.95)} = 29.324. \tag{2.10}$$

In general, we have that

$$LCL = \max\{0, np - 3\sqrt{np(1-p)}\}. \tag{2.11}$$

A control chart for the number of defectives, usually referred to as the *np chart*, is defined by the limits (2.6) and (2.11). Equivalently, we can construct a control chart for the proportion defectives. Indeed, in our discussion, we freely switched from one chart to another. Formally, the equivalence between the two types of control charts is given by (2.4). The right hand side of (2.4) refers to the proportion of defectives. Control limits for the chart for the proportion defectives have the form

$$UCL = p + 3\sqrt{\frac{p(1-p)}{n}} \qquad (2.12)$$

and

$$LCL = \max\{0, p - 3\sqrt{\frac{p(1-p)}{n}}\}. \qquad (2.13)$$

This last chart is usually referred to as the *p chart*.

2.4 Testing with Unequal Lot Sizes

There are many situations in which it is not reasonable to suppose that the lot sizes are equal. For example, let us suppose that we are looking at the number of patients (on a monthly basis) in a hospital who experience postoperative infections following hip replacement surgery. We consider such a set of data in Table 2.2 for a period of two years. We note that the average proportion of patients who contract infections during their stay in the hospital following hip replacement surgery is .08, which we are told is normal for the hospital. We would like to be able to use the database to improve the operation in the hospital. As in the case of the production of bolts, such information viewed at the end of the two year period is likely to be of marginal value. A Pareto glitch in anything but the immediate past is typically difficult to associate with an assignable cause of that glitch.

Of much less difficulty is the minor computational problem of computing the UCL when the number of patients varies from month to month. This is easily dealt with via the formula

$$UCL = .08 + 3\sqrt{\frac{.08(1 - .08)}{n}}. \qquad (2.14)$$

In general, the UCL for the p chart assumes now the form

$$UCL = p + 3\sqrt{\frac{p(1-p)}{n_k}}, \qquad (2.15)$$

where k is the lot's number and n_k is the lot's size. If the lot sizes are not equal, the control limit for the proportion defectives has to be calculated for each lot separately.

Table 2.2. Infections Following Hip Replacement.					
Month	Patients	Infections	Prop.	UCL	Prop./UCL
1	50	3	.060	.195	.308
2	42	2	.048	.206	.232
3	37	6	.162	.214	.757
4	71	5	.070	.177	.399
5	55	6	.109	.190	.575
6	44	6	.136	.203	.673
7	38	10	.263	.212	1.241
8	33	2	.061	.222	.273
9	41	4	.098	.207	.471
10	27	1	.037	.237	.157
11	33	1	.030	.222	.137
12	49	3	.061	.196	.312
13	66	8	.121	.180	.673
14	49	5	.102	.196	.520
15	55	4	.073	.190	.383
16	41	2	.049	.207	.236
17	29	0	.000	.231	.000
18	40	3	.075	.209	.359
19	41	2	.049	.207	.236
20	48	5	.104	.197	.527
21	52	4	.077	.193	.399
22	55	6	.109	.190	.575
23	49	5	.102	.196	.520
24	60	2	.033	.185	.180

We note that in Table 2.2 and Figure 2.4, in the seventh month, the proportion of patients developing infections went to 1.241 times the UCL. If we are examining the data very long after that month, we are unlikely to find the assignable cause. In the situation, here, after much retrospective work, we find that there were three teams (composed of surgeons, anesthetists, surgical nurses, etc.) performing this type of surgery during the period. We shall refer to these as teams A, B, and C. The numbers of patients will be noted by n, the number of these who developed infections m. The results of the breakdown are given in Table 2.3.

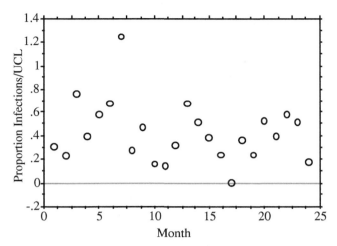

Figure 2.4. Run Chart of Proportion of Infections Divided by UCL.

From Table 2.3, we note that Team C only began operation in month 4. The number of surgeries handled by this team then grew slowly. By the end of the second year, it was handling roughly as many surgeries as the other two teams. Figure 2.5 gives us a fair picture of the progress of Team C. Early on, the proportion of patients developing infections was much higher than that of the other two teams. From month 15 onward, the infection rate of Team C appears comparable to that of the other two teams. Since the data was "cold" by the time we received it, perhaps we shall never know what steps were taken in month 15 which appears to have rid Team C of its earlier comparatively poor performance. Suffice it to say, however, that employment of process control early on might have spared a number of patients the trauma associated with an infection following a joint replacement procedure. Benjamin Franklin's adage that "Experience keeps a hard school, and a fool will learn by no other" is obviously relevant here. Any start-up procedure is likely to have problems associated with it. But, absent an orderly regular measurement procedure, these start-up problems may take a long time in solving. Here, it took almost a year. And that is a better record than that usually associated with surgical procedures whose records are not regularly monitored for "Pareto glitches." Frequently, in a medical setting, poor performance by a surgical team may go unnoticed and uncorrected for years, until a personal injury lawyer brings the matter to the attention of the hospital. When this occurs, the hospital gains a quick and expensive demonstration of the principle that modest expense and effort in implementing a good statistical process control regime generally saves vast sums later on.

Month	n_A	m_A	n_B	m_B	n_C	m_C
1	20	1	30	2	0	0
2	22	2	20	0	0	0
3	20	2	17	4	0	0
4	30	2	35	1	6	2
5	17	2	25	2	13	2
6	20	1	15	2	9	3
7	15	2	10	2	13	6
8	21	1	9	0	3	1
9	19	1	19	2	3	1
10	10	0	15	0	2	1
11	15	1	15	0	3	0
12	25	1	20	1	4	1
13	31	2	20	2	15	4
14	19	1	20	1	10	3
15	25	1	20	2	10	1
16	19	2	15	0	7	0
17	10	0	9	0	10	0
18	14	1	16	1	10	1
19	10	1	10	1	21	0
20	15	1	10	2	23	2
21	20	1	20	2	12	1
22	19	2	17	2	19	2
23	14	1	15	2	20	2
24	20	1	20	1	20	0

Table 2.3. Team Performance Data.

In this discussion, we relied on the p chart. Let us mention that, instead of this chart, one can use a chart for standardized proportion defectives,

$$Z_k = \frac{p_k - p}{\sqrt{\frac{p(1-p)}{n_k}}}, \qquad (2.16)$$

where k is the lot's number, n_k is the lot's size and p_k is the proportion of defectives in lot k. Since the Z_k's are approximately normal with mean 0 and standard deviation 1, the upper and lower control limits become 3 and -3, respectively.

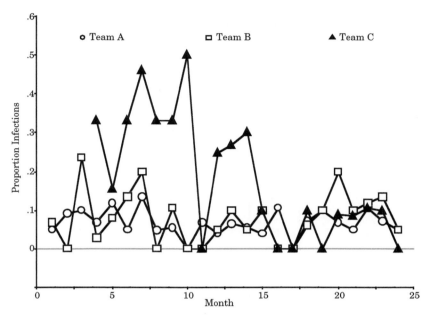

Figure 2.5. Comparison of Team Performances.

2.5   Testing with Open-Ended Count Data

Worse even than feedback from end inspection is that obtainable from items returned due to unsatisfactory performance. Unfortunately, such information is frequently the major source of information to a firm as to the quality of a product. One problem in dealing with such data is that the potential number of returns is essentially infinite.

Under fairly general conditions, e.g., that the probability that there will be an item returned in a small interval of time is proportional to the length of the interval and that the returns in one time interval do not impact the number of returns in a later interval, the number of items X returned per week has, roughly, the Poisson distribution, i.e.,

$$P(X \geq m) = 1 - \sum_{i=0}^{m-1} e^{-\lambda} \frac{\lambda^i}{i!} \qquad (2.17)$$

where λ is the average value of X.

Table 2.4. Weekly Returns	
Week	Number Returned
1	22
2	13
3	28
4	17
5	22
6	29
7	32
8	17
9	19
10	27
11	48
12	53
13	31
14	22
15	31
16	27
17	20
18	24
19	17
20	22
21	29
22	30
23	31
24	22
25	26
26	24

In Figure 2.6, we show weekly numbers of items returned for a six month period. The average number of returns per week is computed to be $\bar{X} = 26.27$. We recall that for the Poisson distribution, the mean and the variance σ^2 are equal. Accordingly, our estimate for the standard deviation $\sigma = 5.12$. (With all observations included, the sample standard deviation is 8.825, much bigger than the 5.12 we get making the Poisson assumption.) The upper control limit is given by

$$UCL = 26.27 + 3(5.12) = 41.63. \qquad (2.18)$$

We note that two weeks have return numbers above the UCL. With these two items removed from the pool, we find $\bar{X} = 24.25$, giving $\sqrt{\bar{X}} = 4.92$.

The sample standard deviation is $5.37 \approx 4.92$.

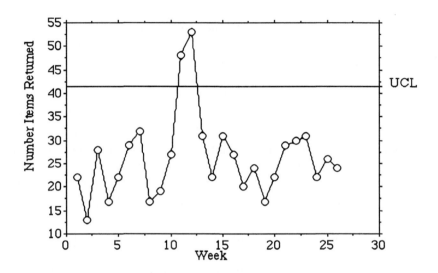

Figure 2.6. Run Chart Items Returned per Week.

What was the difficulty causing the unusually high number of returns in weeks 11 and 12? A backtracking of records indicated that 10 of the actual returns in week 11 and 12 of those in week 12 had been erroneously entered twice. Such a sounding off of a "QC alert" by incorrect measurements or erroneous data rather than by a basic flaw in the production system is quite common in quality control. Here, little harm seems to have been done. But again, we must emphasize the desirability of looking at the data before it has become cold. And we must note that, realistically speaking, anyone who starts a statistical process control paradigm of stepwise optimization will be confronted with the fact that there are reams of "useful data" available which, hot or cold, represent some sort of starting point for analysis.

Indeed, in the United States, SPC consultants frequently are called in to resolve some crisis in which nothing but rough end product acceptance-rejection data is available. If the consultant is clever enough to use such data in order to solve the immediate crisis, he may find that he is thanked, paid (handsomely) and bid farewell to, until the next crisis occurs. It is a continuing tragedy of American managers that they tend to lurch from crisis to crisis, lacking the attention span and perception to obtain the vastly superior increases in quality and productivity if a methodical system of statistical process control were instituted and developed.

Let us conclude this section with a brief discussion of another problem with data whose values cannot be bounded from above. Until now, we have dealt with numbers of defective items. Sometimes, however, we are interested in the number of defects per lot of a fixed size (possibly 1). Depending on the situation, an item can be considered defective if it possesses just one defect of some known sort. In the example of producing bolts, a bolt was determined as being defective if its diameter lay outside tolerance limits. In cases of end inspection of assembled products, an item may usually be defective in many different and well-known ways. Or, say, a molded plastic item can be considered defective if it has at least one scratch on its visible part, or it has at least one flash, or unacceptable flow line, or its bending strength is below a tolerance limit.

As a rule, we can assume that the number of defects per lot is a Poisson random variable, that is, that the number of defects in one "region" is independent of the number of defects in a disjoint region, the probability that there will be a single defect in a small region is proportional to the size of the region, and the probability of more than one defect occurring in a small region is negligible. A particular definition of the region depends, of course, on the case under scrutiny. If the number of defects follows the Poisson distribution (2.17), we obtain essentially the same control chart as discussed earlier in this section. Namely,

$$UCL = \bar{c} + 3\sqrt{\bar{c}} \qquad (2.19)$$

and

$$LCL = \max\{0, \bar{c} - 3\sqrt{\bar{c}}\}, \qquad (2.20)$$

where $\bar{c}$ is either equal to the mean number of defects per lot (λ in (2.17)), assumed known from past experience, or is the mean's estimate,

$$\bar{c} = \frac{1}{N} \sum_{k=1}^{N} c_k, \qquad (2.21)$$

where N is the number of lots, and c_k is the number of defects in lot k. The control chart presented is often called the c *chart*.

Clearly, the c chart should not be used if lots are of unequal sizes. It is natural then to replace the c chart by a control chart for the number of defects per unit, or per item. Denoting the number of defects per unit in lot k of size n_k by u_k,

$$u_k = \frac{c_k}{n_k}, \qquad (2.22)$$

we obtain the following control limits for the kth lot

$$UCL = \bar{u} + 3\sqrt{\frac{\bar{u}}{n_k}} \qquad (2.23)$$

and

$$LCL = \max\{0, \bar{u} - 3\sqrt{\frac{\bar{u}}{n_k}}\}, \qquad (2.24)$$

where

$$\bar{u} = \frac{\sum_{k=1}^{N} c_k}{\sum_{k=1}^{N} n_k}. \qquad (2.25)$$

We note that the limits have to be calculated for each lot separately. The chart obtained is often called the *u chart*.

Obviously, one can use the u chart, instead of the c chart, if lot sizes are equal, $n_k = n$ for all k. It suffices to replace n_k by n in (2.21) to (2.24).

Problems

Problem 2.1. Consider the production of fuel pumps for Diesel engine. In the early stage of the production, before the introduction of the SPC paradigm, 100% end inspection of pumps' cylinders in lots of 100 was being performed.

Lot	Proportion Defectives	Lot	Proportion Defectives
1	.04	17	.13
2	.02	18	.11
3	.03	19	.06
4	.04	20	.04
5	.05	21	.07
6	.04	22	.04
7	.03	23	.07
8	.02	24	.04
9	.03	25	.05
10	.04	26	.09
11	.03	27	.06
12	.04	28	.05
13	.03	29	.07
14	.08	30	.08
15	.11	31	.06
16	.14	32	.08

Construct the run chart for the data and then add the Upper Control
Limit to the chart. Are the data in control? Given the run chart, should
one wait until all 32 lots are inspected and become "cool" or, rather,
should one construct the control chart for the first 20 lots? Does the
control chart for the first 20 lots only exhibit a Pareto glitch?

Problem 2.2. Consider the production process summarized in Figure 2.3. Suppose that from lot 1 onwards, the process is examined in
accordance with the SPC rules. Past experience indicates that 5% of the
items produced are not acceptable. Corresponding Upper Control Limit
is provided by formula (2.7). The limit is crossed on lot 14 and the cause
of the observed Pareto glitch is found and corrected immediately after
the production of lot 15 is completed. The subsequent data are given in
the table below and in Figure 2.3.

Lot	Number Defectives	Proportion Defectives	Lot	Number Defectives	Proportion Defectives
16	3	.03	29	0	.00
17	1	.01	30	0	.00
18	1	.01	31	3	.03
19	0	0	32	1	.01
20	0	0	33	1	.01
21	1	.01	34	1	.01
22	2	.02	35	2	.02
23	0	0	36	0	.00
24	1	.01	37	1	.01
25	1	.01	38	1	.01
26	1	.01	39	0	.00
27	1	.01	40	1	.01
28	2	.02			

a. Delete lots 14 and 15 and compute the sample mean of the proportion of defectives for the first 13 lots. Using the value obtained, compute
the new Upper Control Limit (i.e., that for lots 1 to 13). Use the new
limit as the "trial" limit for verification whether lots 16 onwards are in
control. Give motivation for such an approach.

b. Construct the control chart for lots 16 to 40 without any use of the
past information on the process under scrutiny. Are the lots in control?
Compare the Upper Control Limit obtained with the trial limit.

c. Should one insist on computing trial limits for the future data
(as in **a**) or, rather, wait until sufficiently many data are available for
constructing charts independent of the more distant past? If using the
trial limits is to be recommended, should one recommend relying on the
trial control charts solely?

Problem 2.3. Consider the following data.

Lot	Number Inspected	Number Defectives	Lot	Number Inspected	Number Defectives
1	531	25	14	685	28
2	2000	58	15	2385	89
3	2150	89	16	2150	58
4	1422	61	17	2198	86
5	2331	75	18	1948	41
6	1500	73	19	2271	67
7	2417	115	20	848	30
8	850	27	21	2214	68
9	1700	49	22	1197	56
10	2009	81	23	2150	77
11	1393	62	24	2394	82
12	1250	46	25	850	33
13	2549	115			

Comment on whether the system is in control.

Problem 2.4. The following data pertain to the inspection of pressure relief valves for autoclaves for food processing. The company producing autoclaves is furnished with the valves by two subcontractors.

Lot	Lot Size	Number Defectives	Lot	Lot Size	Number Defectives
1	100	6	16	150	6
2	100	5	17	120	6
3	90	3	18	120	5
4	95	4	19	110	5
5	100	6	20	110	5
6	105	5	21	115	9
7	200	8	22	100	11
8	210	11	23	100	9
9	155	7	24	90	7
10	155	7	25	95	9
11	120	6	26	110	12
12	120	6	27	120	9
13	110	5	28	120	12
14	115	5	29	95	7
15	120	5	30	95	8

It is known that the first, more expensive, subcontractor provided the company with the valves for lots 1 to 20. The second subcontractor's valves were used for the remaining 10 lots.

a. Construct the control charts for lots 1 to 20.

b. Compare the quality of valves from both subcontractors. In order to do this, construct the control chart for lots 21 to 30 using the sample mean of the proportions of defectives for lots 1 to 20 (i.e., construct the control chart using the mean based on the past experience).

Problem 2.5. A company producing refrigerators keeps records of failures of its products under warranty, reported by authorized dealers and repair workshops. In the following table, numbers of failures in consecutive 26 weeks are given for a particular type of refrigerator.

Week	Number Repaired	Week	Number Repaired
1	22	14	38
2	16	15	29
3	29	16	25
4	24	17	16
5	18	18	22
6	21	19	26
7	28	20	27
8	24	21	14
9	29	22	19
10	21	23	29
11	22	24	17
12	16	25	27
13	20	26	25

Determine whether the system is in control. Even if any Pareto glitches are detected, do they help in finding their assignable causes?

Problem 2.6. After the production of a new item has been set up, it was decided that the number of defects in the items should be examined. Three different types of defects are possible. It was decided that lots of size 5 be taken. In the following table, numbers of defects (of whatever type) per lot are given for the first 30 lots.

Lot	Number Defects	Lot	Number Defects
1	11	16	12
2	16	17	14
3	8	18	9
4	12	19	7
5	6	20	5
6	6	21	9
7	5	22	12
8	12	23	11
9	14	24	13
10	11	25	20
11	9	26	14
12	11	27	12
13	12	28	7
14	7	29	8
15	6	30	6

    a. Verify that the system is out of control.

    b. An investigation was performed which revealed that defects of the first type constituted the majority of all defects. The following are the numbers of defects of the first type per lot.

Lot	Number Defects	Lot	Number Defects
1	6	16	8
2	9	17	7
3	5	18	6
4	10	19	4
5	5	20	2
6	4	21	7
7	2	22	9
8	9	23	8
9	10	24	9
10	7	25	16
11	8	26	10
12	7	27	8
13	9	28	6
14	3	29	5
15	2	30	3

Do the data reveal any Pareto glitches?

Chapter 3

The Development of Mean and Standard Deviation Control Charts

3.1 Introduction

It is generally a good idea to approach a pragmatic paradigm from the standpoint of a conceptual model. Even though flawed and incomplete, if some care is given to its formulation, a model based approach is likely to give us a more useful frame of reference than a paradigm of the "just do it" variety. Statistical process control is no exception to this rule.

Let us first of all assume that there is a "best of all possible worlds" mechanism at the heart of the process. For example, if we are turning out bolts of 10 cm diameter, we can assume that there will be, in any lot of measurements of diameters, a variable, say X_0, with mean 10 and a variance equal to an acceptably small number. When we actually observe a diameter, however, we may not be seeing only X_0 but a sum of X_0 plus some other variables which are a consequence of flaws in the production process. These are not simply measurement errors but actual parts of the total diameter measurements which depart from the "best of all possible worlds" distribution of diameter as a consequence of imperfections in the production process. One of these imperfections might be excessive lubricant temperature, another bearing vibration, another nonstandard raw materials, etc. These add-on variables will generally be intermittent in time. A major task in SPC is to find measurements which appear to show "contamination" of the basic production process.

For simplicity's sake, let us assume that the random variables are added. In any lot, indexed by the time, t, of sampling, of samples, we will assume that the measured variable can be written as

$$Y(t) = X_0 + \sum_{i=1}^{k} I_i(t) X_i, \qquad (3.1)$$

where

X_i comes from distribution F_i having mean μ_i and variance σ_i^2

and indicator

$$
\begin{aligned}
I_i(t) &= \ 1 \text{ with probability } p_i \\
&= \ 0 \text{ with probability } 1 - p_i.
\end{aligned}
\qquad (3.2)
$$

If such a model is appropriate, then, with k assignable causes, there may be in any lot 2^k possible combinations of random variables contributing to Y. We assume that there is sufficient temporal separation from lot to lot that the parameters driving the Y process are independent from lot to lot. Further, we assume that an indicator variable I_i maintains its value (0 or 1) throughout a lot. Let $\mathcal{I}$ be a collection from $i \in 1, 2, \ldots, k$. Then

$$Y(t) = X_0 + \sum_{i \in \mathcal{I}} X_i \text{ with probability } [\prod_{i \in \mathcal{I}} p_i][\prod_{i \in \mathcal{I}^c} (1 - p_i)]. \qquad (3.3)$$

In the special case where each distribution is normal, then the observed variable $Y(t)$ is given by

$$Y(t) = \mathcal{N}(\sum_{i \in \mathcal{I}} \mu_i, \sum_{i \in \mathcal{I}} \sigma_i^2), \qquad (3.4)$$

$$\text{with probability } [\prod_{i \in \mathcal{I}} p_i][\prod_{i \in \mathcal{I}^c} (1 - p_i)].$$

Of course, in the real world, we will not know what the assignable causes (and hence the X_i) are, let alone their means and variances nor their probability of being present as factors in a given sampled lot. A major task of SPC is to identify the variables, other than X_0, making up $Y(t)$ and to take steps which remove them.

3.2   A Contaminated Production Process

In the matter of the manufacturing of bolts of 10 cm diameter, let us suppose that the underlying process variable X_0 is $\mathcal{N}(\mu = 10, \sigma^2 = .01)$. In addition, we have (unbeknownst to us) a variable due to intermittent lubricant heating, say X_1 which is $\mathcal{N}(.4, .02)$ and another due to bearing vibration, say X_2 which is $\mathcal{N}(-.2, .08)$ with probabilities of occurrence, respectively, $p_1 = .01$ and $p_2 = .005$. So, for a sampled lot, $Y(t)$ will have the following possibilities:

$$
\begin{aligned}
Y(t) &= X_0; \text{ probability } (1 - .01)(1 - .005) = .98505 & (3.5) \\
Y(t) &= X_0 + X_1; \text{ probability } (.01)(1 - .005) = .00995 & (3.6) \\
Y(t) &= X_0 + X_2; \text{ probability } (1 - .01)(.005) = .00495 & (3.7) \\
Y(t) &= X_0 + X_1 + X_2; \text{ probability } (.01).005 = .00005. & (3.8)
\end{aligned}
$$

Recalling that (see Appendix B) for the sum of normal variates, the resulting variate is also normal with means and variances the sum of the means and variances of the added variables, we have

$$
\begin{aligned}
Y(t) &= \mathcal{N}(10, .01); \text{ probability } .98505 & (3.9) \\
Y(t) &= \mathcal{N}(10.4, .03); \text{ probability } .00995 & (3.10) \\
Y(t) &= \mathcal{N}(9.8, .09); \text{ probability } .00495 & (3.11) \\
Y(t) &= \mathcal{N}(10.2, .11); \text{ probability } .00005. & (3.12)
\end{aligned}
$$

For (3.1) (which is not itself normal), the reader can verify that

$$
\begin{aligned}
E(Y) &= \mu_0 + p_1\mu_1 + p_2\mu_2 & (3.13) \\
Var(Y) &= \sigma_0^2 + p_1\sigma_1^2 + p_2\sigma_2^2 + p_1(1 - p_1)\mu_1^2 + p_2(1 - p_2)\mu_2^2 .
\end{aligned}
$$

For the p_1 and p_2 of .01 and .005, respectively, this gives

$$
\begin{aligned}
E(Y) &= 10.003 \\
Var(Y) &= .0124 . & (3.14)
\end{aligned}
$$

But if we increase p_1 to .10 and p_2 to .05, then we have

$$
\begin{aligned}
E(Y) &= 10.03 \\
Var(Y) &= .0323 . & (3.15)
\end{aligned}
$$

We note how even a modest amount of contamination can seriously inflate the variance. Now, in actuality, we really need to estimate σ_0^2, not

$Var(Y)$. We cannot precisely achieve this goal, since we do not know a priori which lots are contaminated and which are not. However, we can elect to use the following strategy: instead of estimating variability about the mean of the mean of all the lots, rather obtain an unbiased estimate for the variance (or, more commonly of the standard deviation) of the distribution (here there are four possible distributions) of each lot and take the average over all base lots. If we do this, then for the estimate $\hat{\sigma}^2$, we will have

$$E(\hat{\sigma}^2) = \sigma_0^2 + p_1\sigma_1^2 + p_2\sigma_2^2 . \qquad (3.16)$$

We shall follow the convention of basing our estimate for σ_0^2 (or of σ_0) on the average of lot variability estimates throughout this book (except when driven by lot sizes of 1 item per lot). This is our first example of a robust procedure, and, happily, it is followed generally by SPC professionals. Then, with $p_1 = .1$, and $p_2 = .05$, if we use the mean of lot sample means and the mean of lot sample variances, we would obtain estimates for μ_0 and σ_0^2 very close to

$$\mu = 10 + .1(.4) + .05(-.2) = 10.03$$
$$\sigma^2 = .01 + .1(.02) + .05(.08) = .016 .$$

Suppose we made a test using the (still somewhat flawed by contamination) estimates for μ_0 and σ_0^2. We shall assume the lot size is 5. Then for the upper control limit, we would have:

$$\frac{UCL - 10.03}{\sqrt{.016/5}} = 3, \qquad (3.17)$$

giving

$$UCL = 10.03 + 3(.05659) = 10.2 \qquad (3.18)$$
$$LCL = 10.03 - 3(.05659) = 9.86 . \qquad (3.19)$$

We note that the means of the three contaminated possible distributions, 10.4, 9.8 and 10.2, each lies outside (or on the boundary) of the acceptable interval $9.86 < \bar{x} < 10.2$. Hence, even for the very high contamination probabilities $p_1 = .1$ and $p_2 = .05$, using the contaminated values 10.03 and .016 for the mean and variance (as opposed to the uncontaminated values 10 and .01), we would identify each of the contaminated lots, as atypical, with probability .5 or greater using the modest lot size of 5. The contaminated distribution $\mathcal{N}(10.2, .13)$ will be the most difficult to pick as "atypical," but even here, half the area of the density is to the right of the upper control limit, 10.2. We recall that our goal in statistical process control is to find and correct problems. If we have a testing procedure which rings the alarm with at least 50% chance every time we have an atypical lot, then we are doing well. Contamination becomes more of a problem in the case of tests based on the sample variances of lots. We recall that the definitions of sample mean and variance, based on a sample of size n, are given by

$$\bar{x} = \frac{1}{n} \sum_{i=1}^{n} x_i \tag{3.20}$$

and

$$s^2 = \frac{1}{n-1} \sum_{i=1}^{n} (x_i - \bar{x})^2. \tag{3.21}$$

And, if the sample is from a normal distribution with mean μ and variance σ^2, we recall that

$$\chi^2 = \frac{(n-1)s^2}{\sigma^2} \tag{3.22}$$

has a Chi-square distribution with $n-1$ degrees of freedom. Let us suppose we have estimated the variance to be .016 (as opposed to the variance of the uncontaminated population .01). Let us take a contaminated lot with a variance as small (and therefore as hard to distinguish from the norm as possible). In the bolt manufacturing example given, this would be the case where the underlying variance is actually .03. Given the model of contamination used here, a one tailed test on σ^2 is clearly indicated. For moment, let us use a one tailed test with chance one in 500 of a false alarm. Using the tables of the Chi-square distribution, we have:

$$P(\frac{(n-1)s^2}{\sigma^2} \geq 16.924) = .002. \tag{3.23}$$

Now then, for the estimated value of $\sigma^2 = .016$, we shall reject the typicality of a lot if

$$(n-1)s^2 \geq 16.924(.016) = .2708. \tag{3.24}$$

Now, in the event the lot comes from the distribution with variance equal to .03, the critical statistic becomes

$$\chi^2 = \frac{(n-1)s^2}{.03} = \frac{.2708}{.03} = 9.026. \tag{3.25}$$

And, with $n - 1 = 4$ degrees of freedom,

$$P(\chi^2 \geq 9.03) = .0605. \tag{3.26}$$

Thus, we could have a chance of as little as 6% of finding a bad lot. Had we used the true norm variance value of .01, the rejection region would be given by

$$(n - 1)s^2 \geq 16.924(.01) = .16924. \tag{3.27}$$

So for an estimated variance of .03, we would have

$$\chi^2 = \frac{.16924}{.03} = 5.6413, \tag{3.28}$$

giving

$$P(\chi^2 > 5.6413) = .23. \tag{3.29}$$

Contamination is generally more of a problem in the estimation of the standard for the variance than it is for that of the mean. There are many ways of reducing the problem of contamination in the estimation of the normative mean and variance, and we shall discuss some of these later.

In most SPC programs, little allowance is made for contamination in the estimation of the mean and variance of the uncontaminated process. Generally, all the observations are left in for the computation of the overall mean and variance, as long as the assignable cause has not been discovered and removed. We recall that our task in SPC is not to find every atypical lot. Rather, we try to find bad lots (sooner, hopefully, but later will do), track down the assignable causes, and remove them. Shortly, we shall show how the use of median based estimates of the norm process can sometimes enhance our ability to find atypical lots. In Figure 3.1, we see a graphical view of our contamination model. The four different types of epochs, three of them "contaminated," are shown. We have assumed that the sample we take is sufficiently small that it is unlikely that a sample will be drawn from two different epochs. Once we draw a sample from one of the contaminated epochs of production, we have a chance of running a test that will tell us about the presence of contamination. This, hopefully, will be the clue we need to find an assignable cause of the contamination and remove it.

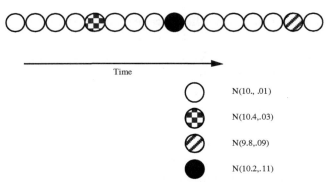

Figure 3.1. Diagram of Contamination Model.

3.3 Estimation of Parameters of the "Norm" Process

Let us suppose we have been sampling lots for some time since the last assignable cause was found and corrected. Thus, we have at our disposal for estimating the mean and variance of the uncontaminated process N sample means and variances. Useful estimates are obtained by simply averaging these statistics:

$$\bar{\bar{x}} = \frac{1}{N}\sum_{j=1}^{N}\bar{x}_j; \tag{3.30}$$

and

$$\bar{s^2} = \frac{1}{N}\sum_{j=1}^{N}s_j^2. \tag{3.31}$$

Now, as is shown in Appendix B, the expectation of $\bar{x}$ is the population mean μ, and the expectation of s^2 is the population variance σ^2. Thus, both estimators are *unbiased*. These results hold regardless of the underlying distribution of the data. Let us suppose we have an estimator $\bar{\hat{\theta}}$ which is composed of an average of unbiased estimators $\{\hat{\theta}_j\}$ of a parameter θ. Then, we immediately have that

$$E(\bar{\hat{\theta}}) = \frac{1}{N}\sum_{j=1}^{N}E(\hat{\theta}_j) = \frac{1}{N}(N\theta) = \theta. \tag{3.32}$$

To develop an estimator for a parameter in statistical process control, it is customary to find an unbiased estimator for the parameter using data from a lot, and then take the average of such estimators across many lots. This procedure tends to work well if the basic unbiased lot estimator has reasonable properties. As we know from the Central Limit Theorem in Appendix B, the asymptotic distribution of the mean of unbiased parameter estimates is normal with mean equal to the parameter, and variance equal to that of a lot estimator divided by N, the number of lots.

To obtain an estimator for the process standard deviation, we can simply take the square root of $\bar{s}^2$. The expectation of this estimator is, indeed, the population variance. A more commonly used estimator for the process standard deviation is the average of the sample standard deviations (i.e., the square roots of the sample variances)

$$\bar{s} = \frac{1}{N}\sum_{j=1}^{N} s_j \tag{3.33}$$

$$= \frac{1}{N}\sum_{j=1}^{N} \sqrt{s^2_j} \; .$$

But we must note that the expected value of s is not quite σ. To see that this is true, we need only recall that for normal data,

$$\frac{(n-1)s^2}{\sigma^2} = z = \chi^2(n-1). \tag{3.34}$$

Thus,

$$E(s) = \frac{\sigma}{2\sqrt{n-1}}\frac{1}{\Gamma(\frac{n-1}{2})}\int_0^\infty e^{-z/2}(\frac{z}{2})^{\frac{n-1}{2}-1}z^{\frac{1}{2}}dz$$

$$= \sigma\sqrt{\frac{1}{2(n-1)}}\frac{1}{\Gamma(\frac{n-1}{2})}\int_0^\infty e^{-z/2}(\frac{z}{2})^{\frac{n}{2}-1}dz \tag{3.35}$$

$$= \frac{\sigma}{\sqrt{n-1}}\frac{\Gamma(\frac{n}{2})}{\Gamma(\frac{n-1}{2})}\sqrt{2}.$$

As an unbiased estimator for σ, then we have (as tabulated in Table 3.1)

$$\hat{\sigma} = \sqrt{n-1}\frac{\Gamma(\frac{n-1}{2})}{\Gamma(\frac{n}{2})}\frac{1}{\sqrt{2}}\bar{s}$$

$$= a(n)\bar{s}. \tag{3.36}$$

Traditionally, because it was not so easy to compute sample deviations on the line before the time of cheap hand held calculators, it used to be common to use a multiple of the average of the sample ranges as an estimate for the population standard deviation. For the j'th sample of size 5, the range R_j is given by

$$R_j = \max\{x_{j1}, x_{j2}, x_{j3}, x_{j4}, x_{j5}\} - \min\{x_{j1}, x_{j2}, x_{j3}, x_{j4}, x_{j5}\}. \quad (3.37)$$

The average of the N ranges is given by

$$\bar{R} = \frac{1}{N} \sum_{j=1}^{N} R_j. \quad (3.38)$$

It is proved in Appendix B, that for normal data, the expected value of the range is a multiple of the standard deviation. Accordingly, when applying $\bar{R}$ to estimate standard deviation σ, we use the formula

$$\hat{\sigma} = b(n)\bar{R}, \quad (3.39)$$

where b_n is given by (B.209).

Table 3.1			
sample size	$a(n)$	$b(n)$	$c(n)$
2	1.253	.8865	1.1829
3	1.1284	.5907	1.0646
4	1.0854	.4857	1.0374
5	1.0638	.4300	1.0260
6	1.0510	.3946	1.0201
7	1.0423	.3698	1.0161
8	1.0363	..3512	1.0136
9	1.0317	.3367	1.0116
10	1.0281	.3249	1.0103
15	1.0180	.2880	1.0063
20	1.0133	.2677	1.0046

In the case where there is no contamination (no removable assignable causes), averages of lots' sample means, $\bar{\bar{x}}$, converge almost surely to μ_0. Such is not the case where we have contamination. Accordingly, we might decide to employ estimates which would not be affected by sample means far away from the underlying population mean.

For example, we might choose as our estimate for μ_0 the median of the sample means:

$$\tilde{\bar{x}} = med\{\bar{x}_j; j = 1, 2, \ldots, N\}. \quad (3.40)$$

This estimator is more "robust" than that based on taking the average of the sample means. In other words, it will be influenced less by information from contaminating distributions, and reflect better the "typical" uncontaminated distribution. For normal data, the sample means are also normal, hence symmetrical, so that the median of their distribution is also equal to the mean. That means that for the uncontaminated case,

$$E(\text{ sample median}) = \mu_0. \tag{3.41}$$

For the contaminated case, the sample median will generally be closer to μ_0 than will $\bar{\bar{x}}$. Regardless of the lot size, the $\bar{\bar{x}}$ estimator for the mean of the uncontaminated process will not converge to that mean. In the Appendix, we show that the $\tilde{x}$ estimator does converge stochastically (as both n and N go to infinity) to the process mean if the proportion of lots from the uncontaminated process is greater than 50%. More importantly, even for small lot sizes the median of means estimator will tend to be closer to the mean of the uncontaminated distribution than will the mean of means estimator. Clearly, unusually large sample means and unusually small means may be the result of contamination of the production process. In fact, it is our purpose to identify these "outliers" as Pareto glitches and remove their source. If we include them in the computation of our estimate for "the norm" it is possible that they may so distort it that we will be impaired in identifying the contaminants.

Similarly, we may find it useful to use a multiple of the median of sample standard deviations to estimate σ_0

$$\tilde{s} = med\{s_j; j = 1, 2, \ldots, N\}. \tag{3.42}$$

Given (3.34), we can solve for the median of the distribution of s_j by solving the equation

$$.5 = \frac{1}{\Gamma(\frac{n-1}{2})} \int_0^{\frac{(n-1)s^2}{\sigma^2}} e^{-z} z^{\frac{n-1}{2}-1} dz. \tag{3.43}$$

We have included in Table 3.1, a column of constants for the use in obtaining the ratio of $E(s)$ and $med(s)$ via

$$\frac{E(s)}{\tilde{s}} = c(n).$$

Thus, the estimator for σ based on the median of sample standard deviations assumes the form

$$\tilde{\sigma} = a(n)c(n)\tilde{s}. \tag{3.44}$$

The disadvantages associated with the use of the median estimation procedure are a slight loss of efficiency in the uncontaminated normal case and a difficulty in having workers on the line enter a long list of sample mean data into a calculator which can then carry out the sorting of the data in order to obtain sample median (i.e., the middle observation in a sort from smallest to largest). The first of these objections is of little practical consequence. Typically, we have rather a long list of sample means. For practical purposes, we can consider it to be infinite. Moreover, a small error in estimating the mean and variance of the uncontaminated process is unlikely to be important. The second objection, concerning the difficulty of requesting line workers to enter data into calculators for sorting, has more validity. However, as a practical matter, it is not a particularly good idea to bother line workers with this degree of technical trivia; they have other more pressing concerns. The estimates for the uncontaminated mean and standard deviation, and hence for the upper and lower control limits, need not be updated at every lot sampling. Daily or less frequently will generally do nicely. The computation of the control limits should be carried out according to a regular standard protocol by a more central computation center than the line. At one of the factories where we have installed an SPC system, the head of the testing laboratories has run data entered into a spreadsheet, which feeds into a standard protocol for plotting and updating control limit estimates. Any crossing of the control limits is a matter of immediate concern and action by the line workers. If they find the cause of a "Pareto glitch" and remove it, this is noted for possible use in deleting data prior to the fix (not a high priority item if the robust median estimation protocol is followed). Each morning, the newest versions of the control charts are put on the desk of the head of the testing laboratories, who then examines them and passes them on to the lines, frequently carrying them himself so that discussions with foremen are possible. Most of the companies with whom we have worked in America and Poland had already some sort of "quality control" or "quality assurance" group. Some used control charts and had strong feelings about using the one rule or another for constructing the mean control chart. Each of these was more or less oriented to the "plus or minus three standard deviations" rule. In showing workers how easy it is to enter data and automatically obtain sample mean and standard deviation data with one of the sturdy, background light powered calculators, such as the Texas Instruments TI-36 Solar (cost of around $15 in 1992), most are willing to begin using direct estimates of the standard deviation rather than the range. But if one prefers to use a range based estimate or some other similar statistic for the standard deviation, that is a matter of small concern. Let us consider now 30 days of simulated data (one sample per shift) from the bolt production line discussed earlier.

Lot	x_1	x_2	x_3	x_4	x_5	$\bar{x}$	s	R	s^2
1	9.927	9.920	10.170	9.976	9.899	9.978	0.111	0.271	0.012
2	9.862	10.003	9.829	9.824	10.077	9.919	0.114	0.253	0.013
3	10.061	10.089	9.950	9.929	9.935	9.993	0.076	0.160	0.006
4	9.820	10.066	10.062	9.897	10.013	9.972	0.109	0.246	0.012
5	9.737	9.937	9.928	10.144	9.965	9.942	0.145	0.406	0.021
6	9.876	9.957	9.845	9.913	9.941	9.906	0.046	0.112	0.002
7	9.898	9.959	9.924	9.989	9.987	9.951	0.040	0.092	0.002
8	10.001	10.050	10.263	9.982	10.076	10.074	0.112	0.281	0.013
9	9.928	10.234	9.832	10.027	10.121	10.028	0.158	0.402	0.025
10	9.896	9.994	10.009	9.835	10.162	9.979	0.125	0.327	0.016
11	10.011	10.011	10.090	10.095	10.120	10.065	0.051	0.108	0.003
12	9.983 9	.974	10.071	10.099	9.992	10.024	0.057	0.125	0.003
13	10.127	9.935	9.979	10.014	9.876	9.986	0.094	0.251	0.009
14	10.025	9.890	10.002	9.999	9.937	9.971	0.056	0.136	0.003
15	9.953	10.000	10.141	10.130	10.154	10.076	0.092	0.201	0.009
16	10.007	10.005	9.883	9.941	9.990	9.965	0.053	0.124	0.003
17	10.062	10.005	10.070	10.270	10.071	10.096	0.101	0.266	0.010
18	10.168	10.045	10.140	9.918	9.789	10.012	0.158	0.379	0.025
19	9.986	10.041	9.998	9.992	9.961	9.996	0.029	0.080	0.001
20	9.786	10.145	10.012	10.110	9.819	9.974	0.165	0.359	0.027
21	9.957	9.984	10.273	10.142	10.190	10.109	0.135	0.316	0.018
22	9.965	10.011	9.810	10.057	9.737	9.916	0.137	0.321	0.019
23	9.989	10.063	10.148	9.826	10.041	10.013	0.119	0.322	0.014
24	9.983	9.974	9.883	10.153	10.092	10.017	0.106	0.270	0.011
25	10.063	10.075	9.988	10.071	10.096	10.059	0.041	0.108	0.002
26	9.767	9.994	9.935	10.114	9.964	9.955	0.125	0.347	0.016
27	9.933	9.974	10.026	9.937	10.165	10.007	0.096	0.232	0.009
28	10.227	10.517	10.583	10.501	10.293	10.424	0.154	0.356	0.024
29	10.022	9.986	10.152	9.922	10.101	10.034	0.091	0.124	0.002
30	9.845	9.901	10.020	9.751	10.088	9.921	0.135	0.337	0.018
31	9.956	9.921	10.132	10.016	10.109	10.027	0.092	0.212	0.009
32	9.876	10.114	9.938	10.195	10.010	10.027	0.129	0.318	0.017
33	9.932	9.856	10.085	10.207	10.146	10.045	0.147	0.352	0.022
34	10.016	9.990	10.106	10.039	9.948	10.020	0.059	0.158	0.003
35	9.927	10.066	10.038	9.896	9.871	9.960	0.087	0.195	0.008
36	9.952	10.056	9.948	9.802	9.947	9.941	0.090	0.254	0.008
37	9.941	9.964	9.943	10.085	10.049	9.996	0.066	0.144	0.004
38	10.010	9.841	10.031	9.975	9.880	9.947	0.083	0.190	0.007
39	9.848	9.944	9.828	9.834	10.091	9.909	0.112	0.262	0.013
40	10.002	9.452	9.921	9.602	9.995	9.794	0.252	0.550	0.064
41	10.031	10.061	9.943	9.997	9.952	9.997	0.050	0.118	0.003
42	9.990	9.972	10.068	9.930	10.113	10.015	0.074	0.183	0.006
43	9.995	10.056	10.061	10.016	10.044	10.034	0.028	0.066	0.001
44	9.980	10.094	9.988	9.961	10.140	10.033	0.079	0.179	0.006
45	10.058	9.979	9.917	9.881	9.966	9.960	0.067	0.176	0.004
46	10.006	10.221	9.841	10.115	9.964	10.029	0.145	0.380	0.021
47	10.132	9.920	10.094	9.935	9.975	10.011	0.096	0.212	0.009
48	10.012	10.043	9.932	10.072	9.892	9.990	0.076	0.179	0.006
49	10.097	9.894	10.101	9.959	10.040	10.018	0.090	0.207	0.008
50	10.007	9.789	10.015	9.941	10.013	9.953	0.097	0.226	0.009
51	9.967	9.947	10.037	9.824	9.938	9.943	0.077	0.213	0.006
52	9.981	10.053	9.762	9.920	10.107	9.965	0.134	0.346	0.018
53	9.841	9.926	9.892	10.152	9.965	9.955	0.119	0.311	0.014
54	9.992	9.924	9.972	9.755	9.925	9.914	0.093	0.236	0.009
55	9.908	9.894	10.043	9.903	9.842	9.918	0.075	0.201	0.006
56	10.011	9.967	10.204	9.939	10.077	10.040	0.106	0.265	0.011
57	10.064	10.036	9.733	9.985	9.972	9.958	0.131	0.330	0.017
58	9.891	10.055	10.235	10.064	10.092	10.067	0.122	0.345	0.015
59	9.869	9.934	10.216	9.962	10.012	9.999	0.132	0.346	0.017
60	10.016	9.996	10.095	10.029	10.080	10.043	0.042	0.099	0.002
61	10.008	10.157	9.988	9.926	10.008	10.017	0.085	0.231	0.007
62	10.100	9.853	10.067	9.739	10.092	9.970	0.165	0.361	0.027
63	9.904	9.848	9.949	9.929	9.904	9.907	0.038	0.101	0.001
64	9.979	10.008	9.963	10.132	9.924	10.001	0.079	0.208	0.006
65	9.982	9.963	10.061	9.970	9.937	9.983	0.047	0.124	0.002

Lot	x_1	x_2	x_3	x_4	x_5	$\bar{x}$	s	R	s^2
66	10.028	10.079	9.970	10.087	10.094	10.052	0.052	0.123	0.003
67	9.995	10.029	9.991	10.232	10.189	10.087	0.115	0.241	0.013
68	9.936	10.022	9.940	10.248	9.948	10.019	0.133	0.312	0.018
69	10.014	10.070	9.890	10.137	9.901	10.002	0.107	0.247	0.011
70	10.005	10.044	10.016	10.188	10.116	10.074	0.077	0.183	0.006
71	10.116	10.028	10.152	10.047	10.040	10.077	0.054	0.124	0.003
72	9.934	10.025	10.129	10.054	10.124	10.053	0.080	0.195	0.006
73	9.972	9.855	9.931	9.785	9.846	9.878	0.074	0.187	0.005
74	10.014	10.000	9.978	10.133	10.100	10.045	0.068	0.155	0.005
75	10.093	9.994	10.090	10.079	9.998	10.051	0.050	0.098	0.003
76	9.927	9.832	9.806	10.042	9.914	9.904	0.093	0.236	0.009
77	10.177	9.884	10.070	9.980	10.089	10.040	0.112	0.293	0.013
78	9.825	10.106	9.959	9.901	9.964	9.951	0.103	0.281	0.011
79	10.333	10.280	10.509	10.631	10.444	10.439	0.140	0.204	0.006
80	9.972	10.116	10.084	10.059	9.914	10.029	0.084	0.202	0.007
81	10.059	9.992	9.981	9.800	9.950	9.956	0.096	0.259	0.009
82	9.832	10.075	10.111	9.954	9.946	9.984	0.112	0.279	0.012
83	9.958	9.884	9.986	10.008	10.113	9.990	0.083	0.229	0.007
84	10.087	9.994	9.915	10.023	9.883	9.980	0.082	0.205	0.007
85	10.232	9.966	9.991	10.021	9.965	10.035	0.112	0.324	0.014
86	10.066	9.948	9.769	10.102	9.932	9.963	0.131	0.333	0.017
87	10.041	10.044	10.091	10.031	9.958	10.033	0.048	0.133	0.002
88	9.868	9.955	9.769	10.023	9.921	9.907	0.096	0.254	0.009
89	10.084	10.018	9.941	10.052	10.026	10.024	0.053	0.143	0.003
90	10.063	10.055	10.104	10.080	10.064	10.073	0.019	0.049	0.001

Table 3.2 (continued)

Let us note what one obtains from this data set as estimates of the uncontaminated mean and standard deviation using the various estimates mentioned. The mean of the 90 sample means is seen to be 10.004. The median of the sample means is 10.000. The mean of the sample standard deviations is seen to be .095, giving (using Table 3.1) as estimate for the standard deviation of the uncontaminated process

$$\hat{\sigma} = a(5)\bar{s} = (1.0638)(.095) = .1011. \qquad (3.45)$$

The median of the sample standard deviations is .093 giving as a robust estimate of the standard deviation of the uncontaminated process

$$\tilde{\sigma} = a(5)c(5)\tilde{s} = (1.0638)(1.026).093 = .1015. \qquad (3.46)$$

The mean of the sample ranges is .232, giving the estimate (using Table 3.1) for the population standard deviation

$$\hat{\sigma}_R = (.232)(.430) = .0998. \qquad (3.47)$$

In constructing our control chart, we shall use the most common estimate for the mean, namely the average of the sample means, as our center for the mean control chart, i.e., 10.004. As our estimate for the standard deviation, we will use the most common of the estimates, namely the average of the sample standard deviations multiplied by the appropriate $a(n)$ from Table 3.1. Thus, our control chart for the mean is given by

$$UCL = \bar{\bar{x}} + 3\frac{a(5)\bar{s}}{\sqrt{5}} \qquad (3.48)$$

$$= 10.004 + 3\frac{.1011}{\sqrt{5}}$$

$$= 10.139; \tag{3.49}$$

$$LCL = \overline{\overline{x}} - 3\frac{a(5)\bar{s}}{\sqrt{5}} \tag{3.50}$$

$$= 10.004 - 3\frac{.1011}{\sqrt{5}}$$

$$= 9.868. \tag{3.51}$$

We note that lots 28 and 79 are above the upper control limit. Hopefully, we would be able to use this information to identify the source of the $\mathcal{N}(10.4,.03)$ contamination and remove it. Similarly, lot 40 has a mean below the lower control limit. Hopefully, we would be able to use this information to find the source of the $\mathcal{N}(9.8,.09)$ contamination and remove it.

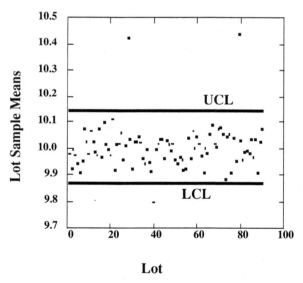

Figure 3.2. Mean Control Chart.

To summarize, the usual method for obtaining the upper and lower control limits for an SPC sample mean control chart is to find the sample mean of the sample means of the lots of size n, hence the expression $\overline{\overline{x}}$. We then find the sample mean of the sample standard deviations of the lots, hence the expression $\bar{s}$. This latter statistic is not quite an unbiased estimator for σ. We correct this via the formula

$$\hat{\sigma} = a(n)\bar{s}. \tag{3.52}$$

The interval in which we take a lot sample mean to be "typical" and hence the production process to be "in control" is given by

$$\bar{\bar{x}} - 3\frac{\hat{\sigma}}{\sqrt{n}} \le \bar{x} \le \bar{\bar{x}} + 3\frac{\hat{\sigma}}{\sqrt{n}}. \tag{3.53}$$

Thus, we have

$$UCL = \bar{\bar{x}} + 3a(n)\frac{\bar{s}}{\sqrt{n}} \tag{3.54}$$

and

$$LCL = \bar{\bar{x}} - 3a(n)\frac{\bar{s}}{\sqrt{n}}. \tag{3.55}$$

Let us now go on to examine the control chart for the standard deviation. We shall follow the convention of estimating the standard deviation using

$$\hat{\sigma} = a(n)\bar{s} = a(n)\frac{1}{N}\sum_{j=1}^{N}s_j. \tag{3.56}$$

We need an estimator of the standard deviation of the lot standard deviations. Now, for each lot

$$
\begin{aligned}
Var(s) &= E(s^2) - [E(s)]^2 & (3.57)\\
&= \sigma^2 - [\frac{\sigma}{a(n)}]^2 & (3.58)\\
&= \sigma^2(\frac{a(n)^2 - 1}{a(n)^2}). & (3.59)
\end{aligned}
$$

Recalling that our customary estimator for σ is $a(n)\bar{s}$, we have

$$UCL = \bar{s} + 3\bar{s}[a(n)^2 - 1]^{\frac{1}{2}}. \tag{3.60}$$

$$LCL = \bar{s} - 3\bar{s}[a(n)^2 - 1]^{\frac{1}{2}}. \tag{3.61}$$

For the data set at hand, this gives us

$$UCL = .095 + 3[1.06381^2 - 1]^{\frac{1}{2}}.100 = .2039. \tag{3.62}$$

Similarly, we have

$$LCL = .095 - 3[1.06381^2 - 1]^{\frac{1}{2}}.100 = -.0139. \tag{3.63}$$

Naturally, this latter control limit is truncated to 0.

Here, we note that we do pick up only one alarm (lot 40) from the 90 standard deviations. Typically, we expect to discover more Pareto glitches from the mean rather than the standard deviation control charts. In the real world, if we are doing our job properly, we will not wait for 30 days of data before intervening to improve the system. Indeed, if we do not investigate the causes for glitches immediately, we shall generally not be able to determine an assignable cause.

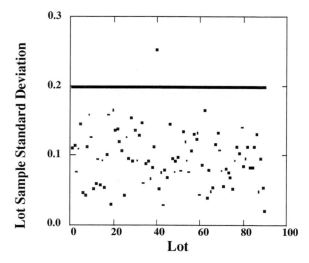

Figure 3.3. Standard Deviation Control Chart.

Generally, whenever we examine variability of measurements within lots, we ourselves rely on standard deviation control charts. In Section 3.7, however, we show how to construct control charts for the lots' ranges.

3.4   Robust Estimators for Uncontaminated Process Parameters

The contamination of the estimators for the "typical" process by data points from the contaminating distribution(s) can pose problems. Let us consider, for example, the situation where the "typical" process is Gaussian with mean 10 and standard deviation .1. The contaminating distribution is Gaussian with mean 9.8 and standard deviation .3. We will assume that with probability .70 a lot comes from $\mathcal{N}(10,.01)$ and with probability .3 it comes from $\mathcal{N}(9.8,.09)$. Thus, we are considering a case where the contamination of the "typical" process is very high indeed. As previously, we shall assume the lot sample size is 5.

Lot	x_1	x_2	x_3	x_4	x_5	$\bar{x}$	s	R	s^2
1	9.987	9.921	9.969	10.239	9.981	10.020	.125	.318	.016
2	9.629	9.798	9.342	9.644	10.465	9.775	0.419	1.123	0.176
3	10.122	9.705	9.910	9.537	9.968	9.848	0.229	0.585	0.053
4	9.861	10.033	9.806	9.930	9.923	9.910	0.085	0.227	0.007
5	9.816	10.061	9.753	10.435	10.165	10.046	0.276	0.682	0.076
6	9.925	9.830	9.556	9.448	9.654	9.682	0.195	0.476	0.038
7	10.103	9.921	10.066	9.970	10.083	10.029	0.079	0.181	0.006
8	10.057	10.078	10.107	10.024	9.993	10.052	0.044	0.113	0.002
9	9.956	10.106	10.349	9.818	10.146	10.075	0.201	0.532	0.040
10	9.591	9.996	10.047	9.544	9.787	9.793	0.228	0.502	0.052
11	10.185	9.809	9.367	10.179	9.791	9.866	0.338	0.818	0.115
12	10.051	9.998	9.998	9.957	9.886	9.978	0.061	0.165	0.004
13	9.966	9.748	9.501	10.086	10.123	9.885	0.260	0.622	0.067
14	9.987	9.863	10.121	10.125	10.093	10.038	0.113	0.262	0.013
15	9.926	9.964	10.137	10.116	9.925	10.014	0.105	0.212	0.011
16	9.734	9.267	9.803	9.903	9.884	9.718	0.261	0.636	0.068
17	9.980	9.980	10.005	9.957	9.927	9.970	0.029	0.078	0.001
18	10.012	9.936	10.080	10.023	10.025	10.015	0.052	0.144	0.003
19	10.263	9.040	10.058	9.670	9.918	9.790	0.471	1.223	0.222
20	9.954	10.025	10.212	9.954	9.933	10.016	0.115	0.278	0.013
21	9.762	10.100	9.928	9.986	10.156	9.987	0.154	0.394	0.024
22	10.008	9.884	10.067	10.002	9.994	9.991	0.066	0.183	0.004
23	10.091	9.936	10.040	10.085	10.012	10.033	0.063	0.155	0.004
24	9.795	9.874	9.491	9.472	9.702	9.667	0.180	0.402	0.032
25	9.968	10.140	9.855	10.030	10.131	10.024	0.119	0.285	0.014
26	9.948	10.008	9.967	10.010	9.890	9.965	0.049	0.119	0.002
27	9.864	9.937	9.990	9.991	9.980	9.952	0.054	0.127	0.003
28	10.162	9.925	9.986	9.963	9.834	9.974	0.120	0.328	0.014
29	9.941	10.064	10.012	9.957	9.981	9.991	0.048	0.122	0.002
30	9.791	9.695	9.415	10.155	9.540	9.719	0.283	0.740	0.080
31	10.052	9.993	9.970	10.040	10.017	10.014	0.034	0.083	0.001
32	10.058	9.982	10.031	10.078	10.104	10.051	0.047	0.122	0.002
33	10.144	10.052	10.077	9.932	9.999	10.041	0.080	0.213	0.006
34	10.080	10.047	10.113	9.874	10.116	10.046	0.100	0.243	0.010
35	9.912	9.991	9.933	9.935	10.099	9.974	0.076	0.187	0.006
36	9.973	9.863	9.785	10.015	10.007	9.929	0.101	0.230	0.010
37	10.003	10.087	10.032	10.068	10.039	10.046	0.033	0.084	0.001
38	10.096	9.977	10.009	10.193	9.951	10.045	0.099	0.242	0.010
39	10.023	10.109	9.893	9.881	9.873	9.956	0.105	0.236	0.011
40	10.188	10.173	10.057	9.927	9.963	10.062	0.119	0.261	0.014
41	10.005	9.973	10.034	10.029	9.903	9.989	0.054	0.131	0.003
42	10.018	9.927	10.002	9.829	9.910	9.937	0.076	0.189	0.006
43	9.961	10.021	10.014	10.147	9.927	10.014	0.084	0.220	0.007
44	9.914	9.870	9.824	10.057	9.965	9.926	0.090	0.233	0.008
45	9.969	9.946	10.317	10.043	10.163	10.088	0.154	0.371	0.024
46	9.976	10.005	10.019	10.001	9.986	9.997	0.017	0.043	0.000
47	9.927	10.052	9.987	10.060	10.003	10.006	0.054	0.132	0.003
48	9.934	10.087	9.919	9.567	9.893	9.880	0.191	0.520	0.036
49	9.956	9.824	10.003	10.044	10.072	9.980	0.098	0.248	0.010
50	10.001	9.935	10.033	10.019	10.137	10.025	0.073	0.202	0.005
51	10.097	9.896	9.977	10.040	10.021	10.006	0.075	0.201	0.006
52	10.179	10.018	10.126	9.711	10.027	10.012	0.181	0.468	0.033
53	9.941	10.316	9.315	9.708	9.518	9.760	0.388	1.001	0.150
54	10.089	9.846	9.949	9.421	9.395	9.740	0.315	0.693	0.099
55	10.006	9.883	10.116	10.033	10.028	10.013	0.084	0.233	0.007
56	9.976	10.010	10.049	10.045	10.081	10.032	0.040	0.104	0.002
57	10.003	9.927	10.009	10.012	9.961	9.982	0.037	0.085	0.001
58	9.875	10.093	10.116	9.971	9.968	10.005	0.099	0.241	0.010
59	9.978	10.101	10.061	9.951	9.884	9.995	0.087	0.217	0.008
60	10.057	9.911	9.752	9.879	10.196	9.959	0.171	0.444	0.029
61	9.853	9.595	9.751	9.765	9.864	0.291	0.763	0.085	
62	9.893	9.574	9.342	9.759	10.011	9.716	0.265	0.669	0.070
63	10.105	10.064	10.166	10.101	9.885	10.064	0.107	0.281	0.011
64	9.963	9.562	9.337	9.381	9.498	9.548	0.249	0.626	0.062
65	9.593	9.945	10.264	9.918	9.925	9.929	0.237	0.671	0.056

Table 3.3

Lot	x_1	x_2	x_3	x_4	x_5	$\bar{x}$	s	R	s^2
\multicolumn{10}{c}{Table 3.3 (continued)}
66	9.580	9.389	10.070	9.324	10.301	9.733	0.432	0.978	0.186
67	9.975	10.103	10.064	10.078	10.057	10.056	0.048	0.128	0.002
68	10.140	9.905	9.709	10.386	9.835	9.995	0.269	0.677	0.072
69	10.181	9.860	10.027	10.051	9.965	10.017	0.118	0.321	0.014
70	9.414	9.791	9.625	9.876	9.704	9.682	0.177	0.462	0.031
71	9.592	9.892	9.587	9.872	10.116	9.812	0.224	0.529	0.050
72	9.952	10.171	9.971	9.992	10.009	10.019	0.088	0.219	0.008
73	9.839	9.888	10.191	9.983	10.024	9.985	0.137	0.352	0.019
74	9.964	10.154	9.855	9.962	10.009	9.989	0.108	0.299	0.012
75	9.770	9.729	9.829	10.088	9.187	9.721	0.329	0.900	0.108
76	9.503	9.682	9.527	9.817	9.964	9.698	0.195	0.461	0.038
77	9.676	9.879	9.909	9.969	10.194	9.925	0.186	0.518	0.035
78	10.015	10.177	9.756	9.997	9.962	9.981	0.151	0.421	0.023
79	9.799	10.026	9.906	10.153	10.014	9.980	0.134	0.354	0.018
80	10.147	9.948	10.014	9.734	9.711	9.911	0.186	0.435	0.035
81	10.104	10.060	9.989	9.813	9.930	9.979	0.114	0.291	0.013
82	10.082	10.047	10.088	9.963	9.932	10.023	0.071	0.156	0.005
83	10.174	10.076	10.159	9.961	9.965	10.067	0.102	0.213	0.010
84	9.922	9.957	10.053	10.075	10.049	10.011	0.067	0.153	0.005
85	10.208	10.008	10.064	10.164	9.889	10.067	0.127	0.319	0.016
86	9.868	9.964	9.799	10.004	9.972	9.921	0.085	0.205	0.007
87	10.029	9.512	9.608	9.784	10.071	9.801	0.248	0.560	0.062
88	10.053	10.152	10.191	9.952	10.045	10.079	0.095	0.239	0.009
89	9.585	10.098	9.538	9.441	9.719	9.676	0.256	0.657	0.066
90	9.956	9.899	9.947	9.909	10.023	9.947	0.049	0.124	0.002

Let us now consider using the standard procedure of creating a mean
control chart. The sample mean of the 90 lot sample means is 9.939.
The sample mean of the 90 lot standard deviations is .146. Thus, our
customary control limits would be given by

$$9.939 - \frac{3(.146)(1.0638)}{\sqrt{5}} \le \bar{\bar{x}} \le 9.939 + \frac{3(.146)(1.0638)}{\sqrt{5}}. \tag{3.64}$$

The resulting control chart is shown in Figure 3.4.

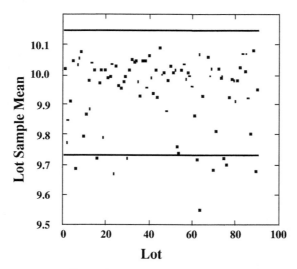

**Figure 3.4. Standard Mean Control Chart With 30%
Contamination.**

Here the control limits seem to bear little relation to the obvious two clusters revealed by the graph. Nevertheless, using the naive sample mean control chart blindly, we would be able to pick 10 of the 29 lots which come from the contaminating distribution.

Now, let us consider, based on the data from Table 3.3, the mean control chart we obtain when we take the median of the lot sample means to estimate the mean of the uncontaminated distribution and the median of the sample standard deviations, appropriately adjusted, to estimate the standard deviation of the uncontaminated distribution. Sorting the 90 lot sample means from smallest to largest, we find that the median of the lot sample means is given by

$$\tilde{x} = 9.984. \tag{3.65}$$

The median of the lot standard deviations is given by .110, giving as the estimate for the uncontaminated standard deviation

$$\tilde{\sigma} = c(5)a(5)\tilde{s} = 1.0638(1.026).110 = .1203. \tag{3.66}$$

This gives as a control interval, using robust estimates,

$$9.984 - \frac{3(.110)(1.0638)}{\sqrt{5}} \leq \bar{\bar{x}} \leq 9.984 + \frac{3(.110)(1.0638)}{\sqrt{5}}. \tag{3.67}$$

In Figure 3.5, we show that the lot mean control chart, using median estimates for the uncontaminated mean and standard deviation, does a much better job of separating $\mathcal{N}(9.8,.09)$ lots from $\mathcal{N}(10.0,.01)$ lots.

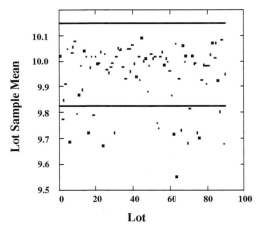

Figure 3.5. Robust Lot Mean Control Chart With 30% Contamination.

Next, we consider in Figure 3.6 the two UCL values obtained for s, depending on whether one uses a mean of the lot standard deviations or a median of the lot standard deviation based estimates. For the 30% contamination example in Table 3.3, the average of the sample standard deviations is $\bar{s} = .146$. The median of the sample standard deviations is $\tilde{s} = .110$. We recall from (3.60) and (3.61) that the upper and lower control limits are given by

$$UCL = \bar{s} + 3\bar{s}[a(n)^2 - 1]^{\frac{1}{2}} = .3049 \qquad (3.68)$$

and

$$LCL = \bar{s} - 3\bar{s}[a(n)^2 - 1]^{\frac{1}{2}} = -.013. \qquad (3.69)$$

For the median of the lot sample standard deviations, we replace $\bar{s}$ by $c(5)\tilde{s} = 1.0260(.110) = .1129$. This gives us

$$UCL = c(5)\tilde{s}(1. + 3[a(n)^2 - 1]^{\frac{1}{2}}) = .2358 \qquad (3.70)$$

and

$$LCL = c(5)\tilde{s}(1. - 3[a(n)^2 - 1]^{\frac{1}{2}}) = -.01. \qquad (3.71)$$

Again, as with the $\bar{s}$ based chart, the lower control limit will be truncated to zero.

We note the dashed upper control limit, representing that corresponding to $\bar{s}$. This limit identifies only 7 of the 29 points from the contaminating distribution. On the other hand, the solid, $\tilde{s}$ based upper control limit splits the two clusters $\mathcal{N}(10,.01)$ and $\mathcal{N}(9.8,.09)$ of points much more effectively, successfully identifying 18 of the 29 points from $\mathcal{N}(9.8,.09)$. Of course, we stated early on that in statistical process control, it is not generally necessary that we identify all the bad lots, since our goal is not removal of bad lots, but identification and correction of the production problem causing the bad lots. Nevertheless, there would seem to be little reason not to avail ourselves of the greater robustness of $\tilde{x}$ and $\bar{s}$, other than the fact that, in so doing, we will be using a procedure slightly different from the standard approaches, based on $\bar{\bar{x}}$ and $\tilde{s}$. We ourselves generally default to more standard approaches with which a client is more familiar.

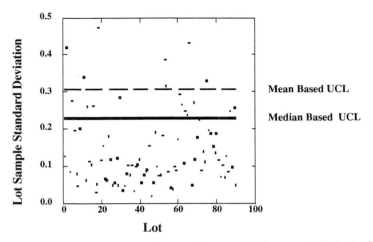

**Figure 3.6. Standard and Robust Control Charts With 30%
Contamination.**

In the Remark in Section 3.6 we show how to use ranges to construct
control charts for the ranges. Thus, if one prefers ranges to standard
deviations, one can not only use ranges to obtain a control chart for the
mean but also to add to it a control chart for the ranges.

3.5   A Process with Mean Drift

Although the contamination process model is probably the most useful
for improvement via statistical process control, various other models may
prove important in various situations. Let us consider, for example, a
model which is related to the case of the juggler in Chapter 1. We
recall that the tiring juggler is confronted with the increasingly difficult
problem of keeping three balls moving smoothly through the air. It
is this sort of anthropomorphic model that most people, not familiar
with quality improvement, assume must be close to production reality.
Happily, this is not generally the case. Nevertheless, let us try to model
such decaying quality in our example of the production of bolts. Here,
we shall assume that a lot sampled at regular time t gives bolts with
measured diameters distributed normally with mean μ_t and variance σ^2.
But we shall assume that the mean drifts according to the so-called
Markov (one step memory) model:

$$\mu_t = \mu_{t-1} + \epsilon_t, \tag{3.72}$$

where ϵ_t is normally distributed with mean 0 and variance τ^2. In our simulation of 90 lots, $\mu_1 = 10$, $\sigma = .1$, and $\tau = .01$.

					Table 3.4			
Lot	x_1	x_2	x_3	x_4	x_5	$\bar{x}$	s	Δ
1	9.927	10.019	9.963	10.073	9.911	9.979	0.067	-0.021
2	9.856	9.985	10.074	9.885	9.970	9.954	0.086	-0.025
3	10.031	9.957	10.133	9.877	10.078	10.015	0.101	0.061
4	9.850	10.071	10.000	10.115	10.011	10.009	0.101	-0.006
5	9.858	10.159	10.039	10.141	10.114	10.062	0.123	0.053
6	10.127	10.006	9.852	10.095	10.090	10.034	0.111	-0.028
7	10.131	10.144	9.964	10.027	9.995	10.052	0.081	0.018
8	9.924	10.008	10.043	10.155	9.939	10.014	0.093	-0.038
9	9.869	9.902	9.739	10.171	10.111	9.958	0.179	-0.055
10	10.132	9.974	9.993	9.967	10.058	10.025	0.070	0.066
11	9.877	9.970	9.980	10.031	9.987	9.969	0.057	-0.056
12	10.089	9.803	9.901	9.925	9.890	9.922	0.104	-0.047
13	10.074	9.954	9.957	9.819	9.943	9.949	0.090	0.027
14	9.977	10.011	9.966	9.876	9.890	9.944	0.058	-0.005
15	10.040	9.961	9.841	10.251	10.125	10.044	0.156	0.100
16	10.049	10.086	10.056	10.045	10.113	10.070	0.029	0.026
17	10.040	10.039	10.014	9.992	9.836	9.984	0.085	-0.085
18	9.964	10.058	9.995	9.940	10.060	10.003	0.054	0.019
19	10.058	9.885	10.081	10.030	10.011	10.013	0.076	0.010
20	10.133	10.049	10.023	10.147	9.904	10.051	0.098	0.038
21	10.063	10.033	10.050	10.065	9.983	10.039	0.034	-0.012
22	10.040	9.954	9.955	10.031	9.940	9.984	0.047	-0.055
23	9.889	10.049	10.117	10.157	9.962	10.035	0.110	0.051
24	10.092	9.889	9.947	9.975	10.203	10.021	0.125	-0.014
25	9.948	10.188	10.014	9.970	9.975	10.019	10.019	-0.002
26	10.119	9.949	10.210	10.114	9.835	10.045	0.150	0.026
27	10.012	10.194	9.980	10.057	10.014	10.051	0.084	0.006
28	9.943	9.835	10.036	9.882	10.160	9.971	0.130	-0.080
29	10.072	10.019	9.946	10.119	10.026	10.036	0.065	0.065
30	10.062	10.280	9.955	10.028	10.131	10.091	0.123	0.055
31	9.988	10.101	10.034	9.951	10.150	10.045	0.081	-0.047
32	10.137	10.031	10.045	10.061	9.936	10.042	0.072	-0.003
33	10.020	10.031	10.060	9.899	9.979	9.998	0.062	-0.044
34	9.870	10.156	10.166	10.052	10.022	10.053	0.120	0.055
35	10.107	10.053	9.785	9.902	9.980	9.965	0.127	-0.088
36	10.118	10.059	10.145	10.149	9.968	10.088	0.076	0.122
37	9.928	9.974	10.120	10.021	9.959	10.001	0.075	-0.087
38	9.934	9.870	10.068	9.828	9.745	9.889	0.121	-0.112
39	9.893	9.997	9.923	9.891	10.007	9.942	0.056	0.053
40	10.120	10.019	9.946	10.001	10.055	10.028	0.065	0.086
41	9.996	9.796	9.967	9.976	9.999	9.947	0.085	-0.081
42	10.071	9.776	9.941	9.904	10.077	9.954	0.126	0.007
43	9.949	9.869	9.925	10.009	9.920	9.934	0.051	-0.019
44	9.930	9.998	9.836	10.195	10.020	9.996	0.132	0.061
45	9.983	9.967	10.045	9.993	10.119	10.021	0.062	0.026
46	10.086	9.995	9.955	9.969	9.830	9.967	0.092	-0.054
47	9.896	9.999	10.051	9.878	9.857	9.936	0.084	-0.031
48	9.997	10.089	9.893	9.968	9.971	9.984	0.070	0.047
49	9.987	9.928	10.006	9.997	10.108	10.005	0.065	0.021
50	9.958	9.999	10.078	9.912	10.163	10.022	0.099	0.017
51	10.075	9.940	10.027	9.800	9.926	9.954	0.106	-0.068
52	10.099	10.042	10.017	9.845	10.061	10.013	0.098	0.059
53	10.097	10.133	9.997	9.911	9.890	10.006	0.108	-0.007
54	10.117	10.020	9.813	9.934	10.070	9.991	0.121	-0.015
55	9.779	10.133	10.086	9.823	10.103	9.985	0.169	-0.006
56	10.015	9.885	10.034	10.013	10.008	9.991	0.060	0.006
57	9.849	10.153	9.996	9.896	9.842	9.947	0.131	-0.043
58	10.010	9.825	9.883	9.967	9.930	9.923	0.072	-0.025
59	9.897	9.975	9.914	10.081	9.842	9.942	0.091	0.019
60	9.857	10.244	9.961	9.940	9.896	9.980	0.153	0.038
61	9.887	9.913	10.021	9.924	9.897	9.929	0.054	-0.051
62	9.920	9.996	10.120	9.957	10.060	10.010	0.080	0.082
63	9.978	9.944	9.921	9.864	10.007	9.943	0.055	-0.068
64	10.126	9.950	10.107	10.027	10.020	10.046	0.071	0.103
65	9.951	9.964	9.883	10.113	9.994	9.981	0.084	-0.065
66	10.083	10.051	10.023	9.918	10.111	10.037	0.074	0.056
67	9.869	10.165	10.095	9.910	9.978	10.003	0.125	-0.034
68	10.196	10.018	9.828	10.037	9.958	10.007	0.133	0.004
69	10.142	10.097	9.957	10.062	10.085	10.069	0.069	0.061
70	9.988	10.043	9.953	10.104	10.018	10.021	0.057	-0.048

Table 3.4(continued)							
Lot	x_1	x_2	x_3	x_4	x_5	$\bar{x}$	s	Δ
71	10.082	9.855	9.943	10.026	10.112	10.004	0.105	-0.017
72	9.967	9.987	9.988	10.080	10.264	10.057	0.123	0.053
73	10.040	10.003	9.964	9.919	10.029	9.991	0.050	-0.066
74	10.035	10.075	10.071	10.175	10.014	10.074	0.062	0.083
75	10.093	10.005	10.106	10.025	10.004	10.047	0.049	-0.027
76	10.066	9.952	9.972	10.166	9.907	10.013	0.103	-0.034
77	10.117	10.107	10.015	9.891	9.994	10.025	0.092	0.012
78	10.119	10.030	10.114	10.152	10.037	10.090	0.054	0.066
79	10.000	10.123	9.974	9.869	10.027	9.999	0.092	-0.092
80	10.012	10.010	9.980	10.052	10.121	10.035	0.055	0.036
81	10.025	10.087	10.248	9.943	9.939	10.048	0.127	0.013
82	10.055	10.171	10.121	10.113	10.162	10.124	0.046	0.076
83	10.124	10.121	9.951	10.096	10.090	10.076	0.072	-0.048
84	9.964	10.185	10.202	10.195	10.071	10.123	0.104	0.047
85	9.971	10.081	10.152	10.035	10.131	10.074	0.074	-0.049
86	10.100	10.137	10.115	10.063	10.041	10.091	0.039	0.017
87	10.153	10.104	10.184	10.080	10.047	10.114	0.055	0.022
88	9.893	9.968	10.133	9.875	10.160	10.005	0.133	-0.108
89	9.932	10.168	10.191	10.116	9.977	10.077	0.116	0.071
90	9.982	10.133	9.916	10.155	9.862	10.010	0.130	-0.067

In Figure 3.7, we note that even though the control limits have not been crossed, it is easy to observe that the data are tending to drift. The control limits are most effective in picking out data from a stationary contaminating distribution. However, we discussed in Chapter 1 the "run" chart developed early on by Ford workers where lot means were simply plotted without any attention to control limits. These charts captured most of the power of the control chart procedure. The reality is that the human visual system can frequently detect pathologies on a control chart even where the conditions are so highly nonstandard that the built-in alarm system given by the control limits is not particularly useful. In Figure 3.8, based on the further continuation of Table 3.4 for another 30 lots, we observe that the means do indeed cross the upper control limit around lot 100. However, the prudent production worker will note trouble in the system long before that.

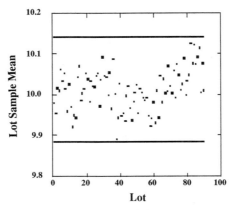

Figure 3.7. Lot Mean Control Chart For Data With Mean Drift.

Table 3.4(continued)							
Lot	x_1	x_2	x_3	x_4	x_5	$\bar{x}$	s	Δ
91	10.226	10.210	10.292	10.350	10.339	10.284	0.064	0.274
92	10.199	9.951	10.032	10.239	10.076	10.100	0.119	-0.184
93	10.264	10.186	10.188	10.264	10.032	10.187	0.095	0.087
94	10.112	10.026	10.104	10.111	10.217	10.114	0.068	-0.073
95	10.126	10.011	10.066	10.147	10.131	10.096	0.057	-0.018
96	10.158	10.155	10.206	10.101	10.260	10.176	0.060	0.079
97	10.156	10.253	10.114	10.176	10.150	10.170	0.052	-0.006
98	10.257	10.067	10.167	10.030	10.177	10.140	0.091	-0.030
99	9.956	10.203	10.146	10.044	10.124	10.095	0.096	-0.045
100	10.057	10.158	10.165	10.220	10.144	10.149	0.059	0.054
101	10.161	10.307	10.210	10.066	10.083	10.165	0.098	0.016
102	10.331	10.109	10.070	10.228	10.060	10.160	0.117	-0.006
103	10.299	10.040	10.051	10.093	10.074	10.111	0.107	-0.048
104	10.209	10.306	10.104	10.153	10.296	10.214	0.088	0.102
105	10.059	10.171	10.037	9.954	10.033	10.051	0.078	-0.163
106	9.967	10.144	10.226	10.207	10.114	10.132	0.103	0.081
107	10.276	10.018	10.206	10.275	10.133	10.182	0.109	0.050
108	10.410	9.947	10.285	10.020	10.125	10.157	0.190	-0.025
109	10.156	10.084	10.119	10.150	10.263	10.154	0.067	-0.003
110	10.209	10.230	10.162	10.072	10.068	10.148	0.076	-0.006
111	9.955	10.052	10.070	10.195	10.188	10.092	0.101	-0.056
112	10.043	9.967	10.220	10.100	10.171	10.069	0.117	-0.034
114	9.881	10.383	10.363	10.243	10.136	10.202	0.205	0.133
115	10.103	10.118	10.215	10.125	10.146	10.141	0.044	-0.060
116	10.039	10.164	10.301	10.278	10.080	10.172	0.116	0.031
117	10.237	9.953	10.216	10.105	10.253	10.153	0.126	-0.020
118	10.136	10.265	10.141	10.070	10.384	10.199	0.125	0.047
119	10.223	9.901	10.073	10.236	10.096	10.106	0.136	-0.094
120	10.096	10.041	9.961	10.132	10.110	10.068	0.068	-0.038

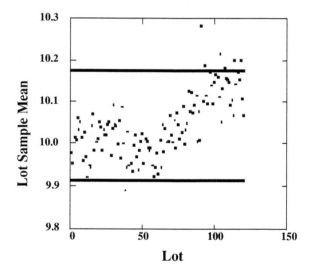

Figure 3.8. Lot Mean Control Chart For Data With Mean Drift.

It is always a good idea, if possible, to come up with a plausible model which explains why the data appears to exhibit "nontypical" lots. In the most common of cases, that of contamination of the means, we could, for example, take the points below the median based UCL in Figure 3.6, and from them compute a control chart. Then we could take the points above the UCL and from these compute another control chart. If the resulting two charts showed all points to be "in control," then we might take as a working hypothesis the notion of a mixture of two different distributions as representing the process.

In the case of the mean drift data, such a model does not appear to be very useful. But let us note that if we have

$$x_{t,j} = \mu_t + \eta_{t,j}, \tag{3.73}$$

where j goes from 1 to 5, $\eta_{t,j}$ is $\mathcal{N}(0,\sigma^2)$, and

$$\mu_{t+1} = \mu_t + \epsilon_{t+1} \tag{3.74}$$

where ϵ_t is $\mathcal{N}(0,\tau^2)$ then we note that the first difference of the $t+1$'th and t'th mean is given by

$$\Delta = \bar{x}_{t+1} - \bar{x}_t = \epsilon_{t+1} + \frac{1}{5}\sum_{j=1}^{5} \eta_{t+1,j} - \frac{1}{5}\sum_{j=1}^{5} \eta_{t,j}. \tag{3.75}$$

Thus, by the independence of all the variables making up Δ we have

$$\sigma_\Delta{}^2 = \tau^2 + \frac{2}{5}\sigma^2, \tag{3.76}$$

and, of course,

$$E(\Delta) = 0. \tag{3.77}$$

For the 119 sample differences of means from lot to lot in Table 3.4, we find a mean of differences equal to .0006, and a variance of .0041. A sample estimate for σ is available by noting that the average of the sample standard deviations is .0921, giving us

$$\hat{\sigma} = a(5)\bar{s} = 1.0638(.0921) = .0978. \tag{3.78}$$

Then we have an estimate of τ^2 via

$$\hat{\tau}^2 = .0041 - \frac{2}{5}(.0978)^2 = .00026. \tag{3.79}$$

We compare the values of .0978 and .00026 with the actual values of .1000 and .0001, respectively. Naturally, the true values will be unknown to us in the real world. However, we can construct a control chart for the Δ_t. The "in control" interval is given by

$$.0006 - 3(.0640) = -.1914 \leq \Delta_t \leq .0006 + 3(.0640) = .1926. \tag{3.80}$$

We plot the 119 sample mean differences in Figure 3.9.

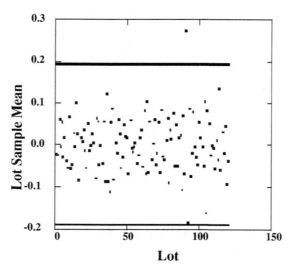

Figure 3.9. Lot Mean Difference Control Chart For Data With Mean Drift.

All the differences, except for that between lots 91 and 90, appear to be "in control." The difference "out of control" appears to be due to chance.

The drift we discussed is, of course, of a more complex nature than those arising from tear and wear of, say, positioning devices in lathes, rolling machines, etc. Such drifts are almost deterministic and monotone. While it is a very good idea to deal with them using control charts, the process is rather routine and obvious, and we shall not dwell on it.

3.6   A Process with Upward Drift in Variance

Another kind of anthropomorphized process is that in which the variability of output randomly increases as time progresses. Again, this is consistent with the performance of an increasingly fatigued juggler, trying to keep the balls moving smoothly through the air. It seldom makes much sense to think of a process which, without any intervention, exhibits declining variance.

The Nashua case study mentioned in Chapter 1 appears to give a counterexample in which the process was "brought under control" by simply leaving it alone until the time delayed servomechanisms had done their job. And there are many other examples of variability diminishing by "breaking in," by workers becoming clever in "tweaking" the process, etc. Nevertheless, it is no doubt true that wear and tear can cause an increase in variability. Let us consider an example of such a situation. We are still attempting to produce bolts of 10 cm diameter. Here, we shall assume that the output from lot to lot follows a normal distribution with mean 10 and starting standard deviation 0.1. As we begin sampling from each lot, however, there is a probability of 10% that the standard deviation will increase by .01 cm. In Table 3.5, we demonstrate 90 simulated lots of five bolts each.

Table 3.5							
Lot	x_1	x_2	x_3	x_4	x_5	$\bar{x}$	s
1	10.036	9.982	9.996	10.157	10.013	10.037	0.070
2	9.851	10.259	10.132	10.038	9.930	10.042	0.162
3	9.916	10.019	9.920	10.140	10.064	10.012	0.096
4	9.829	9.751	9.955	9.847	10.048	9.886	0.116
5	10.194	9.921	10.159	10.003	9.886	10.032	0.138
6	10.057	9.972	10.038	9.864	9.908	9.968	0.083
7	9.875	10.468	10.017	9.997	9.834	10.038	0.252
8	10.006	9.853	10.054	10.009	9.913	9.967	0.081
9	10.013	10.015	10.073	10.184	10.048	10.067	0.070
10	9.957	10.009	10.057	9.948	9.942	9.983	0.049
11	9.902	9.925	9.904	9.913	9.991	9.927	0.037
12	9.851	10.248	10.172	10.006	10.065	10.068	0.153
13	10.022	9.976	10.104	9.997	10.051	10.030	0.050
14	9.990	10.084	10.028	9.973	10.184	10.052	0.085
15	10.125	10.161	10.029	9.811	9.710	9.967	0.198
16	9.858	10.079	9.976	9.967	10.204	10.017	0.130
17	9.837	10.305	9.794	9.868	9.948	9.950	0.206
18	10.201	9.925	9.859	10.127	9.857	9.994	0.160
19	10.052	9.850	9.840	9.951	9.906	9.920	0.087
20	10.017	9.932	10.194	9.954	10.143	10.048	0.116
21	9.778	10.144	10.110	10.118	10.117	10.053	0.154
22	10.056	10.008	10.027	10.054	10.107	10.050	0.038
23	10.058	9.834	9.980	9.895	10.082	9.970	0.106
24	9.955	9.946	10.041	9.928	9.998	9.974	0.046
25	10.001	10.027	9.932	9.953	9.936	9.970	0.042
26	10.016	10.091	9.934	9.948	9.946	9.987	0.067
27	10.052	9.959	9.967	10.050	10.023	10.010	0.045
28	10.005	9.977	10.092	9.829	10.094	9.999	0.109
29	9.881	10.159	9.986	10.090	10.093	10.042	0.109
30	9.775	10.003	9.971	9.819	9.968	9.907	0.103
31	10.042	10.098	9.988	10.108	9.918	10.031	0.079
32	9.917	10.011	10.116	9.970	10.078	10.018	0.080
33	9.728	9.886	9.884	10.139	9.955	9.918	0.149
34	10.032	10.102	9.997	9.986	9.991	10.022	0.049
35	10.071	9.949	9.972	9.976	10.223	10.038	0.113
36	10.004	10.102	10.194	10.022	10.052	10.075	0.076
37	9.859	10.286	9.963	10.000	9.930	10.007	0.164

Lot	x_1	x_2	x_3	x_4	x_5	$\bar{x}$	s
38	10.348	9.967	9.874	10.057	10.000	10.049	0.180
39	10.000	10.044	9.811	9.811	10.144	9.962	0.147
40	9.926	10.305	10.036	9.978	10.039	10.057	0.146
41	9.881	10.002	9.781	9.872	10.197	9.946	0.161
42	10.252	10.203	9.845	10.207	10.012	10.104	0.172
43	10.162	9.779	10.150	9.954	9.950	9.999	0.160
44	10.044	10.013	10.030	9.850	10.014	9.990	0.080
45	9.897	9.994	9.857	9.860	9.902	9.902	0.055
46	9.916	9.924	9.855	9.912	9.980	9.917	0.045
47	10.088	9.754	10.122	9.951	10.013	9.986	0.145
48	10.371	10.031	10.203	10.197	10.109	10.182	0.127
49	10.192	10.076	9.915	10.035	10.021	10.048	0.100
50	9.725	10.001	10.127	10.051	10.053	9.992	0.155
51	10.064	9.888	10.082	9.915	9.940	9.978	0.089
52	10.056	10.131	9.870	9.835	9.894	9.957	0.129
53	9.917	9.973	10.170	9.919	9.997	9.995	0.104
54	9.993	9.910	9.938	10.071	9.921	9.967	0.066
55	10.015	9.867	9.836	9.789	9.748	9.851	0.102
56	10.012	10.040	10.134	9.926	10.165	10.055	0.096
57	9.846	9.807	9.824	10.171	9.837	9.897	0.154
58	10.016	10.046	9.862	9.936	10.040	9.980	0.079
59	9.853	9.989	10.066	10.455	9.813	10.035	0.256
60	9.884	10.012	9.894	9.923	10.152	9.973	0.112
61	10.018	9.661	10.019	9.872	9.685	9.851	0.173
62	10.015	10.034	9.781	9.829	10.042	9.940	0.125
63	9.826	10.200	10.029	10.135	9.930	10.024	0.151
64	10.121	9.863	10.008	10.161	9.894	10.010	0.133
65	10.132	9.913	10.268	10.070	10.148	10.106	0.130
66	9.816	9.880	9.733	9.809	9.947	9.837	0.081
67	10.160	9.843	9.848	10.401	10.123	10.075	0.235
68	9.854	10.031	9.947	9.980	9.981	9.959	0.066
69	9.915	9.931	9.982	9.928	10.105	9.972	0.079
70	10.062	10.147	9.705	10.053	9.869	9.967	0.178
71	10.095	10.088	10.340	10.014	9.965	10.100	0.144
72	10.185	9.986	10.038	10.405	10.117	10.146	0.163
73	9.963	10.162	10.186	9.700	9.751	9.952	0.225
74	10.289	9.940	9.963	10.200	9.904	10.059	0.173
75	10.467	9.884	9.952	10.255	10.190	10.150	0.236
76	9.995	9.811	9.781	10.059	10.277	9.985	0.202
77	9.952	9.943	9.844	10.159	9.882	9.956	0.122
78	9.822	9.795	10.041	9.945	10.098	9.940	0.132
79	9.869	10.136	9.638	9.858	9.591	9.818	0.217
80	9.988	10.016	9.943	10.167	9.772	9.977	0.142
81	9.837	9.840	10.184	10.148	10.076	10.017	0.168
82	9.877	10.260	10.197	9.909	10.044	10.057	0.170
83	10.283	9.827	9.959	10.172	10.256	10.099	0.198
84	10.040	9.970	10.165	10.076	9.888	10.028	0.105
85	9.761	10.010	9.900	10.092	10.088	9.970	0.140
86	9.937	10.098	10.059	9.709	10.153	9.991	0.177
87	10.128	9.836	10.179	10.145	10.067	10.071	0.137
88	9.962	10.070	10.053	10.160	9.886	10.026	0.105
89	10.017	10.126	10.103	9.807	10.181	10.047	0.147
90	9.942	9.857	9.919	9.988	9.800	9.901	0.074

Table 3.5(continued)

The values of $\bar{\bar{x}}$ and $\bar{s}$ are 9.999 and .125, respectively. This gives an $\bar{x}$ control interval of

$$9.999 - \frac{3(.125)a(5)}{\sqrt{5}} = 9.821 \le \bar{x} \le 9.999 + \frac{3(.125)a(5)}{\sqrt{5}} = 10.178. \quad (3.81)$$

We plot the sample control chart in Figure 3.10. Although two of the points are slightly outside the control interval for the sample means of the process, such an alarm is unlikely to lead us to discover the upward drift in the variance of the production process.

The standard control interval for s is given by

$$(1 - 3[a(5)^2 - 1]^{.5})\bar{s} = -.011 \le s \le (1 + 3[a(5)^2 - 1]^{.5})\bar{s} = .261. \quad (3.82)$$

We show the resulting s control chart in Figure 3.11.

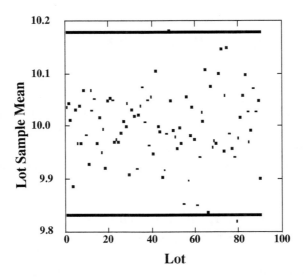

Figure 3.10. Lot Mean Control Chart For Data With Variance Drift.

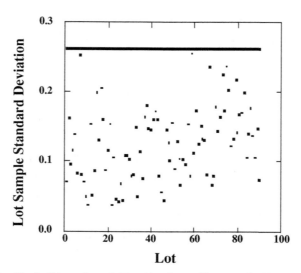

Figure 3.11. Lot Standard Deviation Control Chart For Data With Variance Drift.

We note that none of the s values is "out of control." Nevertheless the trend upward in Figure 3.11 is rather clear. Again, we are essentially left with a "run chart." An experienced control operator would note the trend upward and attempt to find the source of the increasing production variability.

3.7 Charts for Individual Measurements

It is obvious that, whenever possible, measurements taken at some nodal point of a production process should be grouped into statistically homogeneous lots. The greater the size of such a lot, the more accurate are the estimates, $\bar{x}$ and s^2, of the population mean and variance. In turn, the greater the accuracy of our estimates, the greater the chance of detecting a Pareto glitch. It is the limitations in time and cost, as well as the requirement of homogeneity, that make us content with lot size of only 4 or 5.

It may happen, however, that no grouping of the measurements into lots of size greater than one is possible. Typically, the reason is that either the production rate is too slow or the production is performed under precisely the same conditions over short time intervals. In the first case, items appear at a nodal point too rarely to let us assume safely that two or more of them are governed by the same probability distribution. Grouping measurements into lots of size greater than one is then likely to lead to hiding, or "smoothing out," atypical measurements among typical ones. As an example of the second of the two reasons mentioned, let us consider temperature of a chemical reactor. Clearly, the temperature is constant over small time intervals and, therefore, it does not make sense to take more than one measurement of any physical quantity, which is a function of temperature, in one such interval.

In order to see how to construct control charts for individual measurements, let us consider 90 observations coming from $\mathcal{N}(10, .01)$ with probability .855, from $\mathcal{N}(10.4, .03)$ with probability .095, from $\mathcal{N}(9.8, .09)$ with probability .045 and from $\mathcal{N}(10.2, .11)$ with probability .005. The given probabilities have been obtained as in (3.9) to (3.12), using $p_1 = .1$ and $p_2 = .05$. Actually, the 90 observations are the data x_1 of Table 3.2. They are repeated as observations x in Table 3.6.

Table 3.6					
Lot	x	MR	Lot	x	MR
1	9.927		46	10.006	0.052
2	9.862	0.065	47	10.132	0.126
3	10.061	0.199	48	10.012	0.120
4	9.820	0.241	49	10.097	0.085
5	9.737	0.083	50	10.007	0.090
6	9.876	0.139	51	9.967	0.040
7	9.898	0.022	52	9.981	0.014
8	10.001	0.103	53	9.841	0.140
9	9.928	0.073	54	9.992	0.151
10	9.896	0.032	55	9.908	0.084
11	10.011	0.115	56	10.011	0.103
12	9.983	0.028	57	10.064	0.053
13	10.127	0.144	58	9.891	0.173
14	10.025	0.102	59	9.869	0.022
15	9.953	0.072	60	10.016	0.147
16	10.007	0.054	61	10.008	0.008
17	10.062	0.055	62	10.100	0.092
18	10.168	0.106	63	9.904	0.196
19	9.986	0.182	64	9.979	0.075
20	9.786	0.200	65	9.982	0.003
21	9.957	0.171	66	10.028	0.046
22	9.965	0.008	67	9.995	0.033
23	9.989	0.024	68	9.936	0.059
24	9.983	0.006	69	10.014	0.078
25	10.063	0.080	70	10.005	0.009

Table 3.6 (continued)					
Lot	x	MR	Lot	x	MR
26	9.767	0.296	71	10.116	0.111
27	9.933	0.166	72	9.934	0.182
28	10.227	0.294	73	9.972	0.038
29	10.022	0.205	74	10.014	0.042
30	9.845	0.177	75	10.093	0.079
31	9.956	0.111	76	9.927	0.166
32	9.876	0.080	77	10.177	0.250
33	9.932	0.056	78	9.825	0.352
34	10.016	0.084	79	10.333	0.508
35	9.927	0.089	80	9.972	0.361
36	9.952	0.025	81	10.059	0.087
37	9.941	0.011	82	9.832	0.227
38	10.010	0.069	83	9.958	0.126
39	9.848	0.162	84	10.087	0.129
40	10.002	0.154	85	10.232	0.145
41	10.031	0.029	86	10.066	0.166
42	9.990	0.041	87	10.041	0.025
43	9.995	0.005	88	9.868	0.173
44	9.980	0.015	89	10.084	0.216
45	10.058	0.078	90	10.063	0.021

An obvious counterpart of the mean, or $\bar{X}$, control chart, suited to the case of lots of size one, can be obtained in the following way. First, the run chart for $\bar{X}$'s has to be replaced by that for X's themselves. In fact, $X = \bar{X}$ if the sample size is 1. Second, since $X = \bar{X}$ and there is no variability within lots of size 1, we can calculate sample mean and variance, $\bar{X}$ and S^2, for all 90 observations and use as the upper control limit and lower control limit, respectively,

$$\bar{x} + 3a(N)s \qquad (3.83)$$

and

$$\bar{x} - 3a(N)s, \qquad (3.84)$$

where $a(N)$ is given by (3.36) with n replaced by N and $N = 90$. In the example considered,

$$UCL = 9.986 + 3(.101) = 10.289 \qquad (3.85)$$

and

$$LCL = 9.986 - 3(.101) = 9.683, \qquad (3.86)$$

since $a(90) = 1$. The X chart obtained is given in Figure 3.12.

The chart recognizes the Pareto glitch on lot 79, slightly fails to detect atypicality of lot 28 and apparently fails to detect atypicality of lot 40. This rather poor result, in particular when compared with the results for lots of size 5 from the same distributions (see Figure 3.2), should not surprise us. Now, the distance between the UCL and LCL is, approximately, equal to 6 "norm" standard deviations while that for lots of size 5 was reduced by the factor $\sqrt{5}$. Moreover, in Figure 3.2, the "observation" on lot, say 40, was in fact the mean of 5 data from $\mathcal{N}(9.8, .09)$, while in Figure 3.12 it is only the first of those 5 data. So, given the circumstances, we should be happy that the X chart detected the Pareto glitch on lot 79 and hinted at the possibility of another glitch on lot 28.

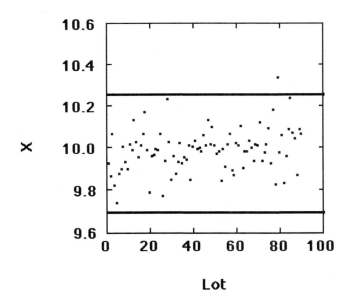

Figure 3.12. X Control Chart.

Table 3.7					
Lot	x	MR	Lot	x	MR
1	9.987		46	9.976	0.007
2	9.629	0.358	47	9.927	0.049
3	10.122	0.493	48	9.934	0.007
4	9.861	0.261	49	9.956	0.022
5	9.816	0.045	50	10.001	0.045
6	9.925	0.109	51	10.097	0.096
7	10.103	0.178	52	10.179	0.082
8	10.057	0.046	53	9.941	0.238
9	9.956	0.101	54	10.089	0.148
10	9.591	0.365	55	10.006	0.083

Table 3.7 (continued)					
Lot	x	MR	Lot	x	MR
11	10.185	0.594	56	9.976	0.030
12	10.051	0.134	57	10.003	0.027
13	9.966	0.085	58	9.875	0.128
14	9.987	0.021	59	9.978	0.103
15	9.926	0.061	60	10.057	0.079
16	9.734	0.192	61	9.853	0.204
17	9.980	0.246	62	9.893	0.040
18	10.012	0.032	63	10.105	0.212
19	10.263	0.251	64	9.963	0.142
20	9.954	0.309	65	9.593	0.370
21	9.762	0.192	66	9.580	0.013
22	10.008	0.246	67	9.975	0.395
23	10.091	0.083	68	10.140	0.165
24	9.795	0.296	69	10.181	0.041
25	9.968	0.173	70	9.414	0.767
26	9.948	0.020	71	9.592	0.178
27	9.864	0.084	72	9.952	0.360
28	10.162	0.298	73	9.839	0.113
29	9.941	0.221	74	9.964	0.125
30	9.791	0.150	75	9.770	0.194
31	10.052	0.261	76	9.503	0.267
32	10.058	0.006	77	9.676	0.173
33	10.144	0.086	78	10.015	0.339
34	10.080	0.064	79	9.799	0.216
35	9.912	0.168	80	10.147	0.348
36	9.973	0.061	81	10.104	0.043
37	10.003	0.030	82	10.082	0.022
38	10.096	0.093	83	10.174	0.092
39	10.023	0.073	84	9.922	0.252
40	10.188	0.165	85	10.208	0.286
41	10.005	0.183	86	9.868	0.340
42	10.018	0.013	87	10.029	0.161
43	9.961	0.057	88	10.053	0.024
44	9.914	0.047	89	9.585	0.468
45	9.969	0.055	90	9.956	0.371

Let us now construct the X chart for observations x_1 of Table 3.3. This time, we have a strong, 30% contamination of the norm $\mathcal{N}(10, .01)$ distribution by the $\mathcal{N}(9.8, .09)$ distribution. The data are repeated in Table 3.7.

We have that

$$UCL = 9.953 + 3(.168) = 10.457 \qquad (3.87)$$

and

$$LCL = 9.953 - 3(.168) = 9.449. \qquad (3.88)$$

The X chart is depicted in Figure 3.13. We note that not only the lower control limit is crossed just once, but the two clusters of the data are not as transparent as they are in Figure 3.4. The reason is similar to that which prevents us from constructing lots of statistically hetero-geneous data. In fact, both $\bar{x}$ and, to the worse, s are now calculated for such a lot of size 90. Since as many as 29 observations come from the contaminating distribution, the value of the norm standard deviation is grossly overestimated. Indeed, $s = .168$ while $\sigma_0 = .1$. Fortunately, in most SPC problems, we are faced with just a few Pareto glitches among many, many more uncontaminated lots (of whatever size).

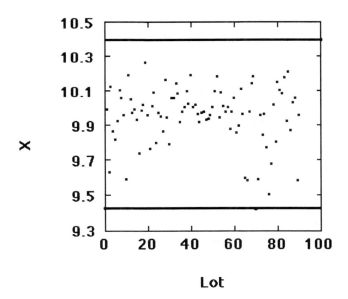

Figure 3.13. X Control Chart With 30% Contamination.

Until now, we have been using rather automatically the "3σ" control limits. We already know that such limits are well justified as long as lots come from a normal distribution. Let us recall that we have then that

$$P(|\frac{X - \mu}{\sigma}| \geq 3) = .0027, \tag{3.89}$$

where X denotes a normally distributed measurement with mean μ and variance σ^2, and, hence, we have chance one in $1/.0027 \simeq 370$ of a false alarm. Moreover, by the Central Limit Theorem, even if individual mea-surements come from another distribution but a lot is of a "reasonable" size, the lot's mean is still approximately normally distributed and the equality like (3.89) approximately holds for $\bar{X}$. This last property allowed us to rely on "normal theory" in justifying the 3σ limits for all $\bar{X}$ charts, regardless of the form of a parent distribution of a production process. But it is no more the case when constructing the X chart for lots of size one. The probabilistic argument preserves its validity only if individual measurements are governed by a normal distribution. Otherwise, although the 3σ limits can still be used, we cannot claim that the probability of getting a false alarm has known value (equal to .0027). It is another matter that the issue of normality should not bother us unless data come from a distribution which is far from normal, for example a heavily skewed distribution.

In practice, we never know whether measurements come from a normal or approximately normal distribution. Accordingly, before the control chart is drawn, a test of normality of the data should be performed. For our purposes, it suffices to construct a *normal probability plot*, which is a very simple and yet efficient device provided by any statistical software. One form of the normal probability plot is the following (see Chambers, Cleveland, Kleiner and Tukey [2]). Given random measurements $X_1, X_2, \ldots, X_N$, we sort them from smallest to largest to obtain order statistics

$$X_{(1)} \leq X_{(2)} \leq \ldots \leq X_{(r)} \leq \ldots \leq X_{(N)}.$$

After standardization, the expected value of the rth order statistic can be estimated by the quantile $z_{(r-.5)/N}$ of order $(r-.5)/N$ of the standard normal random variate Z (arguing similarly as in Subsections B.13.6 and B.13.7 of Appendix B, we may note that constant .5, although quite natural, is somewhat arbitrary). Now, plotting the $X_{(r)}$'s on the horizontal axis against corresponding quantiles $z_{(r-.5)/N}$ on the vertical axis, we should obtain, approximately, a straight line if the measurements follow a normal distribution. The normal probability plot for the measurements of Table 3.6 is given in Figure 3.14, while that for the measurements of Table 3.7 is given in Figure 3.15.

Quite rightly, the normality assumption proves justified for the data of Table 3.6 and it is unwarranted for the data of Table 3.7. In the latter case, we have that a measurement comes from $\mathcal{N}(10, .01)$ with probability .7 and it comes from $\mathcal{N}(9.8, .09)$ with probability .3. Considered unconditionally, each measurement is, in fact, the following sum of random variables

$$Y = X_0 + I_2 X_2 ,$$

where X_0 comes from $\mathcal{N}(10, .01)$, X_2 comes from $\mathcal{N}(-.2, .08)$, and I_2 assumes value 1 with probability .3 and value 0 with probability .7, and all three random variables are mutually independent. Of course, Y is not a normal random variable. The data of Table 3.6 come predominantly from $\mathcal{N}(10, .01)$. Strictly speaking, each datum is a random variable of the form

$$X_0 + I_1 X_1 + I_2 X_2 \ ,$$

where I_1 assumes value 1 with probability .1 and value 0 with probability .9, I_2 assumes value 1 with probability .05 and value 0 with probability .95, X_1 is $\mathcal{N}(.4, .02)$ and X_0 and X_2 are as before. For some purposes, the influence of X_1 and X_2 can be considered to be "almost" negligible, and the data to be approximately normal.

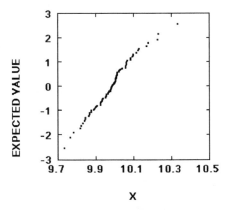

Figure 3.14. Normal Probability Plot.

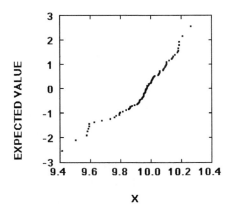

Figure 3.15. Normal Probability Plot for Highly Contaminated Data.

As regards the data set of Table 3.7, the reader experienced in analyzing normal probability plots will find that Figure 3.15 hints at possible skewness of the underlying distribution. The reason may be a strong contamination of a norm process, as Figure 3.13 with its two rather apparent clusters confirms.

Let us now examine the set of measurements x_1 of Table 3.4, tabulated again in Table 3.8.

			Table 3.8		
Lot	x	MR	Lot	x	MR
1	9.927		46	10.086	0.103
2	9.856	0.071	47	9.896	0.190
3	10.031	0.175	48	9.997	0.101
4	9.850	0.181	49	9.987	0.010
5	9.858	0.008	50	9.958	0.029
6	10.127	0.269	51	10.075	0.117
7	10.131	0.004	52	10.099	0.024
8	9.924	0.207	53	10.097	0.002
9	9.869	0.055	54	10.117	0.020
10	10.132	0.263	55	9.779	0.338
11	9.877	0.255	56	10.015	0.236
12	10.089	0.212	57	9.849	0.166
13	10.074	0.015	58	10.010	0.161
14	9.977	0.097	59	9.897	0.113
15	10.040	0.063	60	9.857	0.040
16	10.049	0.009	61	9.887	0.030
17	10.040	0.009	62	9.920	0.033
18	9.964	0.076	63	9.978	0.058
19	10.058	0.094	64	10.126	0.148
20	10.133	0.075	65	9.951	0.175
21	10.063	0.070	66	10.083	0.132
22	10.040	0.023	67	9.869	0.214
23	9.889	0.151	68	10.196	0.327
24	10.092	0.203	69	10.142	0.054
25	9.948	0.144	70	9.988	0.154
26	10.119	0.171	71	10.082	0.094
27	10.012	0.107	72	9.967	0.115
28	9.943	0.069	73	10.040	0.073
29	10.072	0.129	74	10.035	0.005
30	10.062	0.010	75	10.093	0.058
31	9.988	0.074	76	10.066	0.027
32	10.137	0.149	77	10.117	0.051
33	10.020	0.117	78	10.119	0.002
34	9.870	0.150	79	10.000	0.119
35	10.107	0.237	80	10.012	0.012
36	10.118	0.011	81	10.025	0.013
37	9.928	0.190	82	10.055	0.030
38	9.934	0.006	83	10.124	0.069
39	9.893	0.041	84	9.964	0.160
40	10.120	0.227	85	9.971	0.007
41	9.996	0.124	86	10.100	0.129
42	10.071	0.075	87	10.153	0.053
43	9.949	0.122	88	9.893	0.260
44	9.930	0.019	89	9.932	0.039
45	9.983	0.053	90	9.982	0.050

The mean of this data set is 10.01, the standard deviation is .093 and, hence, the control limits are 10.289 and 9.731, respectively. The X chart is plotted in Figure 3.16. The data are not only in control but the chart fails to detect any drift, at least at a first glance. The drift present in the data is too small, relative to the population's standard deviation, to be discovered. The normal probability plot, given in Figure 3.17, reveals that the data can hardly be considered to be normally distributed.

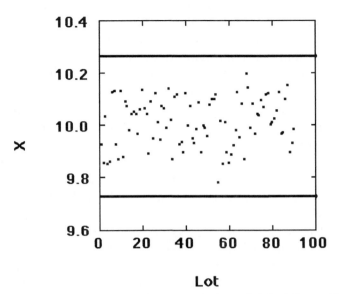

Figure 3.16. X Control Chart For Data With Mean Drift.

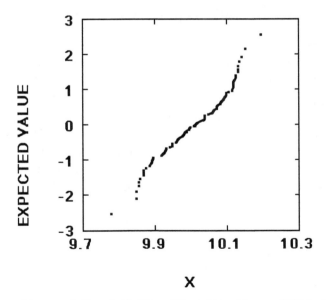

Figure 3.17. Normal Probability Plot For Data With Mean Drift.

As with the normal probability plot, statistical softwares are equipped with similarly constructed probability plots for other distributions. We checked the fit of the data of Table 3.8 to uniform, gamma and Weibull distributions with a variety of possible parameters and we found that no such fit exists. In fact, it follows from the normal probability plot in Figure 3.17 that the underlying distribution is likely to be platykurtic. A conjecture that the data set comes from a contaminated norm process is one possibility. Another, given a weak pattern one can perhaps find in the data upon closer inspection of Figure 3.16, can be that of some type of autocorrelation. As we know, a relatively small drift in the process mean is indeed present.

In order to get a better insight into statistical properties of a data set, it is tempting to construct a chart that would give us some information on the process's variability. To achieve this goal, we can form artificial lots of size 2 or 3 and calculate the so-called *moving ranges*. Given N individual measurements, $X_1, X_2, \ldots, X_N$, we calculate the ith moving range MR_i as the difference between the largest and the smallest value in the ith artificial lot formed from the n measurements $X_i, \ldots, X_{i+n-1}$, where $i = 1, 2, \ldots, N-n+1$ and n is the lot's size, equal to 2 or 3. Thus, if $n = 2$,

$$\mathrm{MR}_1 = |X_2 - X_1|, \mathrm{MR}_2 = |X_3 - X_2|, \ldots, \mathrm{MR}_{N-1} = |X_N - X_{N-1}|.$$

Clearly, proceeding in this way, we violate the requirement that measurements be considered separately and, hence, any analysis of the moving ranges should be undertaken with extreme caution. Hence, also, n should not be greater than 3. It should be noted that the moving ranges are not statistically independent one from another, since the artificial lots are overlapping. Once the moving ranges have been formed, we can plot their run chart and, moreover, we can add the control limits in the following way. Just as we calculated the mean of the range (see (B.208) in Appendix B), we can use the same trick to find that

$$E(R^2) = D^2\sigma^2,$$

where σ^2 is the population variance, D^2 is a suitable constant depending on lot size n and R denotes the range of a lot. Thus,

$$\sigma_R = \sqrt{\mathrm{Var}(R)} = (E(R^2) - [E(R)]^2)^{1/2} = d\sigma, \qquad (3.90)$$

where $d = \sqrt{D^2 - 1/b^2}$ and b is defined by (B.205) and tabulated in Table 3.1 as b_n. Factors $d \equiv d_n$ and again b_n are given in Table 3.9 (for reasons which will be given later, the two factors are given also for lots of sizes greater than 3). Using (3.39), we can estimate the standard deviation of the range as follows

$$\hat{\sigma}_R = b_n d_n \bar{R}. \tag{3.91}$$

Applying (3.91) to the average of $N - n + 1$ moving ranges,

$$\overline{\text{MR}} = \frac{1}{N-n+1} \sum_{i=1}^{N-n+1} \text{MR}_i, \tag{3.92}$$

we obtain the upper and lower control limits for the control chart of moving ranges:

$$UCL = (1 + 3b_n d_n)\overline{\text{MR}} = D_4(n)\overline{\text{MR}} \tag{3.93}$$

and

$$LCL = \max\{0, 1 - 3b_n d_n\}\overline{\text{MR}} = D_3(n)\overline{\text{MR}}. \tag{3.94}$$

Factors $D_3(n)$ and $D_4(n)$ are tabulated in Table 3.9.

Table 3.9				
Lot size	b_n	d_n	$D_3(n)$	$D_4(n)$
2	.8865	.8524	0	3.267
3	.5907	.8888	0	2.575
4	.4857	.8799	0	2.282
5	.4300	.8645	0	2.115
6	.3946	.8480	0	2.004
7	.3698	.8328	.076	1.924
8	.3512	.8199	.136	1.864
9	.3367	.8078	.184	1.816
10	.3249	.7972	.223	1.777
15	.2880	.7546	.348	1.652
20	.2677	.7296	.414	1.586

Remark: Let us note in passing that the above argument enables one to construct the control chart for usual ranges R_j, when lots of size greater than one can be used. Given the run chart for the ranges, it suffices to replace $\overline{\text{MR}}$ in (3.91) and (3.92) by the average of the N ranges, (3.38). Thus, if one prefers ranges to standard deviations, one can use the control chart for the ranges instead of the standard deviation control chart.

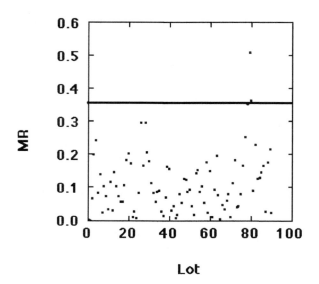

Figure 3.18. MR Control Chart.

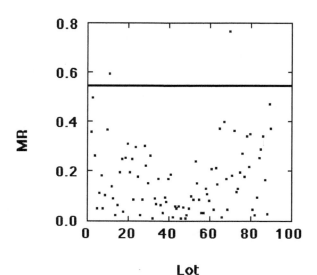

Figure 3.19. MR Control Chart For Data With 30%
Contamination.

The upper control limits for the data sets of Tables 3.6 and 3.7 are
equal to $(.111)(3.267) = .359$ and $(.167)(3.267) = .546$, respectively. The
UCL for the data set of Table 3.8 is $(.103)(3.267) = .3365$. Corresponding
MR charts are given in Figures 3.18, 3.19 and 3.20.

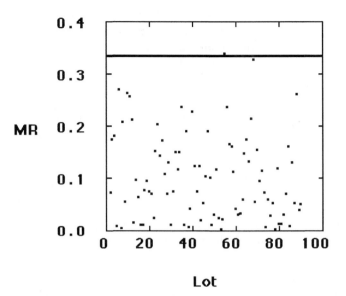

Figure 3.20. MR Control Chart For Data With Mean Drift

For the data of Table 3.6, we note that the UCL is crossed on lots 79 and 80. Along with the X chart, it confirms the existence of a Pareto glitch on lot 79. Note that the detection of this glitch on the MR chart is rather due to the sudden change of the measurements' mean than to the change of the standard deviation. Similarly, the two glitches on the MR chart for the data of Table 3.7 suggest some sort of contamination in the data. Also, the fact that the MR control chart for the data of Table 3.8 reveals a Pareto glitch hints, rather deceptively, at the possibility of some contamination in the data.

Let us turn again to the X charts. We noted that the population standard deviation was considerably overestimated in the case of strongly contaminated data of Table 3.7. We can try to estimate σ using the average of the moving ranges, $\overline{\text{MR}}$, and, thus, to construct the control limits of the following form

$$\bar{x} \pm 3b_2\overline{\text{MR}}. \tag{3.95}$$

With $\overline{\text{MR}} = .167$ and, hence, $b_2\overline{\text{MR}} = .148$ as opposed to $s = .168$, we obtain new control limits, equal to 10.397 and 9.509, respectively. The lower control limit is now crossed twice, on lots 70 and 76. Note that it is the lower control limit which should be likely to be crossed by measurements governed by the contaminated distribution. The lower (and upper) control limit may still be slightly increased by switching from the sample mean $\bar{x}$ to the sample median of all 90 measurements, since the latter is greater than the former by .021. Unfortunately, it does not lead to detecting any more Pareto glitches.

Interestingly, for the data with mean drift (see Table 3.8 and Figure 3.16), the control limits for the X chart provided by the average of the moving ranges are slightly narrower than those based on s. The former are equal to 10.284 and 9.736 while the latter are equal to 10.289 and 9.731. Both types of the limits are very close one to another for the data set of Table 3.6, with those based on $\overline{MR}$ being again minimally narrower. For both sets of data, the sample means are practically equal to sample medians.

These examples show that it is reasonable to base the SPC analysis on all the available control charts as well as on other possible means, such as probability plots. Depending on the case, any chart may prove more informative than another. And it is crucial that charting be accompanied by a thorough visual analysis of the run charts, regardless of what control limits are used.

3.8 Process Capability

In the context of statistical process control, a production or any other process is examined as it is, whatever its technological specifications. One can reasonably claim that, from the methodological point of view, the technological specifications have nothing to do with SPC narrowly understood. What we listen to is the "voice of the process." However, in due time, the "voice of the customer" can hardly be left unheard. It is the purpose of this Section to align the two voices.

There are many ways of doing this, and we shall describe just one of them. We shall refer to just some of a multitude of capability indices. Terminology and approach we shall use seems to be prevalent now, although neither of the two can be claimed already well established. In any case, they are consistent, e.g., with those used by "The Big Three" (Chrysler Corporation, Ford Motor Company and General Motors Corporation; see their *Statistical Process Control (SPC) Reference Manual* [5]).

It is our basic prerequisite that, when studying process capability, we deal only with processes in a state of statistical control. Prior to calculations of any capability indices, a process has to be brought to stability. Indeed, an out of control process is unpredictable and hence any scrutiny of its capability is useless.

In the "best of all possible worlds," the voice of a process is heard via the process's overall mean $\bar{\bar{x}}$ and its inherent variation, measured by an estimate of the population standard deviation, such as $\hat{\sigma}$ given by (3.36) or (3.39). One should note here that the estimates of inherent variation are based on within-lot or, as is said more often, within-group variation. On the other hand, the voice of the customer is heard via nominal value and tolerance (or specification) limits. The two voices are fundamentally different. It is *summary statistics* for lots that are examined for the purpose of controlling a process, while *individual measurements* are compared to specifications. Still, the aforementioned statistics needed for charting sample means enable one to readily assess process capability in relation to technological specifications imposed.

One of the standard measures of process capability is the so-called C_p index,

$$C_p = \frac{USL - LSL}{6\hat{\sigma}}, \qquad (3.96)$$

where USL denotes the upper specification limit, LSL denotes the lower specification limit and $\hat{\sigma}$ is an estimate of the process's inherent variation. The index has a very clear interpretation provided the process is not only in statistical control but also the individual data come from an approximately normal distribution. The estimated 6σ range of the process's inherent variation, which appears in the denominator of (3.96) and which itself is often termed *process capability*, is the range where (in principle) 99.73% of the underlying probability mass lies. It is easy to see that the 6σ range is closely related to the natural "in control" interval for individual data. The C_p index relates the process's inherent variation to the tolerance width, $USL - LSL$. For obvious reasons, C_p should be equal at least to 1 for the process capability to be considered acceptable.

If the individual data of an in control process do not follow normal distribution, sometimes the following modification of the C_p index is postulated

$$C_p = \frac{USL - LSL}{P_{99.865} - P_{0.135}}, \qquad (3.97)$$

where $P_{0.135}$ and $P_{99.865}$ are the 0.135 and 99.865 percentiles of the underlying distribution, respectively. Of course, the two indices coincide for normal distribution. In a non-normal case, it is then suggested to find the underlying distribution and estimate the percentiles needed. However, we agree with Wheeler and Chambers [7] that (3.96) can be used to advantage in most practical situations, even when the process distribution is heavily skewed. Although (3.96) is then not equivalent to (3.97), usually, as Wheeler and Chambers amply show, the great bulk of the probability mass of the underlying distribution still remains within the 6σ range. To put it otherwise, as a rule, whether the process distribution is normal or not, if C_p is greater than 1, the process can be considered capable.

A safer approach has been proposed by Wetherill and Brown [6]. Namely, if we realize that the 6σ range represents the actual process range (minus, approximately, 0.3% of it) uder normality, then – given data from whatever distribution – we can simply calculate their actual range and cut off 0.15% of it at each end.

Process capability, as measured by C_p alone, refers only to the relationship between process's inherent variation and tolerance width as specified by the user, irrespective of the process centering. The capability index which accounts also for process centering is the so-called C_{pk} index,

$$C_{pk} = \min\{CPU, CPL\}, \tag{3.98}$$

where

$$CPU = \frac{USL - \bar{\bar{x}}}{3\hat{\sigma}} \text{ and } CPL = \frac{\bar{\bar{x}} - LSL}{3\hat{\sigma}}.$$

CPU and CPL are referred to, respectively, as upper and lower capability indices. The C_{pk} index relates process capability to the process mean via its estimate $\bar{\bar{x}}$. Clearly, also C_{pk} should be equal at least to 1 for the process capability to be considered acceptable.

For reasons already given for C_p, while it is good to have data distributed normally, the C_{pk} index can either be used unchanged in, or readily adapted to, cases with non-normal data.

Let us consider summary statistics for lots 94 to 125 of size 5 of diameter 3 from Figure 1.22 (the problem is described in the Remark above the Figure), given in Table 3.10.

\multicolumn{6}{c}{Table 3.10}					
Lot	$\bar{x}$	s	Lot	$\bar{x}$	s
94	6.736	.031	110	6.732	.025
95	6.742	.027	111	6.742	.026
96	6.742	.024	112	6.746	.024
97	6.726	.018	113	6.740	.012
98	6.732	.022	114	6.722	.016
99	6.742	.017	115	6.732	.017
100	6.742	.024	116	6.734	.015
101	6.728	.016	117	6.732	.021
102	6.730	.015	118	6.742	.017
103	6.732	.009	119	6.726	.020
104	6.744	.012	120	6.731	.027
105	6.730	.019	121	6.738	.009
106	6.728	.028	122	6.742	.022
107	6.728	.014	123	6.742	.015
108	6.740	.025	124	6.740	.004
109	6.730	.018	125	6.746	.011

We leave it to the Reader to verify that the system is in control (cf. Problem 3.10). The diameter was specified to 6.75 mm with tolerances ± 0.1 mm. Assuming approximate normality, let us calculate C_p and C_{pk}. Clearly, $USL - LSL = .2$. It follows from the Table that $\bar{s} = 0.0187$ and $\bar{\bar{x}} = 6.7356$. Thus

$$C_p = \frac{0.2}{6(1.0638)(0.0187)} = 1.67,$$

$$CPU = \frac{6.85 - 6.7356}{0.06} = 1.91,$$

$$CPL = \frac{6.7356 - 6.65}{0.06} = 1.43$$

and

$$C_{pk} = 1.43.$$

The C_p value shows high capability with respect to variation, while the C_{pk} value can be considered satisfactory. (It is now rather common to treat C_{pk} values greater than 1.33 as satisfactory and C_p values greater than 1.6 as showing high capability.) It is another matter that we should not attach too much meaning to the numbers obtained, as they are in fact random. Obviously, centering the process at the nominal value of 6.65 and leaving inherent variation unchanged would lead to increase of C_{pk} (cf. Problem 3.12).

As given by (3.96) and (3.98), both C_p and C_{pk} apply to cases with two-sided specification limits. The C_p index is meaningless when a one-sided specification limit is given. If only upper specification limit is given, C_{pk} can be defined as equal to CPU. Analogously, for a one-sided, lower specification limit, one can take C_{pk} equal to CPL.

Neither of the two indices should be used alone. It is clear that a process with C_p well above 1, say, 1.6, may be producing non-conforming products from time to time due to being poorly centered. At the same time, high values of C_p show that we can reduce or eliminate the non-conforming product by merely centering the process, and thus increasing C_{pk}. Comparing, however, the costs of decreasing the process's variation and better centering the process it may occur that both actions should be undertaken to suficiently decrease C_{pk} at a minimal cost. Generally speaking, knowing values of both indices may help prioritize the order in which a process should be improved.

We know well that the ultimate aim of SPC is continual improvement. Capability indices help measure that improvement. Repeatedly applying the PDSA cycle to a process results in its changes. After the process was changed, it has to be brought to a state of statistical control and, once that has been achieved, current process's capability indices are measured and compared with those which resulted from the previous turn of the PDSA wheel.

A process with mean drift, considered in Section 3.5, is one example of processes with non-zero between-lot, or between-group, variation. That is, there may be situations when process's total variation is not equal to inherent variation only, as is the case in the best of all possible worlds. Total variation can be a result of the presence of both within-lot or inherent variation and between-lot variation. The latter is usually due to special causes but nevertheles it does not always manifest itself as producing out of control signals, at least for some time. It may therefore be of use to compare C_p and C_{pk} with their variants which account for total variation. Such variants are obtained by replacing in C_p and C_{pk} the 6σ range of a process's inherent variation by the 6σ range of the process's total variation. Accordingly, the following *performance indices* can be introduced:

$$P_p = \frac{USL - LSL}{6\hat{\sigma}_{total}}, \tag{3.99}$$

$$P_{pk} = \min\{\frac{USL - \bar{\bar{x}}}{3\hat{\sigma}_{total}}, \frac{\bar{\bar{x}} - LSL}{3\hat{\sigma}_{total}}\}, \tag{3.100}$$

where $\hat{\sigma}_{total}$ is an estimate of the process's total variation. The most common such estimate is provided by the sample standard deviation based on all lots pooled into one sample of individual observations. The 6σ range of the process's total variation, estimated by $6\hat{\sigma}_{total}$, is itself called *process performance*. Let us emphasize again that the performance indices should be used only to compare their values with those of C_p and C_{pk}, respectively, and in this way to help measure continual improvement and prioritize the order in which to improve processes. Let us also note that differences between corresponding capability and performance indices point to the presence of between-lot variation.

References

[1] Andrews, D.F., Bickel, P.J., Hampel, F.R., Huber, P.J., Rogers, W.H., and Tukey, J.W. (1972). *Robust Estimates of Location.* Princeton: Princeton University Press.

[2] Chambers, J.M., Cleveland, W.S., Kleiner, B., and Tukey, P.A. (1983). *Graphical Methods for Data Analysis.* Boston: Duxbury Press.

[3] Huber, P.J. (1977). *Robust Statistical Procedures.* Philadelphia: Society for Industrial and Applied Mathematics.

[4] Mosteller, F. and Tukey, J.W. (1977). *Data Analysis and Regression.* Reading: Addison-Wesley.

[5] *Statistical Process Control Reference Manual.* (1995) Chrysler Corporation, Ford Motor Company, and General Motors Corporation.

[6] Wetherill, G.B. and Brown, D.W. (1991). *Statistical Process Control.* Chapman & Hall.

[7] Wheeler, D.J. and Chambers, D.S. (1992). *Understanding Statistical Process Control.* SPC Press.

Problems

Remark: Unless otherwise stated, the reader is asked to use a method of his or her choice to estimate unknown population mean and standard deviation.

Problem 3.1. Construct the mean control chart using the sample ranges to estimate the population standard deviation

   a. for the data in Problem 1.1,

   b. for the data in Problem 1.2.

Compare the charts obtained with the corresponding charts from Problem 1.1 and Problem 1.2, respectively.

Problem 3.2. Construct the mean and standard deviation control charts for the following data from lots of size 5.

Lot	$\bar{x}$	s	Lot	$\bar{x}$	s
1	62.028	.0040	13	62.018	.0050
2	62.038	.0030	14	62.026	.0032
3	62.026	.0041	15	62.027	.0031
4	62.025	.0040	16	62.019	.0052
5	62.016	.0030	17	62.025	.0031
6	62.022	.0031	18	62.030	.0020
7	62.027	.0020	19	62.023	.0031
8	62.028	.0040	20	62.025	.0009
9	62.036	.0041	21	62.020	.0030
10	62.026	.0031	22	62.026	.0032
11	62.025	.0040	23	62.023	.0019
12	62.023	.0049	24	62.025	.0030

Assume that assignable causes are found for every Pareto glitch observed and recompute the upper and lower control limits, so that they could be used for examining future data.

Remark: Problems 3.3-3.10 are the continuation of Problems 1.5-1.9 from Chapter 1. The operations under examination are indicated in Figure 1.22 in Chapter 1 and, along with the nominal dimensions, in Figure 3.21 below. More precisely, the two lengths were specified as 3.10 mm with tolerance -.2 mm and 59.20 mm with tolerances $\pm.1$ mm, respectively. The two diameters were specified as 8.1 mm with tolerance -.2 mm and 6.75 mm with tolerances $\pm.1$ mm, respectively. It should be noted that, for obvious reasons, technological specifications may require readjustments/resettings in the production process, even if the process is in control. From the point of view of statistical process control for quality improvement, such readjustments/resettings can cause some minor inconvenience but, of course, have to be taken into account. For instance, in the case of operation 3 (see Problems 3.10 and 3.12 below), a foreman may be said to center the actual diameters at the nominal value, 6.75 mm. If so, suitable caution is needed when analyzing future lots on the basis of control limits obtained in the immediate past.

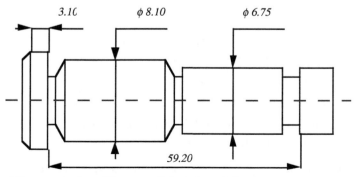

Figure 3.21. Piston Of A Fuel Pump.

Problem 3.3. Consider the (standard) control charts for Problem 1.5. Assume that assignable causes for every Pareto glitch observed are found and recompute the control limits. Use the limits obtained to examine the following data for the next 30 lots of size five.

Lot	$\bar{x}$	s	Lot	$\bar{x}$	s
1	3.034	.032	16	3.022	.018
2	3.028	.017	17	3.028	.028
3	3.046	.018	18	3.050	.010
4	3.010	.012	19	3.022	.019
5	3.010	.028	20	3.032	.021
6	3.030	.016	21	3.012	.024
7	3.018	.028	22	3.038	.024
8	3.048	.024	23	3.014	.016
9	3.028	.019	24	3.040	.024
10	3.012	.010	25	3.030	.024
11	3.016	.020	26	3.036	.013
12	3.036	.013	27	3.030	.020
13	3.012	.008	28	3.022	.021
14	3.020	.018	29	3.036	.021
15	3.050	.024	30	3.028	.014

Comment on whether the given data are in control without any reference to the past experience (use estimators of the population mean and standard deviation of the same type as previously). Compare both analyses.

Problem 3.4. Consider the data in Problem 1.5. Construct the mean and standard deviation control charts using the estimates of the population mean and standard deviation based on the sample medians. Which of the two types of estimates of the population mean and standard deviation seems to be more reliable? Given the new charts for the data examined (instead of the standard control charts), solve Problem 3.3.

Problem 3.5. Consider the (standard) control charts for Problem 1.7. Assume that assignable causes for every Pareto glitch observed are found and recompute the control limits. Use these new limits to examine the data for the next 21 lots. The data mentioned are given in Problem 1.9. Compare the conclusions obtained with the corresponding conclusions for Problem 1.9.

Problem 3.6. Consider the data in Problem 1.7. Construct the mean and standard deviation control charts using the estimates of the population mean and standard deviation based on the sample medians. Which of the two types of estimates of the population mean and standard deviation seems to be more reliable? Given the new charts for the data examined (instead of the standard control charts), solve Problem 3.5.

Problem 3.7. Consider the (standard) control charts for Problem 1.8. Assume that assignable causes for every Pareto glitch observed are found and recompute the control limits. Use these new limits to examine the data for the next 21 lots. The data mentioned are given in Problem 1.9.

Compare the conclusions obtained with the corresponding conclusions for Problem 1.9.

Problem 3.8. Consider the data in Problem 1.8. Construct the mean and standard deviation control charts using the estimates of the population mean and standard deviation based on the sample medians. Which of the two types of estimates of the population mean and standard deviation seems to be more reliable? Given the new charts for the data examined (instead of the standard control charts), solve Problem 3.7.

Problem 3.9. Consider the data in Problem 1.10. Assume that assignable causes for every Pareto glitch observed (if any) are found and recompute the control limits. Use these new limits to examine the data for lots 78 to 101. The data mentioned are summarized in the table below.

Lot	$\bar{x}$	s	Lot	$\bar{x}$	s
78	7.992	.025	90	7.990	.021
79	8.008	.008	91	8.028	.016
80	8.000	.016	92	8.020	.020
81	8.000	.016	93	8.014	.028
82	7.996	.027	94	7.980	.037
83	7.998	.019	95	8.016	.046
84	8.000	.035	96	8.038	.023
85	8.005	.013	97	8.014	.028
86	8.024	.015	98	8.026	.018
87	8.029	.017	99	8.008	.022
88	8.026	.024	100	8.022	.030
89	8.030	.016	101	8.020	.014

Verify also whether the given data are in control without any reference to the past experience. Compare both analyses.

Problem 3.10. In Table 3.10, summary statistics for lots 94 to 125 of size 5 of diameter 3 (see Figure 1.22) are given. Show that the system is in control. Compare the control charts obtained with those from Problems 1.7 and 1.9.

Problem 3.11. The system of statistical process control referred to in Problem 1.11 includes also measurements of another cylinder turned in a metal casting. The inner diameter of the cylinder is specified as $97 - .2$ mm. Lots of size 5 are taken. The head of the plant's testing laboratory verified that the process is in control and that the last control limits are 96.907 and 96.857 for the lot means and .037 for the lot standard deviations. He decided that the given limits be used as trial control limits for subsequent lots. After a few new in-control lots were observed, the lots' sample means suddenly crossed the upper control limit and stayed consistently above it until the production process was stopped. All the time, lots' sample standard deviations were well below the limit .037, in fact never crossing the level of .02. A failure of the turning tool's seat was found. The seat was replaced by a new one, but the

means' values proved first to lie barely below the upper control limit
and soon crossed it. Again, the standard deviations stayed well below
the control limit. The production process had to be stopped once more.
Investigation revealed excessive hardness of castings processed. A cause
of the material's excessive hardness was duly removed. Subsequent lots'
means and standard deviations are given in the following table.

Lot	$\bar{x}$	s	Lot	$\bar{x}$	s
1	96.868	.010	13	96.884	.011
2	96.874	.014	14	96.877	.009
3	96.861	.011	15	96.878	.007
4	96.872	.019	16	96.900	.008
5	96.878	.018	17	96.877	.014
6	96.873	.008	18	96.875	.008
7	96.875	.012	19	96.880	.013
8	96.874	.009	20	96.879	.014
9	96.870	.079	21	96.880	.010
10	96.872	.008	22	96.878	.013
11	96.882	.014	23	96.876	.008
12	96.894	.021	24	96.891	.012

Determine whether the system is in control.

Problem 3.12. Consider the data set in Problem 3.10.

   a. Suppose that the process has been centered at its nominal value
6.75 while its standard deviation has remained unchanged (realistically
speaking, the assumptions not to be fulfilled immediately after resetting
the machine). Calculate the new value of C_{pk}, using the same $\hat{\sigma}$ as in
Section 3.8 (see calculations following Table 3.9) and $\bar{\bar{x}} = 6.75$.

   b. Find the value of $\hat{\sigma}$ such that C_{pk} assumes the same value as in **a**,
while $\bar{\bar{x}}$ is the same as in Section 3.8 (see calculations following Table 3.9).
That is, show that the process capability can be increased not only by
centering a process, but also by reducing its variability.

Problem 3.13. Construct control charts for the ranges

   a. for the data set in Problem 1.1,

   b. for the data set in Problem 1.2.

Are there any Pareto glitches revealed by the charts obtained? Compare the charts obtained with the standard deviation control charts for
the same data.

Problem 3.14. Consider the following data set of simulated individual
measurements.

Lot	x	Lot	x
1	-0.529	21	-2.728
2	0.096	22	0.957
3	-0.869	23	0.734
4	-1.404	24	-1.614
5	2.361	25	-0.224
6	-0.097	26	0.417
7	1.350	27	-0.213
8	-1.378	28	-1.645
9	-0.297	29	-2.298
10	0.490	30	0.436
11	-0.184	31	-0.505
12	-0.784	32	1.003
13	0.469	33	-0.144
14	0.101	34	5.063
15	0.162	35	0.559
16	0.693	36	0.994
17	-0.824	37	0.324
18	1.294	38	-0.449
19	0.287	39	-0.275
20	-0.936	40	1.120

The data come from $\mathcal{N}(0,1)$ with probability .9 and from $\mathcal{N}(3,1)$ with probability .1 (actually, only measurements 5, 7 and 34 come from the latter distribution). Determine whether the system is in control. Perform the investigation thrice: construct X charts using s (given that $a(40) = 1.006$) as well as using moving ranges of artificial lots of size 2 and 3. Interpret the results.

Problem 3.15. Consider the following data set of simulated individual measurements.

Lot	x	Lot	x
1	-0.953	16	4.094
2	-0.880	17	3.261
3	-0.172	18	4.640
4	-0.205	19	4.398
5	2.067	20	1.466
6	-1.016	21	2.754
7	0.398	22	4.267
8	-2.352	23	2.720
9	0.265	24	3.193
10	-1.049	25	4.560
11	4.218	26	4.222
12	2.389	27	3.186
13	3.323	28	3.459
14	2.578	29	2.919
15	2.985	30	1.615

The first 10 measurements come from $\mathcal{N}(0,1)$ while measurements 11 to 30 from $\mathcal{N}(3,1)$. Determine whether the system is in control. Perform the investigation thrice: construct X charts using s (given that $a(30) = 1.009$) as well as using moving ranges of artificial lots of size 2 and 3. Interpret the results.

Chapter 4

Sequential Approaches

4.1 Introduction

Statistical Process Control usually deals with data lots indexed in time. If we look at the mean drift model in Section 3.5 as being close to reality, then we might be tempted to look on the production process as wandering, over time, from a region of satisfactory performance, to one which is not satisfactory. If this were the case, then we might want to develop a test where an alarm sounded as we moved outside the region of satisfactory performance so that we could take corrective action.

In our view, this is generally a poor way to implement quality control. It is very much in the vein of the "quality assurance" philosophy which has worked rather badly both in the Soviet Union and in the United States. It rests largely on the anthropomorphic fallacy of machines and systems operating as so many tiring jugglers.

Nevertheless, since this notion is still so popular in the quality control community, it is appropriate that we address it. In developing the CUSUM testing procedure, we recall the important concept of likelihood ratio tests covered in Appendix B. The argument below follows essentially that of Wilks [7].

4.2 The Sequential Likelihood Ratio Test

If our observations come from a density $f(x; \theta)$ and we have a time ordered data set $(x_1, x_2, \ldots, x_n)$, we may wish to test whether the true parameter is θ_0 or θ_1. The logarithm of the ratio of the two densities gives us a natural criterion for deciding between the two parameters:

$$z_t(\theta) = \ln(\frac{f(x_t, \theta_1)}{f(x_t, \theta_0)}).$$ (4.1)

We wish to find constants k_0 and k_1 such that as long as

$$\ln(k_0) < z_1 + z_2 + \ldots + z_t < \ln(k_1),$$ (4.2)

we are in a "region of uncertainty" and continue sampling (i.e., we are in region G_n).

But when

$$z_1 + z_2 + \ldots + z_t \geq \ln(k_1),$$ (4.3)

we decide for parameter θ_1 (i.e., we are in region $G_n{}^1$).

And when

$$z_1 + z_2 + \ldots + z_t \leq \ln(k_0),$$ (4.4)

we decide for parameter θ_0 (i.e., we are in region $G_n{}^0$).

At the nth stage, assuming we have not previously declared for θ_0 or θ_1, the sample $(x_1, x_2, \ldots, x_n)$ falls in one of three sets:

$$(x_1, x_2, \ldots, x_n) \in G_n, \text{ continue sampling;}$$ (4.5)

$$(x_1, x_2, \ldots, x_n) \in G_n{}^0, \text{ declare for } \theta_0;$$ (4.6)

$$(x_1, x_2, \ldots, x_n) \in G_n{}^1, \text{ declare for } \theta_1.$$ (4.7)

Thus, the probability, if the true parameter is θ, of ever declaring for θ_0 is given by

$$\mathcal{L}(\theta) = P(G_1{}^0) + P(G_2{}^0) + \ldots.$$ (4.8)

By the definition of the likelihood ratio test in (4.2),

$$P(G_n{}^0|\theta_1) \leq k_0 P(G_n{}^1|\theta_0).$$ (4.9)

So

$$\mathcal{L}(\theta_1) = \sum_{n=1}^{\infty} P(G_n{}^0|\theta_1) \leq k_0 \sum_{n=1}^{\infty} P(G_n{}^1|\theta_0).$$ (4.10)

Let us suppose that if θ is truly equal to θ_0, we wish to have

$$\mathcal{L}(\theta_0) = 1 - \alpha.$$ (4.11)

And if θ is truly equal to θ_1, we wish to have

$$\mathcal{L}(\theta_1) = \beta, \tag{4.12}$$

where α and β are customarily referred to as Type I and Type II errors. Then, we must have

$$\beta = \mathcal{L}(\theta_1) \leq k_0 \mathcal{L}(\theta_0) = k_0(1 - \alpha). \tag{4.13}$$

So

$$k_0 \geq \frac{\beta}{1 - \alpha}. \tag{4.14}$$

By a similar argument for $G_n{}^1$, we have

$$k_1 \leq \frac{1 - \beta}{\alpha}. \tag{4.15}$$

Rather than embarking on the rather difficult task of finding precise value for k_0 and k_1, let us choose

$$k_0 = \frac{\beta}{1 - \alpha} \tag{4.16}$$

and

$$k_1 = \frac{1 - \beta}{\alpha}. \tag{4.17}$$

What will the resulting actual Type I and Type II errors (say α^* and β^*) be? Substituting in (4.14) and (4.15), we have

$$\frac{\beta}{1 - \alpha} = k_0 \geq \frac{\beta^*}{1 - \alpha^*} \tag{4.18}$$

and

$$\frac{1 - \beta}{\alpha} = k_1 \leq \frac{1 - \beta^*}{\alpha^*}. \tag{4.19}$$

So

$$\beta^* \leq \frac{\beta}{1 - \alpha} \tag{4.20}$$

and

$$\alpha^* \leq \frac{\alpha}{1 - \beta}. \tag{4.21}$$

In practise, since α and β are usually less than .2, this ad hoc selection of k_0 and k_1 gives values for the actual Type I and Type II errors which are close to the target values.

4.3   CUSUM Test for Shift of the Mean

Let us now consider a sequential test to be used in the detection of a shift of the mean of a production process from μ_0 to some other value μ_1. We will assume that the variance, σ^2, is known and unchanged as we shift from one mean to the other. We carry out the test on the basis of the log likelihood ratio of N sample means, each of size n:

$$
\begin{aligned}
\mathcal{R}_1 &= \ln\left(\frac{\prod_{j=1}^{N} \frac{1}{\sqrt{2\pi}\frac{\sigma}{\sqrt{n}}} \exp(-\frac{(\bar{x}_j - \mu_1)^2}{2\sigma^2/n})}{\prod_{j=1}^{N} \frac{1}{\sqrt{2\pi}\frac{\sigma}{\sqrt{n}}} \exp(-\frac{(\bar{x}_j - \mu_0)^2}{2\sigma^2/n})}\right) \\
&= \frac{1}{2\sigma^2/n} \sum_{j=1}^{N} [(\bar{x}_j - \mu_0)^2 - (\bar{x}_j - \mu_1)^2] \\
&= \frac{\mu_0 - \mu_1}{2\sigma^2/n} [N(\mu_1 + \mu_0) - 2N\bar{\bar{x}}] \\
&= N\frac{\mu_1 - \mu_0}{\sigma^2/n} [\bar{\bar{x}}_N - \frac{\mu_0 + \mu_1}{2}].
\end{aligned}
\tag{4.22}
$$

The test statistic is then clearly based on the difference between $\bar{\bar{x}}_N$ and the average of μ_0 and μ_1. Note that if we considered the case where μ_1 was unspecified, then we would replace μ_1 in (4.22) by the maximum likelihood estimator for μ, namely $\bar{\bar{x}}_N$, and the test statistic would be simply

$$
\mathcal{R}_1 = \frac{N(\bar{\bar{x}}_N - \mu_0)^2}{\sigma^2/n}
\tag{4.23}
$$

as shown in (B.168).

We note that (4.23) indicates a procedure somewhat similar to that of a control chart for the mean. A major difference is that, at any stage j, we base our test on $\bar{\bar{x}}_j$ rather than on $\bar{x}_j$. Clearly such a cumulative procedure (CUSUM) is not so much oriented to detecting "Pareto glitches," but rather to discovering a fundamental and persistent change in the mean.

Now let us return to the case of two levels for μ_0 and μ_1. How shall we adapt such a test to process control? We might take the upper limit for acceptable μ and call it μ_1. Then we could take the lower acceptable limit for μ and call it μ_0. Typically, but not always, the target value is the midpoint

$$\mu^* = \frac{\mu_0 + \mu_1}{2}. \qquad (4.24)$$

Then the analogue of the acceptance interval in the control chart approach becomes the region of uncertainty

$$\ln(k_0) < N\frac{\mu_1 - \mu_0}{\sigma^2/n}[\bar{\bar{x}}_N - \frac{\mu_0 + \mu_1}{2}] < \ln(k_1). \qquad (4.25)$$

Suppose that we observe the first N observations to be equal to μ^*. Then our statistic becomes

$$\mathcal{R}_1 = N\frac{\mu_1 - \mu_0}{\sigma^2/n}[\frac{\mu_0 + \mu_1}{2} - \frac{\mu_0 + \mu_1}{2}] = 0. \qquad (4.26)$$

Then, suppose the next observation is $\mu^* + \delta$. The statistic becomes

$$\begin{aligned}
\mathcal{R}_1 &= (N+1)\frac{\mu_1 - \mu_0}{\sigma^2/n}[\frac{\mu_0 + \mu_1}{2} + \frac{\delta}{N+1} - \frac{\mu_0 + \mu_1}{2}] \qquad (4.27) \\
&= \frac{\mu_1 - \mu_0}{\sigma^2/n}\delta.
\end{aligned}$$

We note that this statistic is independent of N. So, in one sense, the statistic does not have the highly undesirable property that a run of 100 lots, each precisely on target, can mask a 101'st bad lot.

On the other hand, suppose that the 101'st lot gives $\bar{x}_{101} = \mu^* + \delta$, and the 102'nd lot gives $\bar{x}_{102} = \mu^* - \delta$. Then, the pooled test will have returned our statistic to a neutral $\mathcal{R}_1 = 0$. If we believe that Pareto glitches both below and above the mean can occur, we note that the sequential test allows (undesirably) for the cancellation of positive departures from the standard by negative ones.

We note an additional problem with the kind of sequential test developed above. We recall that this test identifies "in control" with the sequential "region of uncertainty." Thus the "in control" period is viewed as a region of instability, a kind of lull before the inevitable storm. In Statistical Process Control, we view Pareto glitches not as evidence of inevitable fatigue in the production system. Rather they are clues to be used to find means of improving the production system. It is the region where items are in control which represents stability.

Let us observe the use of this test on the simulated data set of Table 3.2. We recall that this data came predominantly from the distribution $\mathcal{N}(10,.01)$ (with probability .985) but with data also coming from

$\mathcal{N}(10.4, .03)$ (with probability .00995) and from $\mathcal{N}(9.8, .09)$ (with probability .00495).

We need to ask what are the natural choices for α and β so that we might pick values for $\ln(k_0)$ and $\ln(k_1)$. We note that, as we have posed the problem, neither α nor β is particularly natural. Suppose we use .01 for both α and β. Then

$$\ln(k_0) = \ln(\frac{\beta}{1 - \alpha}) = \ln(\frac{.01}{.99}) = -4.595. \qquad (4.28)$$

Similarly, we have $\ln(k_1) = 4.595$. In Table 4.1, we add to Table 3.2 the indicated sequential test statistic, namely

$$\mathcal{R}_1 = N\frac{\mu_1 - \mu_0}{\sigma^2/n}[\bar{\bar{x}}_N - \frac{\mu_0 + \mu_1}{2}]. \qquad (4.29)$$

The value of $\mu_1 - \mu_0$ which we shall use is σ. (In practice, the values for $-\ln(k_0)$ and $\ln(k_1)$ are generally taken to be in the 4 to 5 range. The value for $\mu_1 - \mu_0$ is commonly taken to be $.5\sigma$ or 1.0σ.)

Table 4.1

Lot	x_1	x_2	x_3	x_4	x_5	$\bar{x}$	$\bar{\bar{x}}_N$	$\mathcal{R}_1$	$\mathcal{R}_2$
1	9.927	9.920	10.170	9.976	9.899	9.978	9.978	-1.100	-0.492
2	9.862	10.003	9.829	9.824	10.077	9.919	9.948	-5.150	-1.629
3	10.061	10.089	9.950	9.929	9.935	9.993	9.963	-5.505	-1.421
4	9.820	10.066	10.062	9.897	10.013	9.972	9.966	-6.900	-1.543
5	9.737	9.937	9.928	10.144	9.965	9.942	9.961	-9.750	-1.950
6	9.876	9.957	9.845	9.913	9.941	9.906	9.952	-14.400	-2.629
7	9.898	9.959	9.924	9.989	9.987	9.951	9.952	-16.800	-2.840
8	10.001	10.050	10.263	9.982	10.076	10.074	9.967	-13.200	-2.087
9	9.928	10.234	9.832	10.027	10.121	10.028	9.974	-11.700	-1.744
10	9.896	9.994	10.009	9.835	10.162	9.979	9.974	-13.000	-1.838
11	10.011	10.011	10.090	10.095	10.120	10.065	9.982	-9.900	-1.335
12	9.983	9.974	10.071	10.099	9.992	10.024	9.986	-8.400	-1.084
13	10.127	9.935	9.979	10.014	9.876	9.986	9.986	-9.100	-1.129
14	10.025	9.890	10.002	9.999	9.937	9.971	9.985	-10.500	-1.255
15	9.953	10.000	10.141	10.130	10.154	10.076	9.991	-6.750	-0.779
16	10.007	10.005	9.883	9.941	9.990	9.965	9.989	-8.800	-0.984
17	10.062	10.005	10.070	10.270	10.071	10.096	9.996	-3.400	-0.369
18	10.168	10.045	10.140	9.918	9.789	10.012	9.996	-3.600	-0.379
19	9.986	10.041	9.998	9.992	9.961	9.996	9.996	-3.800	-0.390
20	9.786	10.145	10.012	10.110	9.819	9.974	9.995	-5.000	-0.500
21	9.957	9.984	10.273	10.142	10.190	10.109	10.001	1.050	0.102
22	9.965	10.011	9.810	10.057	9.737	9.916	9.997	-3.300	-0.315
23	9.989	10.063	10.148	9.826	10.041	10.013	9.998	-2.300	-0.214
24	9.983	9.974	9.883	10.153	10.092	10.017	9.998	-2.400	-0.219
25	10.063	10.075	9.988	10.071	10.096	10.059	10.001	1.250	0.112
26	9.767	9.994	9.935	10.114	9.964	9.955	9.999	-1.300	-0.114
27	9.933	9.974	10.026	9.937	10.165	10.007	9.999	-1.350	-0.116
28	10.227	10.517	10.583	10.501	10.293	10.424	10.014	20.300	1.716
29	10.022	9.986	10.152	9.922	10.101	10.034	10.015	21.750	1.806
30	9.845	9.901	10.020	9.751	10.088	9.921	10.012	18.000	1.470
31	9.956	9.921	10.132	10.016	10.109	10.027	10.013	20.150	1.618
32	9.876	10.114	9.938	10.195	10.010	10.027	10.013	20.800	1.644
33	9.932	9.856	10.085	10.207	10.146	10.045	10.014	23.100	1.798
34	10.016	9.990	10.106	10.039	9.948	10.020	10.014	23.800	1.825
35	9.927	10.066	10.038	9.896	9.871	9.960	10.013	22.050	1.667
36	9.952	10.056	9.948	9.802	9.947	9.941	10.011	19.800	1.476
37	9.941	9.964	9.943	10.085	10.049	9.996	10.010	18.500	1.360
38	10.010	9.841	10.031	9.975	9.880	9.947	10.009	17.100	1.241
39	9.848	9.944	9.828	9.834	10.091	9.909	10.006	11.700	0.838
40	10.002	9.452	9.921	9.602	9.995	9.794	10.000	0.000	0.000
41	10.031	10.061	9.943	9.997	9.952	9.997	10.001	2.050	0.143
42	9.990	9.972	10.068	9.930	10.113	10.015	10.001	2.100	0.145
43	9.995	10.056	10.061	10.016	10.044	10.034	10.002	4.300	0.293

Lot	x_1	x_2	x_3	x_4	x_5	$\bar{x}$	$\bar{\bar{x}}_1$	$\mathcal{R}_1$	$\mathcal{R}_2$
44	9.980	10.094	9.988	9.961	10.140	10.033	10.002	4.400	0.297
45	10.058	9.979	9.917	9.881	9.966	9.960	10.001	2.250	0.150
46	10.006	10.221	9.841	10.115	9.964	10.029	10.002	4.600	0.303
47	10.132	9.920	10.094	9.935	9.975	10.011	10.002	4.700	0.307
48	10.012	10.043	9.932	10.072	9.892	9.990	10.002	4.800	0.310
49	10.097	9.894	10.101	9.959	10.040	10.018	10.002	4.900	0.313
50	10.007	9.789	10.015	9.941	10.013	9.953	10.001	2.500	0.158
51	9.967	9.947	10.037	9.824	9.938	9.943	10.000	0.000	0.000
52	9.981	10.053	9.762	9.920	10.107	9.965	10.000	0.000	0.000
53	9.841	9.926	9.892	10.152	9.965	9.955	9.999	-2.650	-0.163
54	9.992	9.924	9.972	9.755	9.925	9.914	9.997	-8.100	-0.493
55	9.908	9.894	10.043	9.903	9.842	9.918	9.996	-11.000	-0.663
56	10.011	9.967	10.204	9.939	10.077	10.040	9.996	-11.200	-0.669
57	10.064	10.036	9.733	9.985	9.972	9.958	9.996	-11.400	-0.675
58	9.891	10.055	10.235	10.064	10.092	10.067	9.997	-8.700	-0.511
59	9.869	9.934	10.216	9.962	10.012	9.999	9.997	-8.850	-0.515
60	10.016	9.996	10.095	10.029	10.080	10.043	9.998	-6.000	-0.346
61	10.008	10.157	9.988	9.926	10.008	10.017	9.998	-6.100	-0.349
62	10.100	9.853	10.067	9.739	10.092	9.970	9.998	-6.200	-0.352
63	9.904	9.848	9.949	9.929	9.904	9.907	9.996	-12.600	-0.710
64	9.979	10.008	9.963	10.132	9.924	10.001	9.996	-12.800	-0.716
65	9.982	9.963	10.061	9.970	9.937	9.983	9.996	-13.000	-0.721
66	10.028	10.079	9.970	10.087	10.094	10.052	9.997	-9.900	-0.545
67	9.995	10.029	9.991	10.232	10.189	10.087	9.998	-6.700	-0.366
68	9.936	10.022	9.940	10.248	9.948	10.019	9.999	-3.400	-0.184
69	10.014	10.070	9.890	10.137	9.901	10.002	9.999	-3.450	-0.186
70	10.005	10.044	10.016	10.188	10.116	10.074	10.000	0.000	0.000
71	10.116	10.028	10.152	10.047	10.040	10.077	10.001	3.550	0.188
72	9.934	10.025	10.129	10.054	10.124	10.053	10.002	7.200	0.379
73	9.972	9.855	9.931	9.785	9.846	9.878	10.000	0.000	0.000
74	10.014	10.000	9.978	10.133	10.100	10.045	10.000	0.000	0.000
75	10.093	9.994	10.090	10.079	9.998	10.051	10.001	3.750	0.194
76	9.927	9.832	9.806	10.042	9.914	9.904	10.000	0.000	0.000
77	10.177	9.884	10.070	9.980	10.089	10.040	10.000	0.000	0.000
78	9.825	10.106	9.959	9.901	9.964	9.951	10.000	0.000	0.000
79	10.333	10.280	10.509	10.631	10.444	10.439	10.005	19.750	0.994
80	9.972	10.116	10.084	10.059	9.914	10.029	10.006	24.000	1.200
81	10.059	9.992	9.981	9.800	9.950	9.956	10.005	20.250	1.006
82	9.832	10.075	10.111	9.954	9.946	9.984	10.005	20.500	1.012
83	9.958	9.884	9.986	10.008	10.113	9.990	10.005	20.750	1.019
84	10.087	9.994	9.915	10.023	9.883	9.980	10.004	16.800	0.820
85	10.232	9.966	9.991	10.021	9.965	10.035	10.005	21.250	1.031
86	10.066	9.948	9.769	10.102	9.932	9.963	10.004	17.200	0.829
87	10.041	10.044	10.091	10.031	9.958	10.033	10.004	17.400	0.834
88	9.868	9.955	9.769	10.023	9.921	9.907	10.003	13.200	0.629
89	10.084	10.018	9.941	10.052	10.026	10.024	10.003	13.350	0.633
90	10.063	10.055	10.104	10.080	10.064	10.073	10.004	18.000	0.849

Table 4.1 (continued)

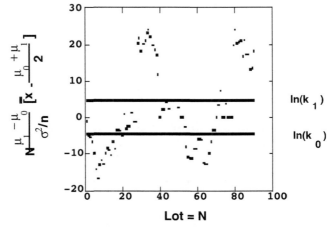

Figure 4.1. Sequential Test (CUSUM) for Mean Shift.

4.4   Shewhart CUSUM Chart

We display the results of our sequential test graphically in Figure 4.1. We recall that of the 90 lot simulations, all but lots 28, 40 and 79 came from $\mathcal{N}(10.0,.01)$. Lots 28 and 79 came from $\mathcal{N}(10.4,.03)$. Lot 40 came from $\mathcal{N}(9.8,.09)$. Unfortunately, the fact that the first eight lot sample means were, by chance, below 10.0 gives us an erroneous false alarm early on. We do see the big jump upward at lot 28, but it is rather clear that this test is not appropriate for detecting Pareto glitches of an intermittent, nonpersistent nature, as with the data set in Table 4.1. We recall, moreover, that each of the three glitches was picked up satisfactorily using the standard mean control chart strategy as employed in Figure 3.2.

A popular empirical alternative to tests based on a formal sequential argument (e.g., the CUSUM Chart) is the "Shewhart CUSUM Chart." Rather than being based on testing each lot mean, as in the case of the control chart for the mean, this test is based on the pooled running mean

$$\bar{\bar{x}}_N = \frac{1}{N} \sum_{j=1}^{N} \bar{x}_j \ . \tag{4.30}$$

If all the lot means are identically and independently distributed with common lot mean μ_0 and lot variance $\sigma_0{}^2/n$, where n is the lot size, then

$$E(\bar{\bar{x}}_N) = \mu \tag{4.31}$$

and

$$Var(\bar{\bar{x}}_N) = \frac{Var(\bar{x}_j)}{N} = \frac{\sigma_0{}^2/n}{N} = \frac{\sigma_0{}^2}{nN} \ . \tag{4.32}$$

Accordingly, in the Shewhart CUSUM procedure for a change from the mean level μ_0, we use as the "in control" region:

$$-3 < \frac{\bar{\bar{x}}_N - \mu_0}{\frac{\sigma_0}{\sqrt{nN}}} < 3 \ . \tag{4.33}$$

There are troubling problems associated with the Shewhart CUSUM which are not encountered with the more orthodox CUSUM chart. To see this, suppose that each of the first N $\bar{x}_j$ values is precisely equal to μ_0. Then, if the $N+1$'st lot mean is equal to $\mu_0 + \delta$, we find that

$$\frac{\bar{\bar{x}}_N - \mu_0}{\frac{\sigma_0}{\sqrt{n(N+1)}}} = \frac{\frac{\delta}{\sqrt{N+1}}}{\frac{\sigma_0}{\sqrt{n}}} \ . \tag{4.34}$$

Thus, the Shewhart CUSUM approach does, unfortunately, have inertia built into it. It will take an ever larger departure from μ_0 to ring the alarm, the longer the chain of satisfactory lots. This problem is indicated in Figure 4.2 where we apply the Shewhart CUSUM to the data in Table 4.1, taking $\mu_0 = 10$ and $\sigma_0 = 0.10$.

We recall that in the paradigm of process control, there is an emphasis on seeing Pareto glitches as soon as they occur. Because of its inertia, the Shewhart CUSUM tends to be a lagging indicator of a Pareto glitch. And an isolated glitch tends to be masked by satisfactory lots. One way to discount the inertial effect somewhat is to use only the last, say, 50 lots in the computation of the running mean.

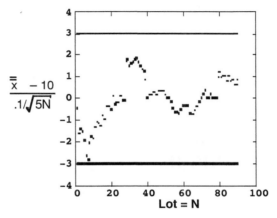

Figure 4.2. Shewhart CUSUM Test for Mean Shift.

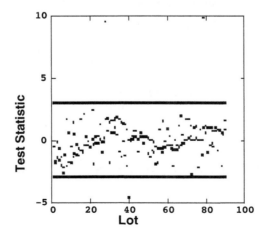

Figure 4.3. Combined Shewhart CUSUM and Mean Control Charts.

Naturally, we could "normalize" the mean test statistic for the ordinary control chart:

$$\frac{\bar{x} - \mu_0}{\sigma_0/\sqrt{n}} \; . \tag{4.35}$$

Then, we could plot both the Shewhart CUSUM and the normalized mean statistics on the same chart, examining any case outside the ± 3 limits. Such a combined chart is demonstrated in Figure 4.3.

We note that the lots 28 and 79 show out of control on the high side, and lot 40 out of control on the low, as we observed in Figure 3.2. Generally speaking, it is probably better to view these charts separately or in parallel rather than superimposed display. We show a parallel graph in Figure 4.4.

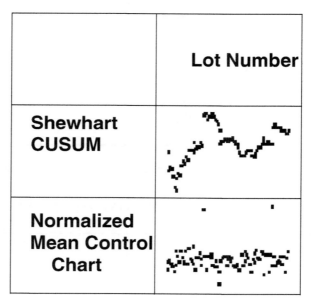

Figure 4.4. Combined Shewhart CUSUM and Mean Control Charts.

4.5   Performance of CUSUM Tests on Data with Mean Drift

We should expect the CUSUM to perform better on the mean drift situation described by (3.75) with 90 lots of simulated data given in Table 3.4. We show this data again in Table 4.2 below. We note the addition of three columns. The first of these $\bar{\bar{x}}_N$ is simply the cumulative or running mean. The second is the CUSUM test statistic, already developed and denoted as $\mathcal{R}_1$,

$$\mathcal{R}_1 = N \frac{\mu_1 - \mu_0}{\sigma^2/n} [\bar{\bar{x}}_N - \frac{\mu_0 + \mu_2}{2}]. \tag{4.36}$$

The third is the "Shewhart CUSUM" test statistic, developed above,

$$\mathcal{R}_2 = \frac{\bar{\bar{x}}_N - \mu_0}{\frac{\sigma_0}{\sqrt{nN}}}. \tag{4.37}$$

We recall that n is the lot size, 5. N is the lot number. In the case of $\mathcal{R}_1$, we shall use $(\mu_0 + \mu_1)/2 = 10$. In the case of $\mathcal{R}_2$, $\mu_0 = 10$.

Table 4.2									
Lot	x_1	x_2	x_3	x_4	x_5	$\bar{x}$	$\bar{\bar{x}}$	$\mathcal{R}_1$	$\mathcal{R}_2$
1	9.927	10.019	9.963	10.073	9.911	9.979	9.979	-1.050	-0.470
2	9.856	9.985	10.074	9.885	9.970	9.954	9.967	-3.300	-1.044
3	10.031	9.957	10.133	9.877	10.078	10.015	9.983	-2.550	-0.658
4	9.850	10.071	10.000	10.115	10.011	10.009	9.989	-2.200	-0.492
5	9.858	10.159	10.039	10.141	10.114	10.062	10.004	1.000	0.200
6	10.127	10.006	9.852	10.095	10.090	10.034	10.009	2.700	0.493
7	10.131	10.144	9.964	10.027	9.995	10.052	10.015	5.250	0.887
8	9.924	10.008	10.043	10.155	9.939	10.014	10.015	6.000	0.949
9	9.869	9.902	9.739	10.171	10.111	9.958	10.009	4.050	0.604
10	10.132	9.974	9.993	9.967	10.058	10.025	10.010	5.000	0.707
11	9.877	9.970	9.980	10.031	9.987	9.969	10.006	3.300	0.445
12	10.089	9.803	9.901	9.925	9.890	9.922	9.999	-0.600	-0.077
13	10.074	9.954	9.957	9.819	9.943	9.949	9.996	-2.600	-0.322
14	9.977	10.011	9.966	9.876	9.890	9.944	9.992	-5.600	-0.669
15	10.040	9.961	9.841	10.251	10.125	10.044	9.995	-3.750	-0.433
16	10.049	10.086	10.056	10.045	10.113	10.070	10.000	0.000	0.000
17	10.040	10.039	10.014	9.992	9.836	9.984	9.999	-0.850	-0.092
18	9.964	10.058	9.995	9.940	10.060	10.003	9.999	-0.900	-0.095
19	10.058	9.885	10.081	10.030	10.011	10.013	10.000	0.000	0.000
20	10.133	10.049	10.023	10.147	9.904	10.051	10.003	3.000	0.300
21	10.063	10.033	10.050	10.065	9.983	10.039	10.004	4.200	0.410
22	10.040	9.954	9.955	10.031	9.940	9.984	10.003	3.300	0.315
23	9.889	10.049	10.117	10.157	9.962	10.035	10.005	5.750	0.536
24	10.092	9.889	9.947	9.975	10.203	10.021	10.005	6.000	0.548
25	9.948	10.188	10.014	9.970	9.975	10.019	10.006	7.500	0.671
26	10.119	9.949	10.210	10.114	9.835	10.045	10.007	9.100	0.798
27	10.012	10.194	9.980	10.057	10.014	10.051	10.009	12.150	1.046
28	9.943	9.835	10.036	9.882	10.160	9.971	10.008	11.200	0.947
29	10.072	10.019	9.946	10.119	10.026	10.036	10.009	13.050	1.084
30	10.062	10.280	9.955	10.028	10.131	10.091	10.011	16.500	1.347
31	9.988	10.101	10.034	9.951	10.150	10.045	10.012	19.375	1.556
32	10.137	10.031	10.045	10.061	9.936	10.042	10.013	20.800	1.644
33	10.020	10.031	10.060	9.899	9.979	9.998	10.013	21.450	1.670
34	9.870	10.156	10.166	10.052	10.022	10.053	10.014	23.800	1.825
35	10.107	10.053	9.785	9.902	9.980	9.965	10.013	22.750	1.720
36	10.118	10.059	10.145	10.149	9.968	10.088	10.015	27.000	2.012
37	9.928	9.974	10.120	10.021	9.959	10.001	10.014	25.900	1.904
38	9.934	9.870	10.068	9.828	9.745	9.889	10.011	20.900	1.516
39	9.893	9.997	9.923	9.891	10.007	9.942	10.009	17.550	1.257
40	10.120	10.019	9.946	10.001	10.055	10.028	10.010	20.000	1.414
41	9.996	9.796	9.967	9.976	9.999	9.947	10.008	16.400	1.145
42	10.071	9.776	9.941	9.904	10.077	9.954	10.007	14.700	1.014
43	9.949	9.869	9.925	10.009	9.920	9.934	10.005	10.750	0.733
44	9.930	9.998	9.836	10.195	10.020	9.996	10.005	11.000	0.742
45	9.983	9.967	10.045	9.993	10.119	10.021	10.005	11.250	0.750
46	10.086	9.995	9.955	9.969	9.830	9.967	10.005	11.500	0.758
47	9.896	9.999	10.051	9.878	9.857	9.936	10.003	7.050	0.460
48	9.997	10.089	9.893	9.968	9.971	9.984	10.003	7.200	0.465
49	9.987	9.928	10.006	9.997	10.108	10.005	10.003	7.350	0.470
50	9.958	9.999	10.078	9.912	10.163	10.022	10.003	7.500	0.474

Lot	x_1	x_2	x_3	x_4	x_5	$\bar{x}$	$\bar{\bar{x}}$	$\mathcal{R}_1$	$\mathcal{R}_2$
51	10.075	9.940	10.027	9.800	9.926	9.954	10.002	5.100	0.319
52	10.099	10.042	10.017	9.845	10.061	10.013	10.002	5.200	0.322
53	10.097	10.133	9.997	9.911	9.890	10.006	10.003	7.950	0.488
54	10.117	10.020	9.813	9.934	10.070	9.991	10.002	5.400	0.329
55	9.779	10.133	10.086	9.823	10.103	9.985	10.002	5.500	0.332
56	10.015	9.885	10.034	10.013	10.008	9.991	10.002	5.600	0.335
57	9.849	10.153	9.996	9.896	9.842	9.947	10.001	2.850	0.169
58	10.010	9.825	9.883	9.967	9.930	9.923	10.000	0.000	0.000
59	9.897	9.975	9.914	10.081	9.842	9.942	9.999	-2.950	-0.172
60	9.857	10.244	9.961	9.940	9.896	9.980	9.998	-6.000	-0.346
61	9.887	9.913	10.021	9.924	9.897	9.929	9.997	-9.150	-0.524
62	9.920	9.996	10.120	9.957	10.060	10.010	9.997	-9.300	-0.528
63	9.978	9.944	9.921	9.864	10.007	9.943	9.996	-12.600	-0.710
64	10.126	9.950	10.107	10.027	10.020	10.046	9.997	-9.600	-0.537
65	9.951	9.964	9.883	10.113	9.994	9.981	9.997	-9.750	-0.541
66	10.083	10.051	10.023	9.918	10.111	10.037	9.998	-6.600	-0.363
67	9.869	10.165	10.095	9.910	9.978	10.003	9.998	-6.700	-0.366
68	10.196	10.018	9.828	10.037	9.958	10.007	9.998	-6.800	-0.369
69	10.142	10.097	9.957	10.062	10.085	10.069	9.999	-3.450	-0.186
70	9.988	10.043	9.953	10.104	10.018	10.021	9.999	-3.500	-0.187
71	10.082	9.855	9.943	10.026	10.112	10.004	9.999	-3.550	-0.188
72	9.967	9.987	9.988	10.080	10.264	10.057	10.000	0.000	0.000
73	10.040	10.003	9.964	9.919	10.029	9.991	10.000	0.000	0.000
74	10.035	10.075	10.071	10.175	10.014	10.074	10.001	3.700	0.192
75	10.093	10.005	10.106	10.025	10.004	10.047	10.001	3.750	0.194
76	10.066	9.952	9.972	10.166	9.907	10.013	10.002	7.600	0.390
77	10.117	10.107	10.015	9.891	9.994	10.025	10.002	7.700	0.392
78	10.119	10.030	10.114	10.152	10.037	10.090	10.003	11.700	0.592
79	10.000	10.123	9.974	9.869	10.027	9.999	10.003	11.850	0.596
80	10.012	10.010	9.980	10.052	10.121	10.035	10.003	12.000	0.600
81	10.025	10.087	10.248	9.943	9.939	10.048	10.004	16.200	0.805
82	10.055	10.171	10.121	10.113	10.162	10.124	10.005	20.500	1.012
83	10.124	10.121	9.951	10.096	10.090	10.076	10.006	24.900	1.222
84	9.964	10.185	10.202	10.195	10.071	10.123	10.008	33.600	1.640
85	9.971	10.081	10.152	10.035	10.131	10.074	10.008	34.000	1.649
86	10.100	10.137	10.115	10.063	10.041	10.091	10.009	38.700	1.866
87	10.153	10.104	10.184	10.080	10.047	10.114	10.011	47.850	2.294
88	9.893	9.968	10.133	9.875	10.160	10.005	10.011	48.400	2.307
89	9.932	10.168	10.191	10.116	9.977	10.077	10.011	48.950	2.320
90	9.982	10.133	9.916	10.155	9.862	10.010	10.011	49.500	2.333

Table 4.2(continued)

We note the results using the CUSUM statistic for mean shifts, $\mathcal{R}_1$, in Figure 4.5.

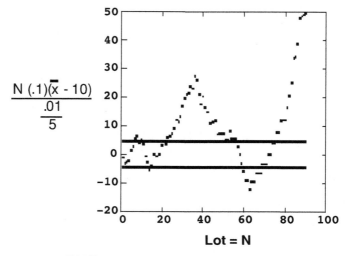

$$\frac{N\,(.1)(\overline{x} - 10)}{\dfrac{.01}{5}}$$

Lot = N

Figure 4.5. CUSUM Test for Data with Mean Drift.

The CUSUM procedure here more or less automatically grasps the shifts upward and downward in the production process which our eye would do if we did our analysis based on Figure 3.7.

In Figure 4.6, we note the performance of the Shewhart CUSUM on the data in Table 4.2. The inertia of the Shewhart CUSUM does not deter us from capturing the directions of the mean trends, but it does tend to suppress the ringing of the alarm. It may be appropriate in some circumstances to change our control limits to $\pm\, 2$.

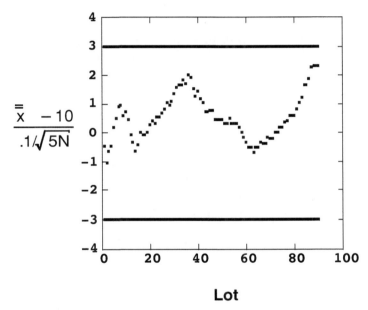

Figure 4.6. Shewhart CUSUM Test for Data with Mean Drift.

4.6   Sequential Tests for Persistent Shift of the Mean

We noted already in Section 4.3 that the CUSUM procedures are oriented to discovering persistent changes in the mean, while they are not well suited to detecting "Pareto glitches." In fact, they are most effective when data reveal a persistent and single shift of the mean from one value to another. Such situations indeed happen and when they do and we know a fair amount a priori about the magnitude of shift to expect, CUSUMs do frequently give us a device for early detection that the shift has occurred. However, as we shall note in the example in Table 4.3,

usually the standard Shewhart control chart will detect the change in
level easily, provided the change is not too small.

One possible example of the situation mentioned is that of a failure of
a cutting tool of a metal-turning lathe, caused by a sudden change of the
hardness of metal elements treated. Only seemingly more difficult is a
similar task of picking up occasional switches of the mean from one value
to another, provided that there are essentially only two possible values
of the mean. Such a situation may happen if elements to be machined
can be preprocessed according to two different technologies.

Let us assume that the mean μ_0 of the production process has the
required value but a shift to some other value μ_1 is possible. Let us
assume also that the variance, σ^2, is known and constant regardless of
whether the shift of the mean occurs or not. As in the previous sections,
we denote the running number of lots taken, each of size n, by N. Let

$$z_N = \frac{\bar{x}_N - \mu_0}{\sigma/\sqrt{n}} \qquad (4.38)$$

be the standardized sample mean of the Nth lot. The most widely used
procedure in the case considered, to be named after E. S. Page who
initiated investigations in the area, consists in calculating the following
two cumulative sums after each Nth lot has been obtained:

$$SH_N = \max[0, (z_N - k) + SH_{N-1}] \qquad (4.39)$$

and

$$SL_N = \max[0, (-z_N - k) + SL_{N-1}], \qquad (4.40)$$

where $SH_0 = SL_0 = 0$ and k is a suitable constant. Note that the upper
cumulative sum, SH_N, is always nonnegative and increases whenever z_N
is greater than k. Analogously, the lower CUSUM, SL_N, is nonnegative
as well and increases whenever z_N is smaller than $-k$. The value of k
is usually selected to be half the assumed value of the (absolute) mean
shift in standardized units. That is, if the magnitude of the mean shift
is believed to be equal to $|\mu_1 - \mu_0|$, we get the shift $\sqrt{n}|\mu_1 - \mu_0|/\sigma$ in
standardized units and we usually set $k = \sqrt{n}|\mu_1 - \mu_0|/(2\sigma)$. Now, if
either SH_N or SL_N becomes greater than some prescribed constant h,
the process is considered to be out of control. More precisely, an upward
shift can be believed to have happened in the former and a downward
shift in the latter case. The value of h is usually selected to be equal to
4 or, less frequently, to 5. We shall see later that there is some rationale
behind the given choices of k and h.

			Table 4.3					
Lot	x_1	x_2	x_3	x_4	x_5	$\bar{x}$	s	EWMA
1	8.798	12.374	10.456	9.419	9.997	10.209	1.361	10.07
2	9.525	7.985	10.211	10.352	9.415	9.498	.940	9.88
3	10.867	11.307	8.249	10.817	11.634	10.575	1.343	10.178
4	11.831	10.996	10.044	9.901	11.402	10.835	.842	10.397
5	9.321	11.073	9.805	10.901	9.722	10.164	.775	10.32
6	8.714	11.066	10.765	9.320	8.949	9.763	1.079	10.134
7	11.155	10.874	8.571	9.170	10.473	10.049	1.123	10.105
8	8.696	10.505	10.783	9.309	9.998	9.858	.858	10.023
9	9.202	10.072	10.278	10.091	8.833	9.695	.637	9.914
10	10.476	10.243	9.005	10.966	9.996	10.137	.727	9.989
11	8.636	10.490	11.323	9.750	10.364	10.13	.998	10.03
12	8.547	10.975	9.885	10.648	8.428	9.697	1.173	9.919
13	9.559	9.868	10.752	8.860	8.428	9.493	.903	9.777
14	9.145	10.528	10.774	7.737	11.201	9.877	1.423	9.81
15	10.068	8.488	9.316	9.050	11.076	9.6	1.002	9.74
16	10.157	9.819	8.125	9.033	9.386	9.304	.785	9.595
17	9.386	9.042	10.025	11.159	12.538	10.43	1.428	9.873
18	11.766	10.748	9.762	9.622	10.298	10.439	.866	10.061
19	9.582	9.905	8.496	9.173	9.817	9.395	.577	9.839
20	11.299	9.588	9.737	10.266	10.324	10.243	.672	9.974
21	11.071	8.509	10.976	9.810	11.762	10.426	1.28	10.125
22	11.355	10.548	9.495	10.105	11.254	10.551	.783	10.267
23	10.824	9.475	10.465	10.320	10.754	10.368	.54	10.3
24	10.957	11.485	11.055	10.123	12.669	11.258	.93	10.619
25	12.572	9.389	9.667	10.206	10.585	10.484	1.256	10.574
26	11.129	12.759	11.488	10.919	12.138	11.687	.757	10.945
27	10.572	10.286	11.077	9.963	12.468	10.873	.981	10.921
28	9.505	12.442	10.212	9.646	11.260	10.613	1.234	10.818
29	13.590	11.024	10.547	13.254	9.279	11.539	1.837	11.058
30	12.319	9.966	11.120	12.565	8.366	10.867	1.741	10.995
31	11.377	9.962	11.133	11.050	11.064	10.917	.55	10.969
32	10.300	11.521	11.668	11.588	9.706	10.957	.897	10.965
33	9.731	10.742	12.674	10.847	11.714	11.142	1.108	11.024
34	10.157	11.343	11.436	11.204	10.785	10.985	.526	11.011
35	11.190	9.765	11.413	10.785	10.704	10.771	.633	10.931
36	10.205	10.161	10.610	11.949	12.851	11.155	1.193	11.006
37	12.398	12.291	11.084	10.974	11.523	11.654	.664	11.222
38	11.640	9.860	11.514	11.460	9.518	10.798	1.022	11.081
39	11.869	15.250	10.530	10.955	13.770	12.475	1.99	11.545
40	9.291	11.154	10.475	10.913	9.125	10.192	.932	11.094

Typical results for the Page CUSUM procedure are given in Tables 4.3 and 4.4. We simulated 40 lots of size 5, the first 20 lots from the distribution $\mathcal{N}(10,1)$ and the second 20 lots from $\mathcal{N}(11,1)$. Thus, the mean shift was equal to $\sqrt{5}$ in standardized units. Simulated data are displayed in Table 4.3 (along with values of $\bar{x}$, s and a statistic to be used later), while their analysis is provided by Table 4.4. In the fourth and fifth columns of Table 4.4 (denoted by $SH^{.5}$ and $SL^{.5}$, respectively), we give the values of SH_N for $k = .5$ and SL_N for $k = .5$, respectively. In the last two columns, we give the values of SH_N for $k = 1$ and for $k = 2$, respectively. As the results for SL_N with $k = .5$ show, there is no need to present the values of SL_N for $k = 1$ and for $k = 2$. The results are displayed graphically in Figures 4.7 to 4.10.

			Table 4.4			
Lot	$\bar{x}$	z	$SH^{.5}$	$SL^{.5}$	SH^1	SH^2
1	10.209	0.467	0.000	0.000	0.000	0.000
2	9.498	-1.123	0.000	0.623	0.000	0.000
3	10.575	1.285	0.785	0.000	0.285	0.000
4	10.835	1.867	2.152	0.000	1.152	0.000
5	10.164	0.368	2.020	0.000	0.520	0.000
6	9.763	-0.530	0.990	0.030	0.000	0.000
7	10.049	0.109	0.599	0.000	0.000	0.000
8	9.858	-0.317	0.000	0.000	0.000	0.000
9	9.695	-0.682	0.000	0.182	0.000	0.000
10	10.137	0.307	0.000	0.000	0.000	0.000
11	10.113	0.252	0.000	0.000	0.000	0.000
12	9.697	-0.678	0.000	0.178	0.000	0.000
13	9.493	-1.133	0.000	0.811	0.000	0.000
14	9.877	-0.275	0.000	0.586	0.000	0.000
15	9.600	-0.895	0.000	0.981	0.000	0.000
16	9.304	-1.556	0.000	2.037	0.000	0.000
17	10.430	0.962	0.462	0.575	0.000	0.000
18	10.439	0.982	0.944	0.000	0.000	0.000
19	9.395	-1.354	0.000	0.854	0.000	0.000
20	10.243	0.543	0.043	0.000	0.000	0.000
21	10.426	0.952	0.495	0.000	0.000	0.000
22	10.551	1.233	1.228	0.000	0.233	0.000
23	10.368	0.822	1.550	0.000	0.055	0.000
24	11.258	2.813	3.863	0.000	1.868	0.813
25	10.484	1.082	4.445	0.000	1.950	0.000
26	11.687	3.771	7.716	0.000	4.721	1.771
27	10.873	1.953	9.169	0.000	5.674	1.724
28	10.613	1.371	10.040	0.000	6.045	1.095
29	11.539	3.441	12.981	0.000	8.486	2.536
30	10.867	1.939	14.420	0.000	9.425	2.475
31	10.917	2.051	15.971	0.000	10.476	2.526
32	10.957	2.139	17.610	0.000	11.615	2.665
33	11.142	2.553	19.663	0.000	13.168	3.218
34	10.985	2.203	21.336	0.000	14.371	3.421
35	10.771	1.725	22.561	0.000	15.096	3.146
36	11.155	2.583	24.644	0.000	16.679	3.729
37	11.654	3.698	27.842	0.000	19.377	5.427
38	10.798	1.785	29.127	0.000	20.162	5.212
39	12.475	5.534	34.161	0.000	24.696	8.746
40	10.192	0.428	34.089	0.000	24.124	7.174

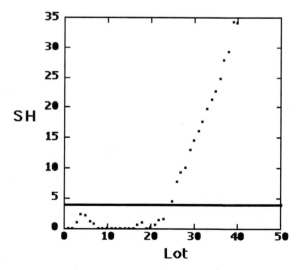

Figure 4.7. Upper CUSUM for Fixed Mean Shift ($k = .5$).

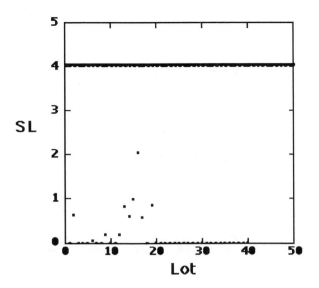

Figure 4.8. Lower CUSUM for Fixed Mean Shift ($k = .5$).

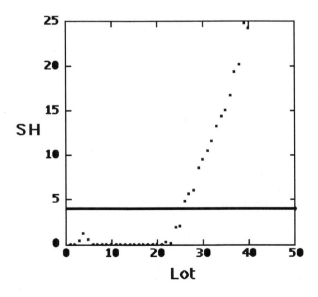

Figure 4.9. Upper CUSUM for Fixed Mean Shift ($k = 1$).

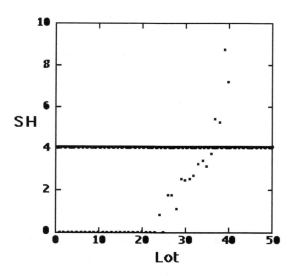

Figure 4.10. Upper CUSUM for Fixed Mean Shift ($k = 2$).

According to the rule given earlier, a reasonable value of k is 1 (there is not much sense in requiring that k equal exactly 1.118). The Page CUSUM test detects then the shift of the mean quite readily. For h equal to 4, the shift is detected on lot 26, while for $h = 5$ on lot 27. The test performs even better if k equal to .5 is used. Clearly, however, the probability of false alarm, when no shift occurred, is a decreasing function of k: the smaller the k the greater the probability of a false alarm. We should not make k smaller than really needed to pick up the shift of a given size. At the same time, as the case with $k = 2$ shows, the test performs poorly for too large k's. We find that, not surprisingly, the Page CUSUM test relies heavily on the proper choice of k (and h) which, in turn, depends on the size of the mean shift, $|\mu_1 - \mu_0|$. The value of $|\mu_1 - \mu_0|$ is usually assumed to be the largest acceptable (absolute) deviation of the mean from its target value μ_0 (note that, unlike in Section 4.3, it is μ_0 which is the target value, while μ_1 is a fixed distance apart from μ_0 in either direction).

Now let us examine the performance of the Page CUSUM test on the data set of Table 4.1. The results are summarized in Table 4.5 and displayed graphically in Figures 4.11 to 4.14. In addition to the Page CUSUM with $k = .5$, we give also the results for $k = .25$.

Lot	$\bar{x}$	z	$SH^{.5}$	$SL^{.5}$	$SH^{.25}$	$SL^{.25}$
						Table 4.5
1	9.978	-0.492	0.000	0.000	0.000	0.242
2	9.919	-1.811	0.000	1.311	0.000	1.803
3	9.993	-0.157	0.000	0.968	0.000	1.701
4	9.972	-0.626	0.000	1.094	0.000	2.086
5	9.942	-1.297	0.000	1.891	0.000	3.133
6	9.906	-2.102	0.000	3.493	0.000	4.985
7	9.951	-1.096	0.000	4.089	0.000	5.831
8	10.074	1.655	1.155	1.934	1.405	3.926
9	10.028	0.626	1.281	0.808	1.781	3.050
10	9.979	-0.470	0.311	0.778	1.061	3.270
11	10.065	1.453	1.264	0.000	2.264	1.567
12	9.992	-0.179	0.585	0.000	1.835	1.496
13	9.986	-0.313	0.000	0.000	1.272	1.559
14	9.971	-0.648	0.000	0.148	0.374	1.957
15	10.076	1.699	1.199	0.000	1.823	0.008
16	9.965	-0.783	0.000	0.283	0.790	0.541
17	10.096	2.147	1.647	0.000	2.687	0.000
18	10.012	0.268	1.415	0.000	2.705	0.000
19	9.996	-0.089	0.826	0.000	2.366	0.000
20	9.974	-0.581	0.000	0.081	1.535	0.331
21	10.109	2.437	1.937	0.000	3.722	0.000
22	9.916	-1.878	0.000	1.378	1.594	1.628
23	10.013	0.291	0.000	0.587	1.635	1.087
24	10.017	0.380	0.000	0.000	1.765	0.457
25	10.059	1.319	0.819	0.000	2.834	0.000
26	9.955	-1.006	0.000	0.506	1.578	0.756
27	10.007	0.157	0.000	0.000	1.485	0.349
28	10.424	9.481	8.981	0.000	10.716	0.000
29	10.034	0.760	9.241	0.000	11.226	0.000
30	9.921	-1.766	6.975	1.266	9.210	1.516
31	10.027	0.604	7.079	0.162	9.564	0.662
32	10.027	0.604	7.183	0.000	9.918	0.000
33	10.045	1.006	7.689	0.000	10.674	0.000
34	10.020	0.447	7.636	0.000	10.871	0.000
35	9.960	-0.894	6.242	0.394	9.727	0.644
36	9.941	-1.319	4.423	1.213	8.158	1.713
37	9.996	-0.089	3.834	0.802	7.819	1.552
38	9.947	-1.185	2.149	1.487	6.384	2.487
39	9.909	-2.035	0.000	3.022	4.099	4.272
40	9.794	-4.606	0.000	7.128	0.000	8.628
41	9.997	-0.067	0.000	6.695	0.000	8.445
42	10.015	0.335	0.000	5.860	0.085	7.860
43	10.034	0.760	0.260	4.600	0.595	6.850
44	10.033	0.738	0.498	3.362	1.083	5.862
45	9.960	-0.894	0.000	3.756	0.000	6.506
46	10.029	0.648	0.148	2.608	0.398	5.608
47	10.011	0.246	0.000	1.862	0.394	5.112
48	9.990	-0.224	0.000	1.586	0.000	5.086
49	10.018	0.402	0.000	0.684	0.152	4.434
50	9.953	-1.051	0.000	1.235	0.000	5.235
51	9.943	-1.275	0.000	2.010	0.000	6.260
52	9.965	-0.783	0.000	2.293	0.000	6.793
53	9.955	-1.006	0.000	2.799	0.000	7.549
54	9.914	-1.923	0.000	4.222	0.000	9.222
55	9.918	-1.834	0.000	5.556	0.000	10.806
56	10.040	0.894	0.394	4.162	0.644	9.662
57	9.958	-0.939	0.000	4.601	0.000	10.351
58	10.067	1.498	0.998	2.603	1.248	8.603
59	9.999	-0.022	0.476	2.125	0.976	8.375
60	10.043	0.962	0.938	0.663	1.688	7.163
61	10.017	0.380	0.818	0.000	1.818	6.533
62	9.970	-0.671	0.000	0.171	0.897	6.954
63	9.907	-2.080	0.000	1.751	0.000	8.784
64	10.001	0.022	0.000	1.229	0.000	8.512
65	9.983	-0.380	0.000	1.109	0.000	8.642
66	10.052	1.163	0.663	0.000	0.913	7.229
67	10.087	1.945	2.108	0.000	2.608	5.034
68	10.019	0.425	2.033	0.000	2.783	4.359
69	10.002	0.045	1.578	0.000	2.578	4.064
70	10.074	1.655	2.733	0.000	3.983	2.159
71	10.077	1.722	3.955	0.000	5.455	0.187
72	10.053	1.185	4.640	0.000	6.390	0.000
73	9.878	-2.728	1.412	2.228	3.412	2.478
74	10.045	1.006	1.918	0.722	4.168	1.222
75	10.051	1.140	2.558	0.000	5.058	0.000

Table 4.5(continued)						
Lot	$\bar{x}$	z	$SH^{.5}$	$SL^{.5}$	$SH^{.25}$	$SL^{.25}$
76	9.904	-2.147	0.000	1.647	2.661	1.897
77	10.040	0.894	0.394	0.253	3.305	0.753
78	9.951	-1.096	0.000	0.849	1.959	1.599
79	10.439	9.816	9.316	0.000	11.525	0.000
80	10.029	0.648	9.464	0.000	11.923	0.000
81	9.956	-0.984	7.980	0.484	10.689	0.734
82	9.984	-0.358	7.122	0.342	10.081	0.842
83	9.990	-0.224	6.398	0.066	9.607	0.816
84	9.980	-0.447	5.451	0.013	8.910	1.013
85	10.035	0.783	5.734	0.000	9.443	0.000
86	9.963	-0.827	4.407	0.327	8.366	0.577
87	10.033	0.738	4.645	0.000	8.854	0.000
88	9.907	-2.080	2.065	1.580	6.424	1.830
89	10.024	0.537	2.102	0.543	6.711	1.043
90	10.073	1.632	3.234	0.000	8.093	0.000

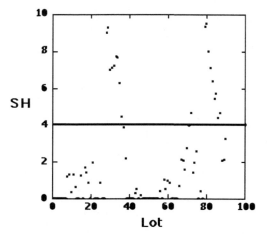

Figure 4.11. Upper CUSUM for Mean Shift ($k = .5$).

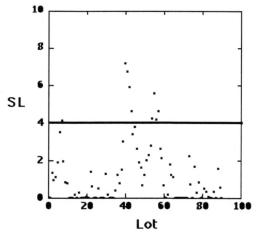

Figure 4.12. Lower CUSUM for Mean Shift ($k = .5$).

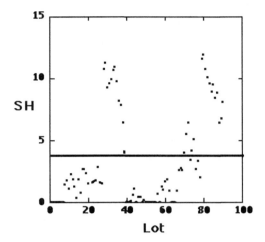

Figure 4.13. Upper CUSUM for Mean Shift ($k = .25$).

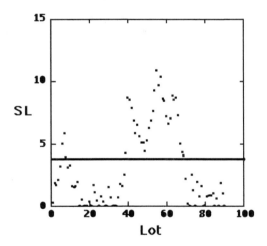

Figure 4.14. Lower CUSUM for Mean Shift ($k = .25$).

Our analysis shows that the Page test is hardly appropriate for detecting Pareto glitches of an intermittent nature, which come from contaminating distributions. We note also that using too small k may lead to particularly erratic behavior of the Page CUSUMs. The results in Table 4.5, when compared with those in Table 4.4, indicate that a persistent shift of the mean should result in a clear upward drift of the upper or lower CUSUM (depending on whether the shift is positive or negative).

As in the case of the original CUSUM Tests, we should expect the

Page procedure to perform better on data with mean drift. For the data
set of Table 4.2, the results are displayed only graphically in Figures 4.15
and 4.16. The value of k was selected to be .5.

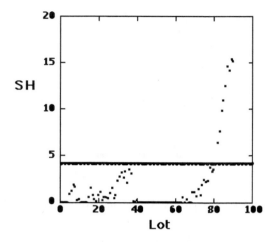

Figure 4.15. Upper CUSUMs for Data with Mean Drift
$(k = .5)$.

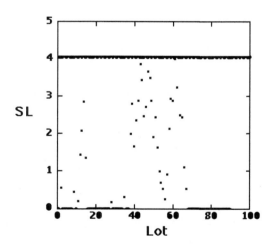

Figure 4.16. Lower CUSUMs for Data with Mean Drift
$(k = .5)$.

Apparently, the Page CUSUM charts are not recommended for use on
data with mean drift either. One can readily verify that the standard
mean control chart picks up the mean shift of the data set of Table 4.3

practically as easily as the Page test. Indeed, the sample mean across lots 1 to 20, $\bar{\bar{x}}$, is equal to 9.9687, while $\bar{s} = .9756$ and hence

$$UCL = 9.9687 + 1.427(.9756) = 11.3609,$$

where the constant 1.427 is provided by Table 1.9 (the lower control limit, which is inessential in the case considered, is equal to 8.5765). Thus, the alarm rings on lot 26, just as it does for the Page test with $k = 1$ and $h = 4$. To be sure, when calculating $\bar{\bar{x}}$ we used our knowledge of the past lots, so that we could estimate μ_0 well. On the other hand, when calculating cumulative sums a good guess at the true value of μ_0 is needed as well. By the way, had we used exact values of μ_0 and σ for the construction of the Shewhart chart, we would have obtained the same result (in standardized z units, we have then simply that $UCL = 3$).

Why, then, should we use the Page test at all? A somewhat vague answer is that the Page test is better in a certain sense for small, single shifts of the mean. A more precise answer requires some additional considerations.

Let us first define the concept of the average run length (ARL) of a test as the expected value of the number of lots until the first out-of-control lot is detected, given the magnitude of the mean shift that occurred. The ARL's can easily be calculated for the mean control chart. Indeed, the ARL corresponds then to the expected value of the number of trials until the first success occurs in a sequence of Bernoulli experiments, where the success consists in producing an out-of-control signal by a single trial, that is, by a single lot. The ARL is, thus, equal to the mean of the geometric distribution, $1/p$, where p is the probability of producing an out-of-control signal by a single lot (see Appendix B). Now, p can be immediately calculated for any given problem. For example, if, in standardized units, the mean shift of absolute value 3 occurred, we get that the new mean lies on one of the control limits (we assume in our analysis that both the mean before the shift and the constant variance of the production process under scrutiny are known). Hence, under the usual normality assumption,

$$
\begin{aligned}
p &= P(z_N + 3 > 3 \text{ or } z_N + 3 < -3) \\
&= P(z_N > 0) + P(z_N < -6) = 1/2
\end{aligned}
$$

and the ARL is equal to 2. Analogously, one obtains that the ARL's for the standardized shifts of values 4, 2, 1 and 0 are $1/.8413 = 1.189$, $1/.1587 = 6.301$, $1/.023 = 43.956$ and $1/.0027 = 370.37$, respectively. We note that the last number is the ARL until a false alarm occurs. Computing the ARL's for the Page CUSUMs is a rather difficult task. This task has been performed by Lucas and Crosier [3] and we shall confine ourselves to citing a few of their results for the Page CUSUMs with $k = .5$ (again for standardized mean shifts) in Table 4.6.

Table 4.6		
Shift	$h = 4$	$h = 5$
0	168	465
1	8.38	10.4
2	3.34	4.01
3	2.19	2.57
4	1.71	2.01

Comparing the ARL's for the standard mean control chart with those for the Page CUSUMs, we find that it may indeed be advantageous to use the latter on data with small, single mean shift. We note also that the Page test with $h = 4$ is likely to ring a false alarm much earlier than the corresponding test with $h = 5$.

A thorough analysis of the ARL's for the Page tests reveals that it is indeed reasonable to use k equal to one-half of the absolute value of the mean shift and to choose h equal to 4 or 5. Lucas and Crosier [4] summarize their numerical work stating that "this k value usually gives a control scheme having the shortest ARL$(2k)$ for a given ARL(0), where ARL(D) is the average run length at a mean shift of D." Moreover, the given values of h have been selected "to give the largest in-control ARL consistent with an adequately small out-of-control ARL."

Until now we have assumed that the variance of a production process is known. In practice, this assumption is neither often fulfilled nor actually needed. Whenever necessary, σ can be replaced by its estimate.

We mentioned at the beginning of this section that the Page test is the most widely used, and misused, tool for detecting persistent shifts of the mean.The upper and lower cumulative sums, (4.39) and (4.40), have an intuitively appealing form and can be used more or less automatically on data with a single mean shift. On the other hand, using automatically the CUSUM test of Section 4.3 on such data is likely to lead to worse results (see Problem 4.1). The latter test is in fact the sequential likelihood ratio test which was developed for choosing the true mean of a population out of its two possible values. For the task mentioned, the sequential likelihood ratio test is the best possible in a certain sense. Namely, for

N denoting the number of observations until a decision is taken, the expected value of N is smaller than the number of observations needed by any fixed-sample-size test which has the same values of Type I and Type II errors. But the problem with detecting the mean shift is that it is a different task and, therefore, no optimality properties of the sequential likelihood ratio test carry over automatically to the case of the CUSUM test of Section 4.3. In order to suggest the way out of the problem, let us turn again to the data set of Table 4.3. The mean of each of the lots is either $\mu_0 = 10$ or $\mu_1 = 11$. Now, instead of automatically calculating $\mathcal{R}_1$ for successive lots, we should stop the test each time a decision is taken (that is, each time $|\mathcal{R}_1|$ becomes greater than 4.595) and start it anew at the next lot. In this way, the question of detecting directly the shift is likely to be replaced by deciding what is the current value of the mean. Results obtained using such a sequence of the CUSUM tests can be hoped to be similar to those obtained by means of the Page test (see Problem 4.1).

The relationship between the CUSUM test of this Section and that of Section 4.3, which is directly based on the likelihood ratio test, has been given for the first time by Johnson [2]. In order to briefly discuss this relationship, let us assume that the in control process has mean μ_0 and we want a test for the mean's shift to μ_1, where $\mu_1 > \mu_0$ (of course, the case with $\mu_1 < \mu_0$ is analogous and does not require separate analysis). For the Page test, (4.39) is used to detect possible upward shift. For the test which is directly based on the likelihood ratio, one can use (4.25) for the shift's detection. The upward shift of the mean to μ_1 is claimed to have occurred when for the first time, $\bar{\bar{x}}_N$ gives

$$N\frac{\mu_1 - \mu_0}{\sigma^2/n}[\bar{\bar{x}}_N - \frac{\mu_0 + \mu_1}{2}] \geq \ln(k_1)$$

for k_1 suitably defined. The left inequality in (4.25) is disregarded since we are interested only in the upward shift (hence, also the current interpretation of μ_0 differs from that in Section 4.3). To simplify notation, let us switch to standardized units. After simple manipulations, the above inequality assumes the form

$$\delta(\sum_{i=1}^{N} z_i - \frac{\delta}{2}N) \geq \ln(k_1),$$

where $\delta = \frac{\mu_1 - \mu_0}{\sigma^2/n}$ is the mean's shift in standardized units. Thus, it follows from (4.17) that the upper limit of uncertainty or continuation region is

$$\sum_{i=1}^{N} z_i = \frac{1}{\delta} \ln \left(\frac{1-\beta}{\alpha} \right) + \frac{\delta}{2} N.$$

The limit given can be given the following equivalent form: flag the process as out of control when, for the first time, the value of $\tilde{SH}_N$ becomes greater than or equal to h, where

$$\tilde{SH}_N = (z_N - k) + \tilde{SH}_{N-1},$$

$$k = \frac{\delta}{2} \ \text{and} \ h = \frac{1}{\delta} \ln \left(\frac{1-\beta}{\alpha} \right),$$

with $\tilde{SH}_0 = 0$. Now, comparing the decision rule obtained with recursive equation (4.39) and recalling that the Page CUSUM test flags detection of the mean's shift when cumulative sum SH_N becomes greater than h, we get the relationship between that test and the one directly based on the likelihood ratio. We find that the Page test is a modification of the other test, well suited to the situation tested. We note also that our comparison provides a new interpretation for parameters k and h of the Page test.

Let us conclude our discussion of the Page test with two remarks. First, Lucas and Crosier [3] have recommended that the upper and lower CUSUMs be given "a head start," that is, that SH_0 and SL_0 in (4.39) and (4.40) be assigned a positive value. They suggest that the head-start value be equal to $h/2$. This head start does not change considerably the average run length until a false alarm, when there is no shift of the mean. On the other hand, it does seemingly reduce the ARL when there is a shift, but *only* if the mean shift and the addition of the head start coincide or almost coincide (see Problem 4.6). The CUSUM test described is known as a FIR CUSUM scheme.

Our second remark is that, in our view, neither the Page test nor FIR CUSUMs should ever be used alone, without a parallel analysis of the standard mean control chart. Some authors, however, claim that sometimes intermittent Pareto glitches can, or even should, be disregarded. If the reader finds a situation when this last claim is justified, he or she may then use a robust CUSUM, developed by Lucas and Crosier [4] in such a way as to make the procedure insensitive to the presence of contaminated data.

We shall close this section with a brief description of the exponentially weighted moving average (EWMA) charts. Although these charts are not based on calculating cumulative sums, they perform as well as CUSUM tests on data with small shifts of the mean of a production process. However, they can hardly be claimed to be superior to the CUSUMs, and we mention them only because they are advocated by some members of the quality control community. The EWMA charts are based on calculating exponentially weighted moving averages of sample means of past lots. Let us assume that a certain number of lots has been observed and that the sample mean across all the lots is $\bar{\bar{x}}$. The exponentially weighted moving average for the Nth lot is computed as

$$\hat{\bar{x}}_N = r\bar{x}_N + (1 - r)\hat{\bar{x}}_{N-1}, \tag{4.41}$$

where $\hat{\bar{x}}_0$ is most often set equal to $\bar{\bar{x}}$. If the means of first lots can be safely assumed to be equal to some μ_0, we can let $\hat{\bar{x}}_0 = \mu_0$. The weighting factor r is positive and not greater than 1.

We note that no averaging takes place, $\hat{\bar{x}}_N = \bar{x}_N$, if $r = 1$. Otherwise, past sample means $\bar{x}_{N-i}$ are included in the moving average $\hat{\bar{x}}_N$ with the exponential weight $r(1 - r)^i$. Clearly, if there is a shift in the lot means, it will likely be reflected in the values of the moving average. Careful analysis shows that, in order that small mean shifts be handily detected, the factor r should lie somewhere between .2 and .5. The most often choices are .25 and .333 (see Robinson and Ho [5], Sweet [6] and Hunter [1]). Control limits for the EWMA chart may be written as

$$UCL = \bar{\bar{x}} + 3\frac{a(n)\bar{s}}{\sqrt{n}}\sqrt{\frac{r}{2-r}} \tag{4.42}$$

and

$$LCL = \bar{\bar{x}} - 3\frac{a(n)\bar{s}}{\sqrt{n}}\sqrt{\frac{r}{2-r}}, \tag{4.43}$$

where $a(n)$ is a constant with values given by (3.37) and Table 3.1. The second term on the right-hand-sides of the above expressions is an approximation of the exact value of three standard deviations of the moving average $\hat{\bar{x}}_N$. In fact, standard deviation of $\hat{\bar{x}}_N$ is equal to

$$\frac{\sigma}{\sqrt{n}}\left[\frac{r(1 - (1 - r)^{2N})}{2 - r}\right]^{1/2},$$

where, as usual, σ is the population standard deviation.

The EWMA chart for the data set of Table 4.3 and $\hat{\bar{x}}_0 = \mu_0 = 10$ is given in Figure 4.17. The values of the exponential moving averages $\hat{\bar{x}}_N$

are given in the last column of Table 4.3. Taking advantage of the fact that we know true values of both the standard deviation σ and initial mean μ_0, we can construct the following control limits:

$$\text{UCL} = 10 + \frac{3}{\sqrt{5}}\sqrt{\frac{.333}{1.667}} = 10.6$$

and

$$\text{LCL} = 10 - \frac{3}{\sqrt{5}}\sqrt{\frac{.333}{1.667}} = 9.4.$$

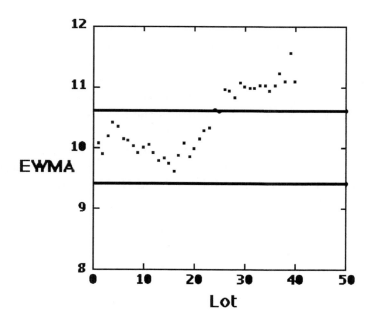

Figure 4.17. The EWMA Chart (μ_0 and σ Known).

If the initial value of the mean is unknown and it is therefore reasonable to set $\hat{\bar{x}}_0 = \bar{\bar{x}}$, we would obviously like that $\bar{\bar{x}}$ be a good estimate of μ_0. Now, if we can expect a single shift of the mean of a production process, we have to proceed cautiously. On the one hand, $\bar{\bar{x}}$ should be based on as many lots as possible. On the other hand, only the lots preceding the mean shift should be taken into account. In practice, the method works reasonably well if we can use at least 40 data to calculate $\bar{\bar{x}}$, that is, given lots of size 5, if we can use at least 8 lots. In our example, $\bar{\bar{x}}_8 = 10.119$ while, say, $\bar{\bar{x}}_{10} = 10.078$. Using either of these cumulative means does not change the UCL or LCL considerably. Moreover, since the influence of past moving averages on the current one decreases exponentially fast, setting $\hat{\bar{x}}_0 = \bar{\bar{x}}_{10}$ or even $\hat{\bar{x}}_0 = \bar{\bar{x}}_8$, instead of setting $\hat{\bar{x}}_0 = \mu_0$, leads to slight changes of only few moving averages (see Problem 4.8). If, however, we used $\bar{\bar{x}}_{40}$, the situation would change dramatically, since $\bar{\bar{x}}_{40}$ assumes then value 10.479 and, hence, both limits would be shifted upward by .479. This situation is depicted in Figure 4.18. Although the EWMA procedure grasps the directions of the mean trends, it rings false alarm on lots 13 through 17.

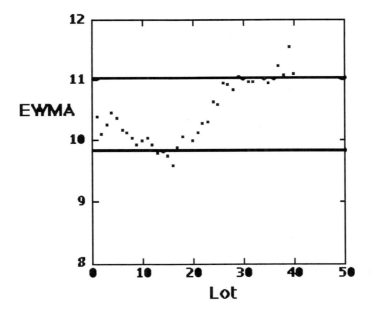

Figure 4.18. The EWMA Chart (Based on $\bar{\bar{x}}_{40}$).

As is the case for the CUSUMs, the EWMA procedure should capture the shifts upward and downward of the means of lots given in Table 4.2. The EWMA chart for this set of data is demonstrated in Figure 4.19. When calculating control limits, we set $\hat{\bar{x}}_0 = \mu_0 = 10$. Given that $\sigma = .1$, one readily obtains that the UCL $= 10.06$ and LCL $= 9.94$. Cumulative means given in Table 4.2 show that the chart would have not been essentially changed had we used practically any of them as $\hat{\bar{x}}_0$.

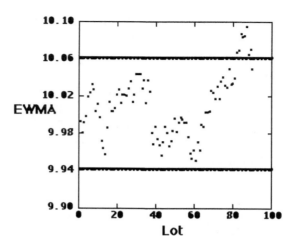

Figure 4.19. The EWMA Chart for Data with Mean Drift.

4.7   CUSUM Performance on Data with Upward Variance Drift

Again, the performance of CUSUM approaches should improve on data where there is a stochastic drift upward in the variance of the output variable. We recall that if lots of observations of size n are drawn from a normal distribution with mean μ and variance σ^2 then letting sample variance be given by

$$s_j{}^2 = \frac{1}{n_j - 1} \sum_{i=1}^{n_j} (x_{j,i} - \bar{x}_j)^2, \tag{4.44}$$

we have the fact (from Appendix B) that

$$z_j = \frac{(n_j - 1)}{\sigma^2} s_j^2 \tag{4.45}$$

has the Chi-square distribution with $n_j - 1$ degrees of freedom. That is

$$f(z_j) = \frac{1}{\Gamma(\frac{n_j-1}{2}) 2^{\frac{n_j-1}{2}}} z_j^{\frac{n_j-1}{2}-1} e^{-\frac{z_j}{2}} . \tag{4.46}$$

Then, assuming that each lot is of the same size, n, following the argument used in (4.22), we have that, after the N'th lot, the likelihood ratio interval corresponding to a continuation of sampling is given by

$$\ln(k_0) < \mathcal{R}_3 < \ln(k_1),$$

where

$$\mathcal{R}_3 = \ln\left(\frac{\left(\frac{1}{\sigma_1^2}\right)^{N\frac{n-3}{2}} \exp\left(-\frac{n-1}{2\sigma_1^2}\sum_{j=1}^{N} s_j^2\right)}{\left(\frac{1}{\sigma_0^2}\right)^{N\frac{n-3}{2}} \exp\left(-\frac{n-1}{2\sigma_0^2}\sum_{j=1}^{N} s_j^2\right)}\right) . \tag{4.47}$$

A straightforward Shewhart CUSUM chart can easily be constructed. We recall that for each sample of size n,

$$s_j^2 = \frac{1}{n-1}\sum_{i=1}^{n}(x_{j,i} - \bar{x}_j)^2 \tag{4.48}$$

is an unbiased estimator for the underlying population variance, σ^2. Thus, the average of N sample variances is also an unbiased estimator for σ^2, i.e.,

$$E\left(\frac{1}{N}\sum_{j=1}^{N} s_j^2\right) = \sigma^2 . \tag{4.49}$$

Recalling the fact that if the $x_{j,i}$ are independently drawn from a normal distribution with mean μ and variance σ^2, then

$$\frac{(n-1)s^2}{\sigma^2} = \chi^2(n-1) . \tag{4.50}$$

For normal data, then, we have, after a little algebra (see Appendix B),

$$Var(s^2) = E(s^2 - \sigma^2)^2 \tag{4.51}$$

$$= \frac{1}{n}\left(\frac{\sigma^4}{4}\frac{4!}{2!} - \frac{n-3}{n-1}\sigma^4\right) \tag{4.52}$$

$$= \frac{2\sigma^4}{n-1} . \tag{4.53}$$

So, then, a natural Shewhart CUSUM statistic is given by

$$\mathcal{R}_4 = \frac{\sum_{j=1}^{N} s_j^2 - N\sigma_0^2}{\frac{\sigma_0^2}{\sqrt{N}}\sqrt{\frac{2}{n-1}}} . \tag{4.54}$$

Here, we shall assume that the output from lot to lot follows a normal distribution with mean 10 and starting standard deviation 0.1. As we begin sampling from each lot, however, there is a probability of 10% that the standard deviation will increase by .01 cm. In Table 4.7, we show the data from Table 3.5.

Lot	x_1	x_2	x_3	x_4	x_5	$\bar{x}$	s	$\sum s_j{}^2$	$\mathcal{R}_3$	$\mathcal{R}_4$
1	10.036	9.982	9.996	10.157	10.013	10.037	0.070	0.005	-0.193	-0.707
2	9.851	10.259	10.132	10.038	9.930	10.042	0.162	0.031	1.713	1.100
3	9.916	10.019	9.920	10.140	10.064	10.012	0.096	0.040	1.920	0.816
4	9.829	9.751	9.955	9.847	10.048	9.886	0.116	0.054	2.628	0.990
5	10.194	9.921	10.159	10.003	9.886	10.032	0.138	0.073	3.835	1.455
6	10.057	9.972	10.038	9.864	9.908	9.968	0.083	0.080	3.841	1.155
7	9.875	10.468	10.017	9.997	9.834	10.038	0.252	0.144	9.548	3.955
8	10.006	9.853	10.054	10.009	9.913	9.967	0.081	0.151	9.555	3.550
9	10.013	10.015	10.073	10.184	10.048	10.067	0.070	0.156	9.361	3.111
10	9.957	10.009	10.057	9.948	9.942	9.983	0.049	0.158	8.868	2.594
11	9.902	9.925	9.904	9.913	9.991	9.927	0.037	0.159	8.276	2.089
12	9.851	10.248	10.172	10.006	10.065	10.068	0.153	0.183	9.982	2.572
13	10.022	9.976	10.104	9.997	10.051	10.030	0.050	0.185	9.489	2.157
14	9.990	10.084	10.028	9.973	10.184	10.052	0.085	0.192	9.496	1.965
15	10.125	10.161	10.029	9.811	9.710	9.967	0.198	0.231	12.703	2.958
16	9.858	10.079	9.976	9.967	10.204	10.017	0.130	0.248	13.709	3.111
17	9.837	10.305	9.794	9.868	9.948	9.950	0.206	0.290	17.216	4.116
18	10.201	9.925	9.859	10.127	9.857	9.994	0.160	0.316	19.124	4.533
19	10.052	9.850	9.840	9.951	9.906	9.920	0.087	0.323	19.130	4.315
20	10.017	9.932	10.194	9.954	10.143	10.048	0.116	0.336	19.737	4.301
21	9.778	10.144	10.110	10.118	10.117	10.053	0.154	0.360	21.444	4.629
22	10.056	10.008	10.027	10.054	10.107	10.050	0.038	0.361	20.851	4.251
23	10.058	9.834	9.980	9.895	10.082	9.970	0.106	0.372	21.257	4.187
24	9.955	9.946	10.041	9.928	9.998	9.974	0.046	0.374	20.765	3.868
25	10.001	10.027	9.932	9.953	9.936	9.970	0.042	0.376	20.272	3.564
26	10.016	10.091	9.934	9.948	9.946	9.987	0.067	0.380	19.978	3.328
27	10.052	9.959	9.967	10.050	10.023	10.010	0.045	0.382	19.485	3.048
28	10.005	9.977	10.092	9.829	10.094	9.999	0.109	0.394	19.992	3.047
29	9.881	10.159	9.986	10.090	10.093	10.042	0.109	0.406	20.498	3.046
30	9.775	10.003	9.971	9.819	9.968	9.907	0.103	0.417	20.905	3.021
31	10.042	10.098	9.988	10.108	9.918	10.031	0.079	0.423	20.813	2.870
32	9.917	10.011	10.116	9.970	10.078	10.018	0.080	0.429	20.720	2.725
33	9.728	9.886	9.884	10.139	9.955	9.918	0.149	0.451	22.226	2.979
34	10.032	10.102	9.997	9.986	9.991	10.022	0.049	0.453	21.733	2.741
35	10.071	9.949	9.972	9.976	10.223	10.038	0.113	0.466	22.340	2.773
36	10.004	10.102	10.194	10.022	10.052	10.075	0.076	0.472	22.246	2.640
37	9.859	10.286	9.963	10.000	9.930	10.007	0.164	0.499	24.253	2.999
38	10.348	9.967	9.874	10.057	10.000	10.049	0.180	0.531	26.761	3.464
39	10.000	10.044	9.811	9.811	10.144	9.962	0.147	0.553	28.268	3.691
40	9.926	10.305	10.036	9.978	10.039	10.057	0.146	0.574	29.674	3.891
41	9.881	10.002	9.781	9.872	10.197	9.946	0.161	0.600	31.581	4.196
42	10.252	10.203	9.845	10.207	10.012	10.104	0.172	0.629	33.788	4.561
43	10.162	9.779	10.150	9.954	9.950	9.999	0.160	0.655	35.694	4.852
44	10.044	10.013	10.030	9.850	10.014	9.990	0.080	0.661	35.601	4.712
45	9.897	9.994	9.857	9.860	9.902	9.902	0.055	0.664	35.209	4.512
46	9.916	9.924	9.855	9.912	9.980	9.917	0.045	0.666	34.715	4.295
47	10.088	9.754	10.122	9.951	10.013	9.986	0.145	0.687	36.122	4.476
48	10.371	10.031	10.203	10.197	10.109	10.182	0.127	0.703	37.029	4.552
49	10.192	10.076	9.915	10.035	10.021	10.048	0.100	0.713	37.336	4.505
50	9.725	10.001	10.127	10.051	10.053	9.992	0.155	0.737	39.042	4.740
51	10.064	9.888	10.082	9.915	9.940	9.978	0.089	0.745	39.149	4.654
52	10.056	10.131	9.870	9.835	9.894	9.957	0.129	0.762	40.157	4.746
53	9.917	9.973	10.170	9.919	9.997	9.995	0.104	0.773	40.563	4.720
54	9.993	9.910	9.938	10.071	9.921	9.967	0.066	0.777	40.270	4.561
55	10.015	9.867	9.836	9.789	9.748	9.851	0.102	0.787	40.577	4.519
56	10.012	10.040	10.134	9.926	10.165	10.055	0.096	0.796	40.784	4.460
57	9.846	9.807	9.824	10.171	9.837	9.897	0.154	0.820	42.490	4.683
58	10.016	10.046	9.862	9.936	10.040	9.980	0.079	0.826	42.398	4.568
59	9.853	9.989	10.066	10.455	9.813	10.035	0.256	0.892	48.305	5.560
60	9.884	10.012	9.894	9.923	10.152	9.973	0.112	0.905	48.911	5.569

			Table 4.7 (continued)							
Lot	x_1	x_2	x_3	x_4	x_5	$\bar{x}$	s	$\sum s_j^2$	$\mathcal{R}_3$	$\mathcal{R}_4$
61	10.018	9.661	10.019	9.872	9.685	9.851	0.173	0.935	51.218	5.885
62	10.015	10.034	9.781	9.829	10.042	9.940	0.125	0.951	52.125	5.945
63	9.826	10.200	10.029	10.135	9.930	10.024	0.151	0.974	53.731	6.129
64	10.121	9.863	10.008	10.161	9.894	10.010	0.133	0.992	54.838	6.223
65	10.132	9.913	10.268	10.070	10.148	10.106	0.130	1.009	55.846	6.297
66	9.816	9.880	9.733	9.809	9.947	9.837	0.081	1.015	55.753	6.180
67	10.160	9.843	9.848	10.401	10.123	10.075	0.235	1.070	60.559	6.911
68	9.854	10.031	9.947	9.980	9.981	9.959	0.066	1.074	60.266	6.757
69	9.915	9.931	9.982	9.928	10.105	9.972	0.079	1.080	60.173	6.640
70	10.062	10.147	9.705	10.053	9.869	9.967	0.178	1.112	62.679	6.964
71	10.095	10.088	10.340	10.014	9.965	10.100	0.144	1.133	64.086	7.099
72	10.185	9.986	10.038	10.405	10.117	10.146	0.163	1.160	66.094	7.333
73	9.963	10.162	10.186	9.700	9.751	9.952	0.225	1.211	70.501	7.962
74	10.289	9.940	9.963	10.200	9.904	10.059	0.173	1.241	72.807	8.236
75	10.467	9.884	9.952	10.255	10.190	10.150	0.236	1.297	77.714	8.932
76	9.995	9.811	9.781	10.059	10.277	9.985	0.202	1.338	81.121	9.376
77	9.952	9.943	9.844	10.159	9.882	9.956	0.122	1.353	81.927	9.396
78	9.822	9.795	10.041	9.945	10.098	9.940	0.132	1.371	83.034	9.464
79	9.869	10.136	9.638	9.858	9.591	9.818	0.217	1.418	87.042	9.992
80	9.988	10.016	9.943	10.167	9.772	9.977	0.142	1.438	88.348	10.088
81	9.837	9.840	10.184	10.148	10.076	10.017	0.168	1.466	90.455	10.308
82	9.877	10.260	10.197	9.909	10.044	10.057	0.170	1.495	92.662	10.542
83	10.283	9.827	9.959	10.172	10.256	10.099	0.198	1.534	95.869	10.928
84	10.040	9.970	10.165	10.076	9.888	10.028	0.105	1.545	96.275	10.878
85	9.761	10.010	9.900	10.092	10.088	9.970	0.140	1.565	97.582	10.968
86	9.937	10.098	10.059	9.709	10.153	9.991	0.177	1.596	99.990	11.224
87	10.128	9.836	10.179	10.145	10.067	10.071	0.137	1.615	101.196	11.296
88	9.962	10.070	10.053	10.160	9.886	10.026	0.105	1.626	101.603	11.246
89	10.017	10.126	10.103	9.807	10.181	10.047	0.147	1.648	103.110	11.363
90	9.942	9.857	9.919	9.988	9.800	9.90	0.074	1.653	40.534	102.917

In Figure 4.20, we display the CUSUM Test with $\sigma_0^2 = .01$ and $\sigma_1^2 = .02$. As with earlier CUSUM Tests, we use $\alpha = \beta = .01$, with the consequential $\ln(k_0) = -4.595$ and $\ln(k_1) = 4.595$. We note how effective the CUSUM procedure is in detecting the kind of random increase in the process variability.

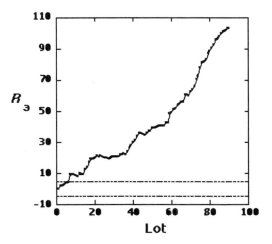

Figure 4.20. CUSUM Test for Data with Variance Drift.

In Figure 4.21, we note the Shewhart CUSUM Test picks up the increase in variance quite handily.

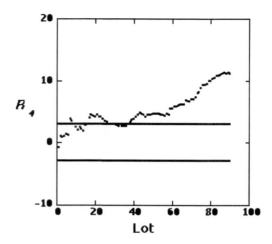

Figure 4.21. Shewhart CUSUM for Data with Variance Drift.

Clearly, statistics $\mathcal{R}_3$ and $\mathcal{R}_4$ can be used for detecting a persistent shift in the variance from one value to another. As in the case of using statistic $\mathcal{R}_1$ to pick up a mean shift, we can replace a single test based on $\mathcal{R}_3$ by a sequence of such tests. Starting the test anew whenever a decision is taken enables us to get rid of the inertia of the single test.

4.8   Acceptance-Rejection CUSUMs

Let us turn now to that area of "quality assurance" which has perhaps made the greatest use of CUSUM procedures. Suppose that we have a production system where the proportion of defective goods at the end of the process is given by p. A manufacturer believes it appropriate to aim for a target of p_0 as the proportion of defective goods. When that proportion rises to p_1 he proposes to intervene and see what is going wrong.

Let us suppose that the size of lot j is n_j. Then the likelihood ratio test is given by

$$\mathcal{R}_5 = \ln\left(\frac{\prod_{j=1}^{N} \frac{n_j!}{x_j!(n-x_j)!} p_1^{x_j}(1-p_1)^{n_j-x_j}}{\prod_{j=1}^{N} \frac{n_j!}{x_j!(n-x_j)!} p_0^{x_j}(1-p_0)^{n_j-x_j}}\right) \tag{4.55}$$

Lot	Defectives	Proportion	Cum Defectives	CUSUM	Shewhart
			Table 4.8		
1	3	.03	3.000	-0.488	0.000
2	2	.02	5.000	-1.509	-0.415
3	5	.05	10.000	-0.934	0.338
4	0	.00	10.000	-3.017	-0.586
5	6	.06	16.000	-1.910	0.262
6	4	.04	20.000	-1.867	0.479
7	2	.02	22.000	-2.887	0.222
8	4	.04	26.000	-2.844	0.415
9	1	.01	27.000	-4.396	0.000
10	2	.02	29.000	-5.416	-0.185
11	7	.07	36.000	-3.778	0.530
12	9	.09	45.000	-1.076	1.523
13	11	.11	56.000	2.689	2.764
14	12	.12	68.000	6.985	4.073
15	14	.14	82.000	12.345	5.600
16	15	.15	97.000	18.236	7.181
17	12	.12	109.000	22.533	8.246
18	10	.10	119.000	25.766	8.981
19	8	.08	127.000	27.936	9.414
20	3	.03	130.000	27.448	9.176
21	5	.05	135.000	28.022	9.210
22	6	.06	141.000	29.129	9.374
23	0	.00	141.000	27.046	8.801
24	1	.01	142.000	25.494	8.376
25	3	.03	145.000	25.005	8.207
26	3	.03	148.000	24.517	8.048
27	4	.04	152.000	24.560	8.010
28	6	.06	158.000	25.667	8.198
29	5	.05	163.000	26.242	8.273
30	5	.05	168.000	26.817	8.348
31	3	.03	171.000	26.328	8.212
32	3	.03	174.000	25.840	8.083
33	7	.07	181.000	27.478	8.368
34	8	.08	189.000	29.648	8.746
35	2	.02	191.000	28.628	8.522
36	0	.00	191.000	26.544	8.109
37	6	.06	197.000	27.651	8.288
38	7	.07	204.000	29.289	8.559
39	4	.04	208.000	29.332	8.542
40	4	.04	212.000	29.376	8.527

This gives, as the "interval of acceptability"

$$\ln(k_0) < \ln(\frac{p_1}{p_0}) \sum_{j=1}^{N} x_j + \ln(\frac{1 - p_1}{1 - p_0}) \sum_{j=1}^{N} (n_j - x_j) < \ln(k_1). \quad (4.56)$$

As has been our practice, we will use here $\alpha = \beta = .01$ giving $\ln(k_0)$

$= -4.596$ and $\ln(k_1) = 4.596$. Let us recall the defect data in Table 2.1 reproduced here as Table 4.8 with additional columns added for CUSUM and Shewhart CUSUM tests. We recall that the lot sizes were all equal to 100. Let us suppose that the manufacturer has a contract assuring that no more than 7% of the items are defective. Then, we might use .05 for p_1. For p_0 we will use .03. We note (Figure 4.22) that we go below the lower control limit here at lot 10. Perhaps it would be appropriate here to see why the performance is better than expected. Generally we will only take interest when we get to lot 14, where the cumulative failure rate becomes unacceptably high. We recall from Figure 2.2 in Chapter 2 that it was at lot 14 that we detected a Pareto glitch.

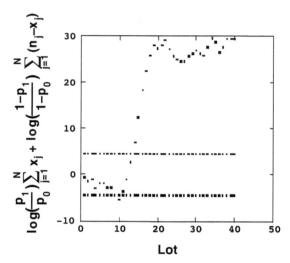

Figure 4.22. CUSUM Test for Defect Data.

A natural Shewhart CUSUM is available by first noting that under the hypothesis that the production proportion of defectives is p_0. If the number of defectives in lot j is equal to x_j, then

$$E(\sum_{j=1}^{N} x_j) = p_0 \sum_{j=1}^{N} n_j . \qquad (4.57)$$

Thus the appropriate Shewhart CUSUM statistic is

$$\mathcal{R}_6 = \frac{\sum_{j=1}^{N} x_j - p_0 \sum_{j=0}^{N} n_j}{\sqrt{\sum_{j=1}^{N} n_j p_0 (1 - p_0)}}. \qquad (4.58)$$

In Figure 4.23, we note that the cumulative defect level arises above 3 on lot 14.

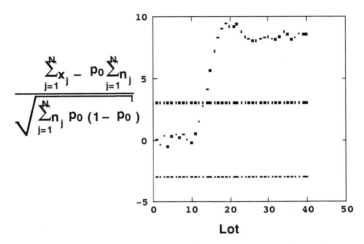

$$\frac{\sum\limits_{j=1}^{N} x_j - p_0 \sum\limits_{j=1}^{N} n_j}{\sqrt{\sum\limits_{j=1}^{N} n_j \, p_0 \, (1 - p_0)}}$$

Figure 4.23. Shewhart CUSUM Test for Defect Data.

References

[1] Hunter, J.S. (1986). "The exponentially weighted moving average," *Journal of Quality Technology, 18,* pp. 203-210.

[2] Johnson, N.L. (1961). "A simple theoretical approach to cumulative sum charts," *Journal of the American Statistical Association, 56,* pp. 835-840.

[3] Lucas, J.M. and Crosier, R.B. (1982a). "Fast initial response for CUSUM quality control schemes: Give your CUSUM a head start," *Technometrics, 24,* pp.199-205.

[4] Lucas, J.M., and Crosier, R.B. (1982b). "Robust CUSUM: A robustness study for CUSUM quality control schemes," *Communications in Statistics — Theory and Methods, 11,* pp. 2669-2687.

[5] Robinson, P.B. and Ho, T.Y. (1978). "Average run lengths of geometric moving average charts by numerical methods," *Technometrics, 20,* pp. 85-93.

[6] Sweet, A.L. (1986). "Control charts using coupled exponentially weighted moving averages," *Transactions of the IIE, 18,* pp.26-33.

[7] Wilks, S.S. (1962). *Mathematical Statistics*. New York: John Wiley & Sons, pp. 472-496.

Problems

Problem 4.1. Perform the standard CUSUM test for shift of the mean for the data set of Table 4.3, setting $\mu_0 = 10$, $\mu_1 = 11$ and $\sigma = 1$. Then, perform a sequence of standard CUSUM tests, starting each test anew whenever a decision is taken (that is, whenever the statistic $\mathcal{R}_1$ becomes greater than 4.595 or smaller than -4.959, stop the test, set $N = 1$ for the next lot, and start the test anew at that lot). Compare the results obtained with those for Page tests with $k = 1$ and $k = .5$ (see Section 4.6 for the latter results).

Problem 4.2. Suppose the simulated data set given in the following table is the set of standardized lot means of 50 lots of size 4. Standardization was performed under the assumption that the population variance is 4 and the mean is equal to a goal value μ_0. It is conjectured, however, that the upward shift by 1 (in stadardized units) is possible. In fact, the first 20 lots come from $\mathcal{N}(0, 1)$ while lots 21 to 50 come from $\mathcal{N}(1, 1)$.

   a. Perform the standard CUSUM test for shift of the mean setting

$$\mathcal{R}_1 = N[\bar{z}_N - .5],$$

since the data is given in standardized units. Then, perform a sequence of standard CUSUM tests, starting each test anew whenever a decision is taken (see Problem 4.1 for explanation).

   b. Construct the Shewhart CUSUM chart for standardized lot means.

   c. Perform four Page CUSUM tests for standardized lot means using $h = 4$ in all cases, and $k = .25, .5, 1$ and 2, respectively.

Comment on the results obtained.

Lot	z	Lot	z
1	0.071	26	2.851
2	0.355	27	1.523
3	-0.176	28	-0.482
4	-0.043	29	3.770
5	1.572	30	-0.875
6	0.129	31	-0.202
7	-0.428	32	0.525
8	-1.495	33	1.867
9	2.590	34	2.831
10	1.319	35	0.321
11	0.377	36	-0.286
12	-0.700	37	2.155
13	-1.269	38	-0.304
14	-0.843	39	0.202
15	0.190	40	1.476
16	-0.795	41	-0.364
17	1.398	42	-0.453
18	0.640	43	0.559
19	0.869	44	0.145
20	-1.709	45	1.068
21	1.064	46	1.157
22	-0.294	47	0.386
23	1.714	48	2.766
24	0.785	49	0.582
25	0.704	50	2.299

Problem 4.3. Suppose the simulated data set given in the following table is the set of 50 lots of size 4 from a population with standard deviation $\sigma = .1$. It is conjectured, however, that the upward shift by 1 (in stadardized units) is possible. In fact, the first 20 lots come from $\mathcal{N}(0, 1)$ while lots 21 to 50 come from $\mathcal{N}(1, 1)$.

a. Perform the standard CUSUM test for shift of the mean setting

$$\mathcal{R}_1 = N[\bar{z}_N - .5],$$

since the data is given in standardized units. Then, perform a sequence of standard CUSUM tests, starting each test anew whenever a decision is taken (see Problem 4.1 for explanation).

b. Construct the Shewhart CUSUM chart for standardized lot means.

c. Perform four Page CUSUM tests for standardized lot means using $h = 4$ in all cases, and $k = .25, .5, 1$ and 2, respectively.

Comment on the results obtained.

Lot	x_1	x_2	x_3	x_4	Lot	x_1	x_2	x_3	x_4
1	4.880	5.085	4.941	5.119	26	5.001	5.175	5.081	4.976
2	4.952	4.998	4.997	5.042	27	5.069	5.011	5.144	5.014
3	5.087	4.988	4.889	5.010	28	5.098	5.009	4.983	5.187
4	5.183	4.992	4.871	4.919	29	5.013	5.121	5.102	5.166
5	4.932	5.112	5.035	4.883	30	5.020	5.087	5.060	4.975
6	4.871	5.060	5.163	4.951	31	5.096	5.061	5.124	5.162
7	5.116	4.957	5.053	5.139	32	5.071	5.075	5.111	5.028
8	4.870	4.976	4.972	5.101	33	4.875	4.925	4.936	4.872
9	4.920	4.980	4.967	5.040	34	5.054	5.061	5.157	5.051
10	5.048	5.186	4.908	5.089	35	5.031	5.100	4.921	5.084
11	4.864	5.065	5.094	5.057	36	5.127	4.964	5.018	5.078
12	4.855	5.003	4.930	5.094	37	4.907	5.045	4.971	5.058
13	4.956	5.085	5.069	5.032	38	5.128	4.931	5.005	5.030
14	4.914	5.162	4.895	5.057	39	5.078	4.933	5.043	5.030
15	5.007	5.018	4.879	5.078	40	4.950	4.969	4.939	5.055
16	5.016	5.043	5.024	5.107	41	5.182	5.014	4.924	5.007
17	4.939	4.961	4.968	5.024	42	5.039	4.980	5.037	4.977
18	5.177	4.882	5.058	4.993	43	5.125	4.986	4.971	5.003
19	4.958	5.238	4.944	5.135	44	5.127	4.929	5.019	5.062
20	5.130	4.969	4.923	5.052	45	4.982	4.976	5.289	4.986
21	5.263	5.222	5.135	4.913	46	4.862	5.034	5.031	5.130
22	5.065	5.060	5.089	5.171	47	5.052	5.007	5.138	5.073
23	5.133	5.056	5.009	5.175	48	5.026	5.123	4.993	5.075
24	4.990	5.185	5.028	5.143	49	4.900	4.845	5.049	5.124
25	5.002	4.865	5.091	5.064	50	5.024	5.152	4.897	5.204

It is believed that the mean, at least for the first lots, is equal to the nominal value 5. It is conjectured, however, that an upward shift of the population mean by $.5\sigma$ to σ is possible. In fact, the first 20 lots come from $\mathcal{N}(5, .01)$ while lots 21 to 50 come from $\mathcal{N}(5.05, .01)$.

   a. Perform the standard CUSUM test for shift of the mean twice, setting first $\mu_0 = 5$, $\mu_1 = 5.05$ and then $\mu_0 = 5$, $\mu_1 = 5.1$.

   b. Perform a sequence of standard CUSUM tests with $\mu_0 = 5$ and $\mu_1 = 5.05$, starting each test anew whenever a decision is taken (see Problem 4.1 for explanation).

   c. Perform a sequence of standard CUSUM tests with $\mu_0 = 5$ and $\mu_1 = 5.1$, starting each test anew whenever a decision is taken.

   d. Construct the Shewhart CUSUM chart for mean shift.

   e. Perform the Page CUSUM test for mean shift thrice, using $h = 4$ in all cases, and $k = .25$, $.5$, and 1, respectively.

   f. Construct the EWMA chart twice, first for $r = .25$ and then for $r = .333$ (use $\hat{\bar{x}} = 5$ in both cases).
Comment on the results obtained.

Problem 4.4. Using statistical tables of normal distribution, compute the average run lengths for standard mean control charts for shifts of values 0, 1, 2 and 4 (in standardized units).

Problem 4.5. In the table below, summary statistics for 20 lots immediately preceding those considered in Problem 3.10 are given.

Lot	$\bar{x}$	s	Lot	$\bar{x}$	s
74	6.740	.023	84	6.772	.021
75	6.731	.017	85	6.751	.027
76	6.771	.018	86	6.778	.025
77	6.731	.022	87	6.782	.018
78	6.748	.020	88	6.755	.023
79	6.752	.016	89	6.742	.024
80	6.744	.023	90	6.736	.022
81	6.736	.019	91	6.742	.017
82	6.741	.021	92	6.748	.018
83	6.781	.026	93	6.734	.021

Use the above data to estimate population mean and standard deviation. Given the estimates obtained, use a CUSUM test of your choice to verify if lots 94 to 125 reveal a $.5\sigma$ mean shift. Interpret your results.

Problem 4.6. Examine the behavior of the FIR CUSUM test on the data set of Table 4.3, assuming $\mu_0 = 10$, $\sigma = 1$, $k = 1$, $h = 4$, $SH_0 = SL_0 = 2$ and
   a. starting the test on lot 1;
   b. starting the test on lot 11;
   c. starting the test on lot 21.
Interpret the results.

Problem 4.7. Consider the data set of Table 3.3. Estimate the population mean and standard deviation, and construct the EWMA chart taking $r = .333$. Repeat the analysis for $r = .25$. Interpret the results.

Problem 4.8. Perform the EWMA test on the data set of Table 4.3, setting $r = .333$ and
   a. $\hat{\bar{x}}_0 = \bar{\bar{x}}_8$;
   b. $\hat{\bar{x}}_0 = \bar{\bar{x}}_{10}$.
Compare the results with corresponding results in Section 4.6.

Problem 4.9. In the following table, simulated lots 1 to 10 come from $\mathcal{N}(5, (.3)^2)$ and lots 11 to 40 from $\mathcal{N}(5, (.6)^2)$.

Lot	x_1	x_2	x_3	x_4	Lot	x_1	x_2	x_3	x_4
1	4.802	5.367	4.725	5.095	21	5.480	4.875	4.687	5.612
2	5.149	4.650	5.017	5.177	22	3.722	4.778	6.330	6.099
3	4.742	4.604	5.323	5.164	23	4.499	5.345	5.360	5.036
4	5.189	5.031	4.789	5.637	24	3.558	5.208	5.644	5.692
5	4.804	5.092	5.243	5.026	25	4.281	3.973	4.810	5.324
6	4.861	4.723	5.265	4.934	26	4.958	5.540	5.219	4.583
7	5.191	4.871	4.826	5.190	27	5.055	4.552	4.474	5.391
8	5.128	5.079	5.536	5.347	28	4.798	5.451	5.336	5.493
9	5.302	4.979	5.141	4.973	29	5.588	4.422	4.766	4.489
10	4.701	4.705	5.297	5.072	30	4.196	5.633	4.164	4.974
11	5.562	4.895	4.246	5.340	31	5.598	5.512	5.197	5.359
12	4.583	5.581	4.924	5.566	32	4.423	5.236	3.835	5.770
13	5.416	5.268	4.527	5.194	33	4.434	4.752	4.578	5.018
14	4.371	4.604	4.813	5.341	34	5.649	4.866	4.540	4.133
15	4.277	4.830	6.435	5.467	35	5.523	5.248	6.410	5.759
16	5.145	4.158	4.887	5.639	36	5.269	5.186	5.032	4.982
17	4.805	5.034	5.529	5.141	37	5.176	5.566	5.523	3.106
18	5.346	4.646	4.658	4.960	38	4.566	4.599	4.906	5.304
19	4.667	5.325	4.996	5.809	39	5.916	5.315	6.270	4.986
20	4.538	5.310	4.084	5.312	40	5.696	5.062	5.729	4.989

Use a suitable CUSUM test (or, if necessary, a sequence of tests) to detect the persistent shift of the variance.

Problem 4.10. Consider the data set of Problem 2.4. Suppose the company aims at signing a contract with a supplier of valves who is able to keep the proportion of defectives always below 5% and to maintain, on the average, a 3% level of defectiveness. Use the CUSUM acceptance-rejection test to verify whether the first of the company's subcontractors satisfies these requirements.

Chapter 5

Exploratory Techniques for Preliminary Analysis

5.1 Introduction

Perhaps the most interesting and challenging stage of a quality control investigation is at the very beginning. It is very unusual for an SPC professional to be presented with neat, chronologically recorded, lot data. And, indeed, it is rather unusual for an SPC program to begin as a well thought-out management decision. Much more common is the situation where the quality consultant is called in "to put out fires." In the United States, these fires may very well be the result of a litigation concerning a real or perceived defect in a production item. It will do no one much good to respond to a potential client whose business is collapsing by offering to implement an orderly program of statistical process control on a production network which is devoid of measuring sensors, which might not even have been charted. A fatal flaw for any consultant is to insist on dealing with the world as he would like to see it rather than as it is. We are accustomed, in both Poland and the United States, to receiving a call for implementing a "comprehensive program of quality control" quickly followed by a statement of the sort "By the way, we have some little problem on which you might care to help." The "little problem" may be a lawsuit or a cancellation by a major client. A situation where the "problem" is actually a business effect rather than the technological cause of the effect requires a certain amount of exploratory work, perhaps to the point of modeling the process which produced the real or perceived defect.

171

5.2   The Schematic Plot

We will first introduce the *schematic plot* of John W. Tukey, the founder of Exploratory Data Analysis [11].[1] Let us suppose we have a set of N measurements of the same variable Y. Let us sort the measurements from smallest to largest

$$y_1 \leq \cdots \leq y_{.25N} \leq \cdots \leq y_{.5N} \leq \cdots \leq y_{.75N} \leq \cdots \leq y_N. \qquad (5.1)$$

Here, $y_{.25N}$ and $y_{.75N}$ denote the lower and upper sample quartiles, respectively. (In the terminology of the founder of Exploratory Data Analysis, John W. Tukey, these are referred to as the upper and lower *hinges*.) Now we compute the *interquartile range*

$$H = y_{.75N} - y_{.25N}. \qquad (5.2)$$

We then compute the size of a

$$\text{Step} = 1.5H. \qquad (5.3)$$

Next we add one step to the upper hinge to obtain the

$$\text{Upper Inner Fence} = \text{Upper Hinge} + \text{Step}. \qquad (5.4)$$

Similarly, we obtain the

$$\text{Lower Inner Fence} = \text{Lower Hinge} - \text{Step}. \qquad (5.5)$$

Finally, we compute the

$$\text{Upper Outer Fence} = \text{Upper Hinge} + 2\text{Step}. \qquad (5.6)$$

$$\text{Lower Outer Fence} = \text{Lower Hinge} - 2\text{Step}. \qquad (5.7)$$

Let us suppose that the data are normally distributed. We will also assume that the sample median falls essentially at the population median (and hence at μ). Further, we will assume that the hinges occur essentially at the population values. Then, expressing all variates in standard form, i.e.,

$$Z = \frac{X - E(X)}{\sqrt{Var(X)}}, \qquad (5.8)$$

---

[1]For a concise model based approach to EDA see [8].

we have the fact that

$$\text{Prob}(Z > \text{Upper Hinge}) = \frac{1}{\sqrt{2\pi}} \int_{.6745}^{\infty} e^{-\frac{z^2}{2}} dz = .25. \qquad (5.9)$$

Next

$$\text{Prob}(Z > \text{Upper Inner Fence}) = \frac{1}{\sqrt{2\pi}} \int_{2.698}^{\infty} e^{-\frac{z^2}{2}} dz = .0035. \qquad (5.10)$$

Then,

$$\text{Prob}(Z > \text{Upper Outer Fence}) = \frac{1}{\sqrt{2\pi}} \int_{4.722}^{\infty} e^{-\frac{z^2}{2}} dz = 1.2 \ 10^{-6}. \qquad (5.11)$$

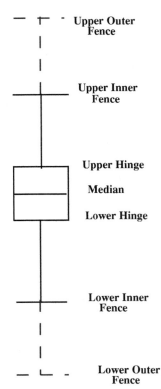

Figure 5.1. Schematic Plot.

Thus, if we are dealing with a data set from a single normal distribution, we would expect that an observation would fall outside the hinges half the time. An observation would fall outside the inner fences 7 times in a thousand. Outside the outer fences, an observation would fall with chances only 2 in a million. In a sense, observations inside the inner fences are not particularly suspect as having been generated by other contaminating distributions. An observation outside the inner fences can begin to point to a contamination, to a Pareto glitch. Outside the outer fences, it is clear that something is pretty much the matter with the assumption that the data came from a dominant normal distribution. Some refer to the schematic plot as a "box plot" or a "box and whiskers plot," but these terms actually refer to slightly less complex, less outlier oriented plots.

We show an idealized schematic plot in Figure 5.1, i.e., one in which the boundaries were determined assuming the data are from a single normal distribution.

Let us now use this approach with the sample means from the data set of Table 3.2. We first sort that data from the lowest $\bar{x}$ to the greatest.

Lot	x_1	x_2	x_3	x_4	x_5	$\bar{x}$	Rank
40	10.002	9.452	9.921	9.602	9.995	9.794	1
73	9.972	9.855	9.785	9.846	9.846	9.878	2
76	9.927	9.832	9.806	10.042	9.914	9.904	3
6	9.876	9.957	9.845	9.913	9.941	9.906	4
63	9.904	9.848	9.949	9.929	9.904	9.907	5
88	9.868	9.955	9.769	10.023	9.921	9.907	6
39	9.848	9.944	9.828	9.834	10.091	9.909	7
54	9.992	9.924	9.972	9.755	9.925	9.914	8
22	9.965	10.011	9.810	10.057	9.737	9.916	9
55	9.908	9.894	10.043	9.903	9.842	9.918	10
2	9.862	10.003	9.829	9.824	10.077	9.919	11
30	9.845	9.901	10.020	9.751	10.088	9.921	12
36	9.952	10.056	9.948	9.802	9.947	9.941	13
5	9.737	9.937	9.928	10.144	9.965	9.942	14
51	9.967	9.947	10.037	9.824	9.938	9.943	15
38	10.010	9.841	10.031	9.975	9.880	9.947	16
7	9.898	9.959	9.924	9.989	9.987	9.951	17
78	9.825	10.106	9.959	9.901	9.964	9.951	18
50	10.007	9.789	10.015	9.941	10.013	9.953	19
26	9.767	9.994	9.935	10.114	9.964	9.955	20
53	9.841	9.926	9.892	10.152	9.965	9.955	21
81	10.059	9.992	9.981	9.800	9.950	9.956	22
57	10.064	10.036	9.733	9.985	9.972	9.958	23
35	9.927	10.066	10.038	9.896	9.871	9.960	24
45	10.058	9.979	9.917	9.881	9.966	9.960	25
86	10.066	9.948	9.769	10.102	9.932	9.963	26
16	10.007	10.005	9.883	9.941	9.990	9.965	27
52	9.981	10.053	9.762	9.920	10.107	9.965	28
62	10.100	9.853	10.067	9.739	10.092	9.970	29
14	10.025	9.890	10.002	9.999	9.937	9.971	30
4	9.820	10.066	10.062	9.897	10.013	9.972	31
20	9.786	10.145	10.012	10.110	9.819	9.974	32
1	9.927	9.920	10.170	9.976	9.899	9.978	33
10	9.896	9.994	10.009	9.835	10.162	9.979	34
84	10.087	9.994	9.915	10.023	9.883	9.980	35
65	9.982	9.963	10.061	9.970	9.937	9.983	36
82	9.832	10.075	10.111	9.954	9.946	9.984	37
13	10.127	9.935	9.979	10.014	9.876	9.986	38
48	10.012	10.043	9.932	10.072	9.892	9.990	39
83	9.958	9.884	9.986	10.008	10.113	9.990	40
3	10.061	10.089	9.950	9.929	9.935	9.993	41
19	9.986	10.041	9.998	9.992	9.961	9.996	42
37	9.941	9.964	9.943	10.085	10.049	9.996	43
41	10.031	10.061	9.943	9.997	9.952	9.997	44
59	9.869	9.934	10.216	9.962	10.012	9.999	45

Table 5.1

Lot	x_1	x_2	x_3	x_4	x_5	$\bar{x}$	Rank
64	9.979	10.008	9.963	10.132	9.924	10.001	46
69	10.014	10.070	9.890	10.137	9.901	10.002	47
27	9.933	9.974	10.026	9.937	10.165	10.007	48
47	10.132	9.920	10.094	9.935	9.975	10.011	49
18	10.168	10.045	10.140	9.918	9.789	10.012	50
23	9.989	10.063	10.148	9.826	10.041	10.013	51
42	9.990	9.972	10.068	9.930	10.113	10.015	52
24	9.983	9.974	9.883	10.153	10.092	10.017	53
61	10.008	10.157	9.988	9.926	10.008	10.017	54
49	10.097	9.894	10.101	9.959	10.040	10.018	55
68	9.936	10.022	9.940	10.248	9.948	10.019	56
34	10.016	9.990	10.106	10.039	9.948	10.020	57
12	9.983	9.974	10.071	10.099	9.992	10.024	58
89	10.084	10.018	9.941	10.052	10.026	10.024	59
31	9.956	9.921	10.132	10.016	10.109	10.027	60
32	9.876	10.114	9.938	10.195	10.010	10.027	61
9	9.928	10.234	9.832	10.027	10.121	10.028	62
46	10.006	10.221	9.841	10.115	9.964	10.029	63
80	9.972	10.116	10.084	10.059	9.914	10.029	64
44	9.980	10.094	9.988	9.961	10.140	10.033	65
87	10.041	10.044	10.091	10.031	9.958	10.033	66
29	10.022	9.986	10.152	9.922	10.101	10.034	67
43	9.995	10.056	10.061	10.016	10.044	10.034	68
85	10.232	9.966	9.991	10.021	9.965	10.035	69
56	10.011	9.967	10.204	9.939	10.077	10.040	70
77	10.177	9.884	10.070	9.980	10.089	10.040	71
60	10.016	9.996	10.095	10.029	10.080	10.043	72
33	9.932	9.856	10.085	10.207	10.100	10.045	73
74	10.014	10.000	9.978	10.133	10.100	10.045	74
75	10.093	9.994	10.090	10.079	9.998	10.051	75
66	10.028	10.079	9.970	10.087	10.094	10.052	76
72	9.934	10.025	10.129	10.054	10.124	10.053	77
25	10.063	10.075	9.988	10.071	10.096	10.059	78
11	10.011	10.011	10.090	10.095	10.120	10.065	79
58	9.891	10.055	10.235	10.064	10.092	10.067	80
90	10.063	10.055	10.104	10.080	10.064	10.073	81
8	10.001	10.050	10.263	9.982	10.076	10.074	82
70	10.005	10.044	10.016	10.188	10.116	10.074	83
15	9.953	10.000	10.141	10.130	10.154	10.076	84
71	10.116	10.028	10.152	10.047	10.040	10.077	85
67	9.995	10.029	9.991	10.232	10.189	10.087	86
17	10.062	10.005	10.070	10.270	10.071	10.096	87
21	9.957	9.984	10.273	10.142	10.190	10.109	88
28	10.227	10.517	10.583	10.501	10.293	10.424	89
79	10.333	10.280	10.509	10.631	10.444	10.439	90

In constructing the Tukey schematic plot, we first compute the median, in this case, the average between the means ranked 45 and 46, respectively,

$$\text{Median} = \frac{9.999 + 10.001}{2}. \qquad (5.12)$$

The lower hinge is obtained by going up one fourth of the ranks. In this case the rank is essentially 23, so the lower hinge is 9.958. Similarly, the upper hinge is the observation having rank 68, or 10.034. This then gives us for the step size

$$\text{Step} = 1.5(10.034 - 9.958) = .114. \qquad (5.13)$$

Then we have

$$\text{Lower Inner Fence} = 9.958 - .114 = 9.844; \qquad (5.14)$$

$$\text{Lower Outer Fence} = 9.958 - 2(.114) = 9.730; \qquad (5.15)$$

$$\text{Upper Inner Fence} = 10.034 + .114 = 10.148; \qquad (5.16)$$

and

$$\text{Upper Outer Fence} = 10.034 + 2(.114) = 10.262. \qquad (5.17)$$

We have one value between the lower outer and inner fences (denoted by an asterisk) and two values outside the upper outer fence (denoted by hollow circles). We recall that our ordinary control chart analysis identified these three points as being "out of control." Why, then, one might well ask, should one bother with the schematic plot? The answer is that in the beginning of an SPC investigation, it is very common to have data not regularly indexed on time. In such a case, the schematic plot provides us a quick feel as to whether there appear to be easily identified Pareto glitches. Once we answer this question affirmatively, we can try and backtrack in time to see whether we might be able to identify the assignable cause of the glitch. Even if we cannot do so, we have "seen gold nuggets in the stream" and we can proceed forward with confidence that a procedure of instituting regular control charting is likely to yield big dividends, quickly.

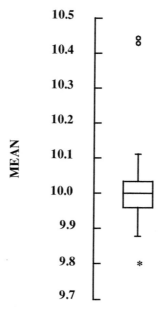

Figure 5.2. Schematic Plot of 90 Bolt Lot Means.

5.3   Smoothing by Threes

It is the basic task of control charting to find the "high frequency" Pareto glitch. Consequently, it might appear that the last thing we would want to consider in this context is smoothing, a device generally oriented to removing the high frequency wiggles in a time indexed data set. A concrete example will show the importance of smoothing to the SPC investigator.

A manufacturer producing flexible construction material had a number of samples that were outside the tolerance limits on a particular strength index. (This was a manufacturer practicing "quality assurance" rather than statistical process control.) There were about a dozen such measurement indices which were measured both by the manufacturer and his client. Not a great deal of attention was paid to these indices, which were perhaps only marginally related to performance of the sheets. Rather the manufacturing workers more or less identified good material visually. Nevertheless, the indices were a part of the state approved "code." Failure to comply with them on the part of the builder might expose him to subsequent lawsuits by an end user.

Some weeks after the material had been used, it was noted by a builder, who was the major client of the manufacturer, that a number of samples were outside stated tolerance levels for one of the strength indices. An on site task force was immediately dispatched to sites where the material had been used. Both the manufacturer and the builder agreed that the material was performing well. Extensive examination over a period of some months failed to find a problem. The head of the quality assurance section left the manufacturing firm "to pursue exciting opportunities elsewhere." Pressure from the builder increased. Veiled threats to changing suppliers were made.

At this time, we were called in to find a solution to the problem. Naturally, the first thing to consider was to take past runs and create a schematic plot. We show the plot in Figure 5.3. None of the observations were outside the inner fences. The schematic plot indicated that the data was skewed to the right, not normally distributed. Several transformations of the data were attempted, namely, the square root, the fourth root, the natural logarithm and the logarithm of the logarithm. None of these brought the chart to symmetry.

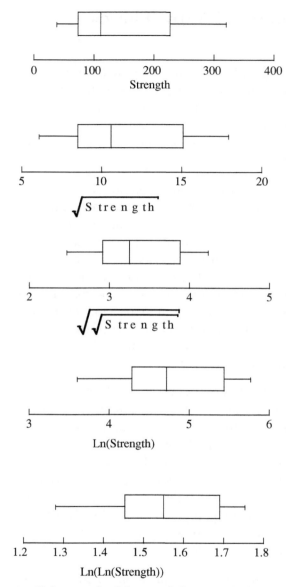

Figure 5.3. Schematic Plots of Strength Data.

Table 5.2					
Lot	x	$\sqrt{x}$	$\sqrt{\sqrt{x}}$	$\ln(x)$	$\ln(\ln(x))$
1	278.000	16.673	4.083	5.628	1.728
2	181.400	13.468	3.670	5.201	1.649
3	265.400	16.291	4.036	5.581	1.719
4	111.000	10.536	3.246	4.710	1.550
5	187.600	13.697	3.701	5.234	1.655
6	74.400	8.626	2.937	4.309	1.461
7	71.840	8.476	2.911	4.274	1.453
8	258.400	16.075	4.009	5.555	1.715
9	225.700	15.023	3.876	5.419	1.690
10	249.200	15.786	3.973	5.518	1.708
11	86.600	9.306	3.051	4.461	1.495
12	94.480	9.720	3.118	4.548	1.515
13	189.000	13.748	3.708	5.242	1.657
14	264.000	16.248	4.031	5.576	1.718
15	72.600	8.521	2.919	4.285	1.455
16	249.000	15.780	3.972	5.517	1.708
17	66.000	8.124	2.850	4.190	1.433
18	226.200	15.040	3.878	5.421	1.690
19	151.400	12.304	3.508	5.020	1.613
20	36.600	6.050	2.460	3.600	1.281
21	85.400	9.241	3.040	4.447	1.492
22	210.800	14.519	3.810	5.351	1.677
23	84.800	9.209	3.035	4.440	1.491
24	57.400	7.576	2.753	4.050	1.399
25	49.800	7.057	2.656	3.908	1.363
26	54.200	7.362	2.713	3.993	1.384
27	204.400	14.297	3.781	5.320	1.671
28	233.000	15.264	3.907	5.451	1.696
29	210.200	14.498	3.808	5.348	1.677
30	77.550	8.806	2.968	4.351	1.470
31	89.900	9.482	3.079	4.499	1.504
32	112.000	10.583	3.253	4.718	1.551
33	64.000	8.000	2.828	4.159	1.425
34	70.600	8.402	2.899	4.257	1.449
35	59.800	7.733	2.781	4.091	1.409
36	66.800	8.173	2.859	4.202	1.435
37	235.600	15.349	3.918	5.462	1.698
38	90.800	9.529	3.087	4.509	1.506
39	216.000	14.697	3.834	5.375	1.682
40	120.500	10.977	3.313	4.792	1.567
41	239.200	15.466	3.933	5.477	1.701
42	87.000	9.327	3.054	4.466	1.496
43	318.800	17.855	4.226	5.765	1.752
44	44.480	6.669	2.583	3.795	1.334
45	91.840	9.583	3.096	4.520	1.509
46	40.800	6.387	2.527	3.709	1.311
47	232.200	15.238	3.904	5.448	1.695

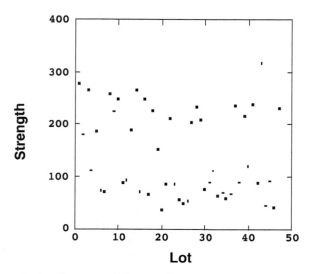

Figure 5.4. Scatter Plots of Strength Data.

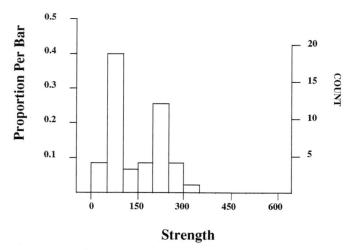

Figure 5.5. Histogram of Strength Data.

Although the data had not been collected in any sort of regular time interval fashion, and we had only one datum per lot, it was possible to order the data by production run as we demonstrate in Table 5.2 and in Figure 5.4. The split of the data into two groupings is made even more apparent by the histogram of the strength data in Figure 5.5. A standard control chart here is obviously inappropriate. The variability induced by the two groupings causes all the lots to fall inside the control limits. We recall that in the case where we have n samples in a lot, to obtain an estimator for the population variance on the basis of N lots, we can use

$$\hat{\sigma}^2 = \frac{1}{N} \sum_{j=1}^{N} s_j^2 \qquad (5.18)$$

where

$$s_j^2 = \frac{1}{n-1} \sum_{i=1}^{n} (x_{j,i} - \bar{x}_j)^2. \qquad (5.19)$$

Note that we could have used as an estimator for σ^2, under the assumption that all the data come from the same normal distribution,

$$\hat{\hat{\sigma}}^2 = \frac{1}{nN} \sum_{j=1}^{N} \sum_{i=1}^{n} (x_{j,i} - \bar{\bar{x}})^2. \qquad (5.20)$$

This estimator is not generally used, because it can be inflated greatly if the mean is not truly constant. But in our example, where there is only one sample per lot, we have little choice except to use $\hat{\hat{\sigma}}^2$ as an estimator for σ^2 .

When we use the estimator on the untransformed "strength" data in Table 5.2, we obtain the inflated estimate for σ^2, 6,881.5. The $\bar{\bar{x}}$ value of 144.398, with three times the square root of our variance estimate, gives, as upper and lower control limits, 393.27 and 0, respectively. A glance at Figure 5.6 shows that all measurements are nominally (and naively) "in control."

It turns out that the data appears to be dividing into a group with "strength" measurements in the proper range, and another very much on the low side. Such a data set is not untypical of the kind confronting statistical process control professionals as soon as they walk in the door. In the United States, in particular, SPC people seem to be called in by desperate managers more frequently than by managers who have cooly decided that this is the year when the company will get serious about quality control. The Deming/Shewhart analysis, which is generally excellent as a procedure once some degree of order has been introduced, frequently needs to be supplemented by nonstandard techniques in the first stages of an SPC implementation.

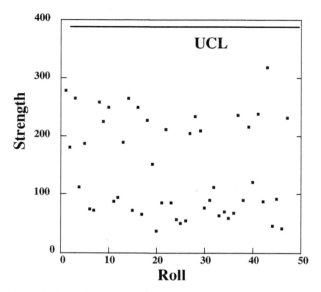

Figure 5.6. Naive Control Chart of Strength Data.

The anecdotal observations of field personnel to the effect that there seemed to be no discernible difference between the material from lots in which the "strength" index was low and those when it was in the satisfactory range could indicate many things. For example, the index could simply have nothing to do with anything. In this situation, that seemed not to be the case. The company engineers were confident that suppressed values of the index should, in some cases, lead to material failures, of which there had, in fact, been none.

Perusing old records, we found cases in which the laboratory had inadvertently made two measurements from the same lot (naturally, they should have been replicating their tests always, but they had not purposely done so). And, in some cases, it turned out that on the same testing machine, with the same material, normal range as well as low values had been found. So, then, one might suppose that the technicians obtaining the low readings were improperly trained or otherwise deficient. But still further investigation found that the same technician obtained normal and low readings for the same material on two different testing machines (there were 3 testing machines). So, then, perhaps there was a problem with one of the machines. But further investigation showed that low values had been observed on each of the machines, with approximately the same frequency.

At this point, we arranged for the chief technician to take samples from three different lots and run them on each of the three machines, each lot replicated three times. For each set of material and for each machine, the technician observed both high and low measurements. A look at the testing curves (which had been automatically suppressed for observation, since an automated read-out software package bundled with the testing device was being utilized) revealed the difficulty.

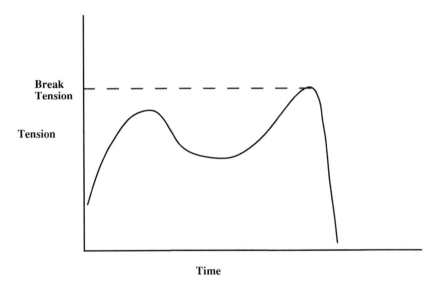

Figure 5.7. Tension Test of Material.

In the case of this material, a sample was stretched by the machine until its elasticity was exceeded. At this time, the tension automatically dropped. But as the material was stretched further, the now deformed material actually increased in strength, until the material broke. The strength measurement which was being recorded was the tension at the time of break of the material. We show such a curve in Figure 5.7.

After retrieving the curves typical of the material under examination (see Figure 5.8) a possible candidate for the cause of the difficulty was apparent. Due to the special characteristics of the material, a high frequency jitter was present. If the software in the measuring instrument was not smoothing the data, and if it was attempting to find the "strength" (tension at failure) by looking for the tension the second time the tension began to decrease, then the kind of problem observed could well occur.

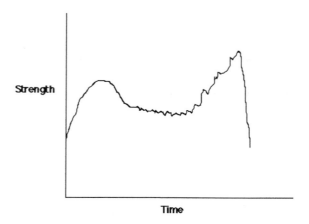

Figure 5.8. Tension Test of Nonstandard Material.

We immediately told the client to disconnect the "automatic yield finder" and reconnect the graphics terminal, so the technician could read the actual value as opposed to a bogus one. Then we called the instrumentation vendor to see if the conjecture about the way they wrote the software to find the "strength" could be substantiated. It was. Next, we indicate a quick fix for such a problem. Again, we rely on an Exploratory Data Analysis algorithm of Tukey's, namely the 3R smooth.

In Table 5.3, the "Tension" column shows the raw measurements (of the sort the software was using to find the second local maximum). In the second column, "Tension3," we have used the "3R" smooth. To show how this works, let us consider the 13'th Tension measurement, which is equal to 93. We look at the observation before and that after to see the triple {101,93,98}. We then replace the 13'th observation by the middle observation of the triple formed by the observation and the one just before and that just after. In other words, we write down 98 to replace 93. We start at the top of the list, and continue all the way through to the end (generally, the first and the last observations are not changed). We denote in boldface those values which have changed from the preceding iteration of the "3" smooth. In this case, three iterations brings us to point where there are no more changes. "3R" means "3 smooth repeated until no further changes take place." Accordingly, here, the "Tension333" column is the "3R" end result. In Figure 5.9, we denote the raw data and the evolution of smoothing as we proceed through the iterations. We note how the smooth tends to eliminate the perception of high frequency jitter as local maxima and minima.

Table 5.3				
Time	Tension	Tension3	Tension33	Tension333
1	100	100	100	100
2	107	107	107	107
3	114	114	114	114
4	121	121	121	121
5	125	125	125	125
6	127	125	125	125
7	125	125	125	125
8	120	120	120	120
9	116	116	116	116
10	111	111	111	111
11	107	107	107	107
12	101	101	101	101
13	93	**98**	98	98
14	98	**93**	**96**	96
15	87	**96**	**93**	**96**
16	96	**91**	96	96
17	91	**96**	96	96
18	105	**100**	100	100
19	100	**105**	**102**	102
20	108	**102**	105	105
21	102	**108**	108	108
22	121	121	121	121
23	125	125	125	125
24	130	**128**	128	128
25	128	**130**	130	130
26	139	139	139	139
27	158	158	158	158
28	168	168	168	168
29	175	175	175	175
30	188	188	188	188
31	195	195	195	195
32	205	205	205	205
33	215	215	215	215
34	220	215	215	215
35	182	182	182	182
36	164	164	164	164
37	140	140	140	140

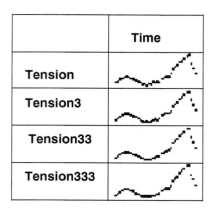

Figure 5.9. Tension Measurements with Smooths.

5.4   Bootstrapping

Some might be curious as to why workers in SPC seldom deal with the t (or "Student's") distribution. After all, it was the late nineteenth century proto quality control expert W.S. Gossett (aka "Student," who used an alias in his early publishing lest his employer, Guiness Breweries, bear the opprobrium of having used statistical techniques in its manufacturing process) who discovered that the distribution of

$$t = \frac{\bar{x} - \mu}{s/\sqrt{n}}$$

was not $\mathcal{N}(0,1)$, as had been supposed, but was something having heavier tails. Why do SPC workers revert to the old assumption that if $x_1, x_2, \ldots, x_n$ are independent and normally distributed, then, indeed:

$$t = z = \frac{\bar{x} - \mu}{s/\sqrt{n}}$$

may be treated as though it were $\mathcal{N}(0,1)$? The reason is clear enough. We do not, in SPC, estimate the underlying variance from the sample variance of a given lot, but from the pooled average of the sample variances from many lots, say M—typically over 50. The estimate[2]

$$\bar{s^2} = \frac{1}{M} \sum_{i=1}^{M} s_i^2$$

---

[2]For lot i of size n_i, $s_i^2 = 1/(n_i - 1) \sum_{j=1}^{n_i} (x_{ij} - \bar{x}_i)^2$.

is very close to σ^2 because by the Strong Law of Large Numbers (see Appendix) the average of many independent sample mean estimates of a parameter tends to the parameter as the number of estimates gets large. For many, perhaps most, situations in SPC we will not be far off the mark if we simply assume that our data consists of draws from a dominant normal distribution plus occasional draws from contaminating normal distributions (as discussed in Chapter 3). However, we have already seen with the example in Section 5.3 that there are significant exceptions to the rule. That case was particularly hard to deal with since the "in control" part was not really dominant. The bad observations were as numerous as the good ones.

Most of the standard testing in Statistical Process Control is based on lot means and standard deviations. Because for almost all realistic cases we are dealing with distributions which produce sample lot means which converge to the Gaussian (normal) distribution as the size of lots gets large (say greater than 20), we frequently assume that the lot means are normally distributed. Yet our lot sizes are generally 10 or less where the convergence of the sample mean to normality has not yet taken place (unless the underlying distribution is itself rather close to Gaussian).

Resampling gives a means for using lot means in such a way that we are not making any assumptions about their being normally distributed. Similarly, we can make tests about lot standard deviations which do not make any assumptions about the underlying normality of the data.

The range of sophistication that can be used in resampling is extensive. But for dealing with low order moments—the mean and variance—the *nonparametric bootstrap* (developed in its full glory by Bradley Efron [1], but in the simple form we use here much earlier by the late Julian Simon [5]) works perfectly well.

Suppose we have a data set of size n and from this data set wish to make a statement about a *confidence interval* of the mean of the distribution from which the data was taken. Then, we can select with replacement n of the original observations and compute the sample mean. If we carry out say 10,000 such resamplings each of size n, we can order the resulting sample means from smallest to largest. Then $[\bar{X}_{250}, \bar{X}_{9750}]$ gives us a 95% confidence interval for the true value of the mean of the distribution from which the sample was taken. We are 95% sure the mean μ lies in this interval.

Similarly, let us take that data set of size n and the same set of 10,000 resamplings and compute s, the sample standard deviation for each of the 10,000 resamplings. Rank order the sample standard deviations. A 95% confidence interval for σ would be $[s_{250}, s_{9750}]$.

Of course, we are pretending that the data we have represents all the data that ever we could have and that further examinations will always yield simply a repeating of our original data set, but with some points missing, and others included more than once. This has a certain intuitive appeal, although mathematically a great deal more work is required to put it on solid ground.

Let us return to the data in Table 3.2. We recall that this data was actually the result of a baseline normal distribution $\mathcal{N}(\mu = 10, \sigma^2 = .01)$, contaminated by two other normals. The variables in each lot were

$$Y(t) = \mathcal{N}(10, .01); \text{ probability } .98505 \qquad (5.21)$$
$$Y(t) = \mathcal{N}(10.4, .03); \text{ probability } .00995 \qquad (5.22)$$
$$Y(t) = \mathcal{N}(9.8, .09); \text{ probability } .00495 \qquad (5.23)$$
$$Y(t) = \mathcal{N}(10.2, .11); \text{ probability } .00005. \qquad (5.24)$$

The proviso was that $Y(t)$ stayed in the same distribution for all items in the same lot.

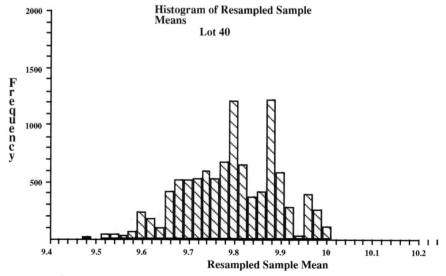

Figure 5.10. Bootstrapped Means from Lot 40.

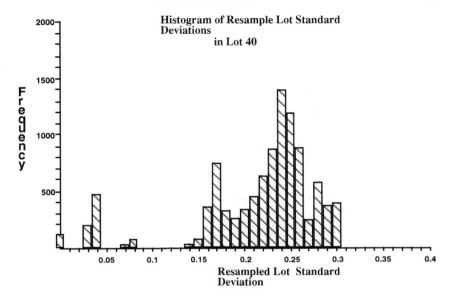

Figure 5.11. Bootstrapped Standard Deviations from Lot 40.

Now we first try a bootstrapping approach where we use the lot histograms to obtain a 99.8% confidence interval for μ and 99.8% confidence interval for σ. Why 99.8%? Because we have been using 3σ level normal theory tests, and that convention corresponds to 99.8% confidence intervals. We note that for lot 40, the 99.8% confidence interval for the standard deviation is $(0, 0.30125)$. For the 90 lots, the average sample variance is .01054, giving an estimate of $\sigma = .1026$. This is well within the confidence interval from lot 40.

The 99.8% confidence interval for the resampled means, based on lot 40, is $(9.482, 10.001)$ The overall average of the lot means from the 90 lots is 10.004. Since 10.004 is outside the 99.8% resampled mean confidence interval from lot 40, we reject lot 40 as being out of control. We recall that when we went through our parametric test, we rejected lot 40 on the basis of standard deviation as well as mean. And we should have done so, since we know that lot 40 comes from $\mathcal{N}(9.8, .09)$.

If we carry through similar testing for lots 28 and 79, we also reject them as being out of control on the basis of the 99.8% mean confidence interval formed from each lot. Again, since we recall that these lots were actually drawn from $\mathcal{N}(10.4, .03)$, we see they should have been rejected as contaminated (out of control). On the other hand, we also reject noncontaminated lots 6, 7, 54, 63 and 73 on the basis of the resampled means confidence interval test. These lots were drawn from the in control distribution $\mathcal{N}(10.0, .01)$, and they were not rejected by the parametric normal theory based procedure in Chapter 3. We know that, in SPC, false alarms can really cause havoc down when they come at such a high rate. It would appear that for very small lot sizes, resampling confidence intervals based on the individual lots is probably not a very good idea.

What else might we try? Let us make the assumption that all the bolts are drawn from the same in control distribution. We put all $90 \times 5 = 450$ observations from Table 3.2 into a pool. From this pool we make 10,000 random draws of size 5. The resulting histogram of the 10,000 sample means is shown in Figure 5.12. The 99.8% confidence interval is (9.8614, 10.139). The three contaminated lots: 28, 40 and 79 are rejected as being out of control. No noncontaminated lots are rejected.

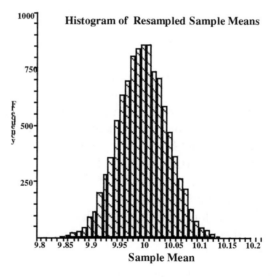

Figure 5.12. Bootstrapped "Lot Means."

The 10,000 sample standard deviations histogram of the 10,000 random draws of size 5 is shown in Figure 5.13. The 98.8% confidence interval is (0.01567, 0.20827). Only lot 40 is rejected on the basis of the standard deviation confidence interval. This is exactly what we gleaned from the Gaussian based test in Section 3.3 for the same data. Note that the pooling of all the data into an urn from which we draw lots of size five will tend to inflate estimates of the lot standard deviation.

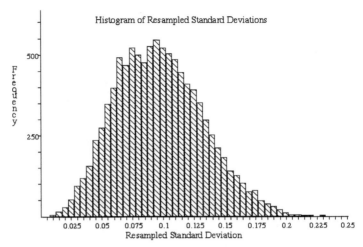

Figure 5.13. Bootstrapped "Lot Standard Deviations."

Next, following [10], we could look at the entire set of 90 lots and obtain resampling confidence intervals there against which we could compare the means and standard deviations of each of the 90 lots. In a way, the procedure we shall suggest is closer to the spirit of the normal theory tests of Chapter 3. Although no explicit assumptions of normality are made, the notion that the distribution of the bootstrapped grand means of all 90 lots of 5 is that of a lot of size 5 but with scale changed by $1/\sqrt{90}$ allows the central limit theorem induced normality of the grand mean be implied for the lot means as well. (Of course, if the underlying distribution of the bolts is not too bad, there will already be some moving toward normality by the averaging of five bolts to get the lot mean.) Let us suppose we make the assumption that all the lots come from the "in control" distribution. Then we can take the overall pooled mean of sample means (10.0044), and record all the means from all the lots as differenced from 10.0044,

$$(diff)_j = \overline{X_j} - 10.0044.$$

Now, it is an easy matter then to use bootstrapping to find a 99.8% confidence interval about the mean of sample means in which the population mean should lie. But, recalling that

$$Var(\overline{\overline{X}}) = Var(\overline{X_j})/90,$$

where j is simply one of the 90 lots, we realize that, under the assumptions given, the standard deviation of the lot means should be taken to be $\sqrt{(90)}$ times that of the mean of lot means. So, to find a confidence interval about 10.0044 in which a lot mean would be expected to fall, with 99.8% chance, if it is truly from the "in control" population, we construct a histogram of the bootstrapped grand differenced (from 10.0044) means multiplied by $\sqrt{90} = 9.4868$ with 10.0044 added. The grand differenced means are obtained from the bootstrapped differences defined above. We show this in Figure 5.14.

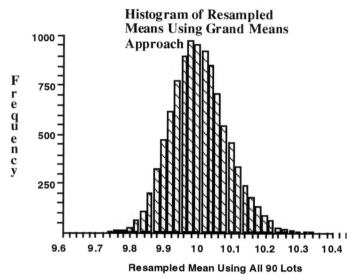

Figure 5.14. Bootstrapped "Lot Means."

The 99.8% confidence interval (a sort of bootstrapped mean control chart) is given by (9.7751,10.294). In Figure 5.14, we note that (contaminated) lots 28 and 79 both are out of control. But (contaminated) lot 40 is not recognized as being out of control from the resampled mean test.

Next, let us carry out a similar approach for the resampled sample standard deviations. The average sample variance over the entire 90 lots is .01054. The 99.8% confidence interval for σ is (.0045941, 0.20639). We note that only lot 40 has a standard deviation outside this confidence interval. We note that between the bootstrapped mean test and the bootstrapped standard deviation test, we have found all three out of control lots.

We note that none of the uncontaminated lots is identified as out of control. Consequently, in this case, the bootstrap control chart worked satisfactorily. By using the deviation of lot means from the grand mean, we have, however, made an assumption which may give us an inflated measure of the in control process variability and causes us to construct confidence intervals which are too wide, thus accepting too many bad lots as being in control. There are other steps we might take. For example, we might use median estimates for the grand mean of sample means and for the overall standard deviation of the in control distribution (see Section 3.3).

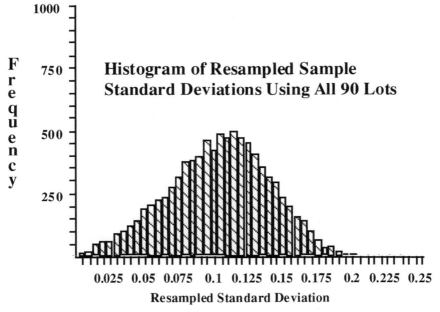

Figure 5.15. Bootstrapped "Lot Standard Deviations."

5.5 Pareto and Ishikawa Diagrams

In free market economies, we are in a different situation than managers in command economies, for there generally is a "bottom line" in terms of profits. The CEO of an automobile company, for example, will need to explain the dividends paid per dollar value of stock. If it turns out that these dividends are not satisfactory, then he can take "dramatic action" such as having his teams of lobbyists demand higher tariffs on foreign automobiles and instructing his advertising department to launch intimidating "buy American" campaigns. Sometimes, he might take even more dramatic action by trying to build better automobiles (but that is unusual). We note that if the decision is made to improve the quality of his product then there is the question of defining what it means for one car to be better than another. It is all very well to say that if profits are good, then we probably are doing OK, but a reasonable manager should look to the reasons why his sales should or should not be expected to rise. Uniformity of product is the measure which we will be using to a very large degree in the development of the statistical process control paradigm. But clearly, this is not the whole story. For example, if a manufacturer was turning out automobiles which had the property that they all ran splendidly for 10,000 miles and then the brake system failed, that really would not be satisfactory as an ultimate end result, even though the uniformity was high. But, as we shall see, such a car design might be very close to good if we were able simply to make appropriate modification of the braking system. A fleet of cars which had an average time to major problems of 10,000 miles but with a wide variety of failure reasons and a large variability of time until failure would usually be more difficult to put right.

The modern automobile is a complex system with tens of thousands of basic parts. As with most real world problems, a good product is distinguished from a bad one according to an implicit criterion function of high dimensionality. A good car has a reasonable price, "looks good," has good fuel efficiency, provides safety for riders in the event of an accident, has comfortable seating in both front and rear seats, has low noise levels, reliably starts without mishap, etc., etc.

Yet, somehow, consumers manage to distill all this information into a decision as to which car to purchase. Certain criteria seem to be more important than others. For example, market analysts for years have noted that Japanese automobiles seem to owe their edge in large measure to the long periods between major repairs. One hears statements such as, "I just changed the oil and filter every five thousand miles, and the thing drove without any problems for 150,000 miles."

Long time intervals between major repairs make up one very important criterion with American car buyers. Fine. So then, an automotive CEO might simply decide that he will increase his market share by making his cars have long times until major repairs. How to accomplish this? First of all, it should be noted that broad spectrum pep talks are of negative utility. Few things are more discouraging to workers than being told that the company has a problem and it is up to them to solve it without any clue as to how this is to be achieved.

A reasonable first step for the CEO would be to examine the relative frequencies of causes of first major repair during a period of, say, three months. The taxonomy of possible causes must first be broken down into the fifty or so groups. We show in Figure 5.16 only the top five. It is fairly clear that management needs to direct a good deal of its attention to improving transmissions. Clearly, in this case, as is generally true, a few causes of difficulty are dominant. The diagram in Figure 5.16 is sometimes referred to as a *Pareto diagram*, inasmuch as it is based on *Pareto's Maxim* to the effect that *the failures in a system are usually the consequence of a few assignable causes rather than the consequence of a general malaise across the system.*

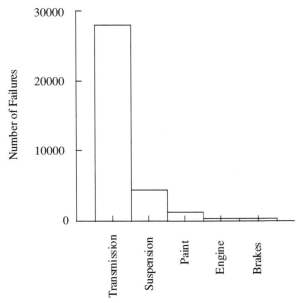

Figure 5.16. Failure Pareto Diagram.

What is the appropriate action of a manager who has seen Figure 5.16? At this point, he could call a meeting of the managers in the Transmission Section and tell them to fix the problem. This would not be inappropriate. Certainly, it is much preferable to a general harangue of the entire factory. At least he will not have assigned equal blame to the Engine Section with 203 failures (or the Undercoating Section with no failures) as to the Transmission Section with 27,955 failures. The use of hierarchies is almost inevitable in management. The Pareto diagram tells top management where it is most appropriate to spend resources in finding (and solving) problems. To a large extent, the ball really is in the court of the Transmission Section (though top management would be well advised to pass through the failure information to the Suspension Section and indeed to all the sections).

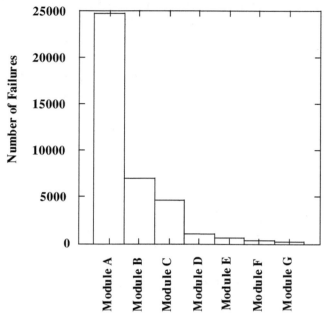

Figure 5.17. Transmission Failure Pareto Diagram.

What should be the approach of management in the Transmission Section? The obvious answer is a Pareto diagram (Figure 5.17) on the 27,955 faulty transmissions. That may not be realistic. It is easier to know that a transmission has failed than what was the proximate cause of that failure. We might hope that the on site mechanics will have correctly diagnosed the problem. Generally speaking, in order to save time, repair diagnostics will be modularized; i.e., there will be a number of subsections of the transmission which will be tested as to whether they are satisfactory or not. Naturally some of the transmissions will have more than one failed module.

Clearly, Module A is causing a great deal of the trouble. It is possible to carry the hierarchy down still another level to find the main difficulty with that module. The problem may be one of poor design, or poor quality of manufacture. Statistical process control generally addresses itself to the second problem.

The "cause and effect" or "fishbone" diagram of Ishikawa is favored by some as a tool for finding the ultimate cause of a system failure. Such a diagram might look like Figure 5.18 for the present problem.

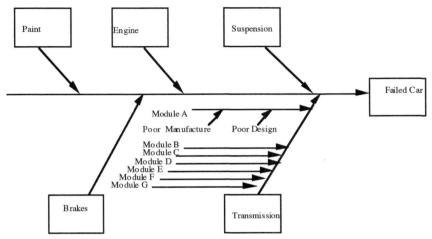

Figure 5.18. Fishbone (Ishikawa) Diagram.

The fishbone diagram should not be thought of as a precise flowchart of production. The chart as shown might lead one to suppose that the transmission is the last major component installed in the car. That is not the case. We note that Figure 5.18 allows for free form expression. For example, fishbone diagrams are frequently the product of a discussion where a number of people are making inputs on the blackboard. Each of the paths starting from a box is really a stand-alone entity. We have here developed only one of the paths in detail. We note that in the case of Transmissions, we go down the next level of hierarchy to the modules and then still one more level to the design and quality of manufacturing. In practice, the fishbone diagram will have a number of such paths developed to a high level of hierarchy. Note that each one of the major branches can simply be stuck onto the main stem of the diagram. This enables people in "brainstorming" sessions to submit their candidates for what the problem seems to be by simply sticking a new hierarchy onto the major stem.

5.6   A Bayesian Pareto Analysis for System Optimization of the Space Station

In 1995, one of us (Thompson) was asked to design a theoretical prototype for implementing Statistical Process Control in the construction and operation of the joint American-Russian Space Station. NASA traditionally has excellent reliability in its design, but, at the time, was not engaged in the operational Statistical Process Control paradigm. (Incredible though this may seem, it is not unusual to find excellent engineering prospective system design unaccompanied by an orderly process optimization of a created system.) It was an interesting opportunity to design a "start from scratch" operation for a system of incredible complexity.

The foregoing industrial examples bear on system optimization for the Space Station. Yet they differ in important aspects. An industrialist might, if he so chooses, simply allocate optimization resources based on customer complaints. We note that we were dealing with nearly 30,000 cases of transmission complaints alone. We have no such leisure when we consider system optimization of the Space Station. We cannot simply wait, calmly, to build up a data base of faulty seals and electrical failures. We must "start running" immediately. Thus, we will require an alternative to a hierarchy of histograms. Yet there are lessons to be learned from the industrial situation.

5.6.1   Hierarchical Structure

First of all, in the case of building a car, we recall that we had a hierarchy of parts of the system to be optimized. We did not simply string out a list of every part in a car. We formed a hierarchy; in the case of a car, we had three levels. Possibly, in the complexity of the Space Station, we will need to extend the hierarchy to a higher number than three, possibly as high as six or seven levels.

A top level might consist, say, of structure, fluid transmission, life support, electromechanical function, kinetic considerations and data collection. Again, we note that modern quality control seldom replaces a bolt or a washer. The irreducible level is generally a "module." We would expect such a practice to be utilized with the Space Station also. If we assume that we have a hierarchy of six levels and that there are roughly seven sublevels for each, then we will be dealing with approximately $7^6 = 117,649$ basic module types for consideration.

Next, in Figure 5.19 we demonstrate the sort of hierarchical structure we advocate through three levels. Even at three levels, using seven categories at each stage, we would be talking about $7^3 = 343$ end stages.

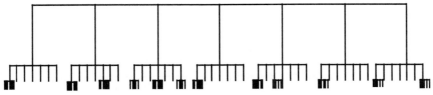

Figure 5.19. Three Levels of Hierarchy.

5.6.2 Pareto's Maxim Still Applies

Again, in the case of the Space Station, it would be folly to assume that at each level of the hierarchy, the probability of less than satisfactory performance in each category is equally likely. We do not have experiential histograms to fall back on. Classical flow charting will not be totally satisfactory, at least in the early days of operation. We need an alternative to the (say) six levels of histograms.

5.6.3 A Bayesian Pareto Model

In this section, we follow arguments in [9] with a look back at [2,3,5,10]. Let us suppose that at a given level of hierarchy, the failures (by this we mean any departures from specified performance) due to the k components are distributed independently according to a homogeneous Poisson process. So, if t is the time interval under consideration, and the rate of failure of the ith component is θ_i, then the number y_i of failures in category i is given (see Section B.14) by

$$f(y_i|\theta_i) = \exp(-\theta_i t)\frac{(\theta_i t)^{y_i}}{y_i!}. \tag{5.25}$$

The expected number of failures in category i during an epoch of time length t is given by

$$E(y_i|\theta_i) = \sum_{y_i=0}^{\infty} y_i e^{-\theta_i t}\frac{(\theta_i t)^{y_i}}{y_i!} = \theta_i t.$$

Similarly, it is an easy matter to show that the variance of the number of failures in category i during an epoch of time length t is also given, in the case of the Poisson process, by $\theta_i t$. Prior to the collection of failure data, the distribution of the ith failure rate is given by the *prior density*:

$$p(\theta_i) = \frac{\theta_i^{\alpha_i-1}\exp(-\frac{\theta_i}{\beta_i})}{\Gamma(\alpha_i)\beta_i^{\alpha_i}}. \tag{5.26}$$

Then, the *joint density* of y_i and θ_i is given by taking the product of $f(y_i|\theta_i)$ and $p(\theta)$:

$$f(y_i, \theta_i) = \exp(-\theta_i t)\frac{(\theta_i t)^{y_i}}{y_i!}\frac{\theta_i^{\alpha_i-1}\exp(-\frac{\theta_i}{\beta_i})}{\Gamma(\alpha_i)\beta_i^{\alpha_i}}. \tag{5.27}$$

Then, the *marginal distribution* of y_i is given by

$$\begin{aligned} f(y_i) &= \frac{t^{y_i}}{y_i!\Gamma(\alpha_i)\beta_i^{\alpha_i}}\int_0^\infty \exp(-\theta_i(t+\frac{1}{\beta_i}))\theta_i^{y_i+\alpha_i-1}d\theta_i \\ &= \frac{t^{y_i}}{y_i!\Gamma(\alpha_i)\beta_i^{\alpha_i}(t+1/\beta_i)^{y_i+\alpha_i}}\Gamma(y_i+\alpha_i). \end{aligned} \tag{5.28}$$

Then the *posterior density* of θ_i given y_i is given by the quotient of $f(y_i, \theta_i)$ divided by $f(y_i)$:

$$g(\theta_i|y_i) = \exp[-\theta_i(t+1/\beta_i)]\theta_i^{y_i+\alpha_i-1}(t+1/\beta_i)^{y_i+\alpha_i}/\Gamma(y_i+\alpha_i). \tag{5.29}$$

Then, looking at all k categories in the level of the hierarchy with which we are currently working, we have for the prior density on the parameters $\theta_1, \theta_2, \ldots, \theta_k$,

$$p(\theta_1, \theta_2, \ldots, \theta_k) = \Pi_{i=1}^k \frac{\theta_i^{\alpha_i-1}\exp(-\frac{\theta_i}{\beta_i})}{\Gamma(\alpha_i)\beta_i^{\alpha_i}}. \tag{5.30}$$

Similarly, after we have recorded over the time interval $[0, t]$, $y_1, y_2, .., y_k$ failures in each of the modules at the particular level of hierarchy, we will have the posterior distribution of the θ_i given the y_i,

$$g(\theta_1, \theta_2, \ldots, \theta_k|y_1, y_2, \ldots, y_k) = \tag{5.31}$$

$$\Pi_{i=1}^k \exp[-\theta_i(t+1/\beta_i)]\theta_i^{y_i+\alpha_i-1}(t+1/\beta_i)^{y_i+\alpha_i}/\Gamma(y_i+\alpha_i).$$

It should be observed in (5.26) that our prior assumptions concerning α had roughly the same effect as adding α_i failures at the beginning of the observation period.

We note that

$$E[\theta_i] = (y_i+\alpha_i)\frac{t}{t+1/\beta_i}. \tag{5.32}$$

Furthermore,

$$Var[\theta_i] = (\frac{t}{t+1/\beta_i})^2(y_i+\alpha_i). \tag{5.33}$$

We note that if we rank the expectations from largest to smallest, we may plot $E[t\theta_i]$ values to obtain a Bayesian Pareto plot very similar to the Pareto plot in Figure 5.16.

How shall one utilize expert opinion to obtain reasonable values of the α_i and β_i? First of all, we note that equations (5.32) and (5.33) have two unknowns. We are very likely to be able to ask an expert the question, "how many failures do you expect in a time interval of length t?" This will give us the left hand side of equation (5.32). An expression for the variance is generally less clearly dealt with by experts, but there are various ways to obtain nearly equivalent "spread" information. For example, we might ask the expert to give us the number of failures which would be exceeded in a time interval of length t only one time in ten.

5.6.4 An Example

Let us suppose that at the top level of hierarchy, we have seven subcategories. At the beginning of the study, expert opinion leads us to believe that for each of the subcategories, the expected "failure rate" per unit time is 2, and the variance is also 2. This gives us, before any data are collected, $\alpha_i = 2$ and $\beta_i = 1$. So, for each of the prior densities on θ_i we have the gamma density shown in Figure 5.20.

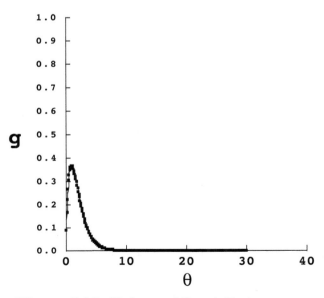

Figure 5.20. Priors without Data.

However, after 5 time units have passed, we discover that there have been
100 "failures" in the first module, and 5 in each of the other modules.
This gives us the posterior distributions shown in Figure 5.21.

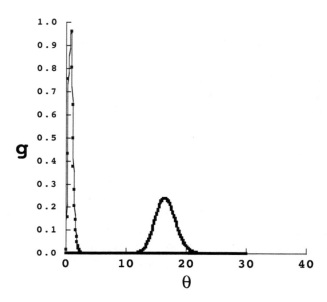

Figure 5.21. Evolving Posterior Distributions.

Clearly, we now have a clear indication that the posterior on the right
(that of the first module) strongly indicates that the major cause of
"failures" is in that first module, and that is where resources should be
allocated until examination of the evolutionary path of the posteriors
in lower levels of the hierarchy gives us the clue to the cause of the
problem(s) in module seven, which we then can solve.

Perhaps of more practical use to most users would be a *Bayesian
Pareto Chart*, which is simply the expected number of failures in a time
epoch of length seven. From (5.32) we note that

$$E[t\theta_i] = (y_i + \alpha_i)\frac{t}{t + 1/\beta_i}.$$

We show such a chart in Figure 5.22.

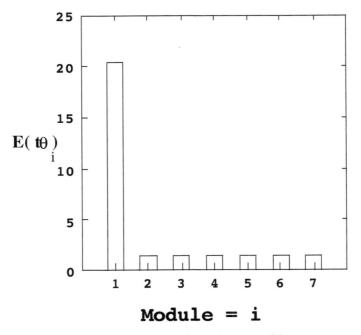

Figure 5.22. Bayesian Pareto Chart.

One very valid criticism might have to do with the inappropriateness of the assumption that the rates of failure in each category at a given level of hierarchy are independent. The introduction of dependency in the prior density will not be addressed here, since the study of the independent case allows us conveniently to address the evolution of posterior densities without unnecessarily venturing into a realm of algebraic complexity.

5.6.5 Allowing for the Effect of Elimination of a Problem

It should be noted that when we solve a problem, it is probably unwise to include all the past observations which include data before the problem was rectified. For example, if we fix the first module in Figure 5.22, then we should discount, in a convenient way, observations which existed prior to the "fix." On the other hand, we need to recognize the possibility that we have not actually repaired the first module. It might be unwise to discount completely those 100 failures in the 5 time units until we are really sure that the problem has been rectified. Even if we did not discount the failures from the time period before the problem has been rectified, eventually the posterior distribution would reflect the fact that less attention needs be given to repairs in the seventh module. But "eventually" might be a long time.

One way to discount records from the remote past is to use an *exponential smoother* such as

$$\hat{z}_i = (1 - r)\hat{z}_{i-1} + rz_i$$

where a typical value for r is 0.25. Let us consider the data in Table 5.4. Here, a malfunction in the first module was discovered and repaired at the end of the fifth time period. z_{ij} represents the number of failures of the ith module in the tth time period. $z_{i0} = \alpha_i$.

Table 5.4

Module	z_{i0}	z_{i1}	z_{i2}	z_{i3}	z_{i4}	z_{i5}	z_{i6}	z_{i7}	z_{i8}	z_{i9}	z_{i10}
1	2	20	18	23	24	15	2	2	0	2	1
2	2	1	1	2	0	1	2	0	1	1	2
3	2	1	2	0	1	1	1	2	1	0	0
4	2	0	2	0	2	1	1	1	1	2	0
5	2	2	1	0	0	2	0	0	2	2	1
6	2	0	2	1	1	1	1	1	1	2	1
7	2	1	1	0	2	1	1	1	1	0	2

Application of the exponential smoother with $r=.25$ gives the values in Table 5.5.

Table 5.5

Module	$\hat{z}_{i0}$	$\hat{z}_{i1}$	$\hat{z}_{i2}$	$\hat{z}_{i3}$	$\hat{z}_{i4}$	$\hat{z}_{i5}$	$\hat{z}_{i6}$	$\hat{z}_{i7}$	$\hat{z}_{i8}$	$\hat{z}_{i9}$	$\hat{z}_{i10}$
1	2	6.5	9.38	12.78	15.59	15.44	12.08	9.56	7.17	3.29	1.57
2	2	1.75	1.56	1.67	1.25	1.19	1.39	1.04	1.03	1.01	1.75
3	2	1.75	1.81	1.36	1.27	1.20	1.15	1.36	1.27	0.32	0.08
4	2	1.5	1.62	1.22	1.41	1.31	1.23	1.17	1.13	1.78	0.45
5	2	2.00	1.75	1.31	0.98	1.24	0.93	0.70	1.02	1.76	1.19
6	2	1.5	1.62	1.47	1.35	1.26	1.20	1.15	1.11	1.78	1.19
7	2	1.75	1.56	1.17	1.38	1.28	1.21	1.16	1.12	0.28	1.57

In Figure 5.23, we show the exponentially weighted Pareto chart at the end of time interval 5 and the exponentially weighted Pareto chart at the end of time interval 10.

In Figure 5.24, we show time lapsed exponentially weighted charts for all ten time intervals. It is clear that by the end of the ninth time interval, we should consider relegating module one to a lower level of risk of failures and reallocating inspection resources accordingly.

Finally, we might ask whether the Bayesian framework might ever be appropriately replaced, as the Space Station matures, by a more classical control chart strategy applied to a network. The answer is obviously in the affirmative. So far, with the manner in which we have presented the Pareto histogram, our orientation has really been more to the old-fashioned "quality assurance" paradigm, which is, in practice, utilized much more than Deming control charts. This is the leisurely approach by which managers may attack problems when they are good and ready to do so, and can use long histories of complaints as input. By the incorporation of expert opinion, we have achieved a Bayesian paradigm which allows us to search for problems from the outset. But, as time progresses, it would clearly be in order to move toward control charting.

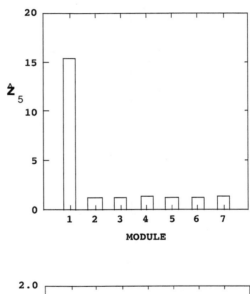

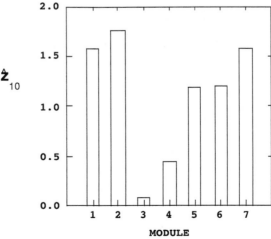

Figure 5.23. Exponentially Weighted Pareto Charts.

Time

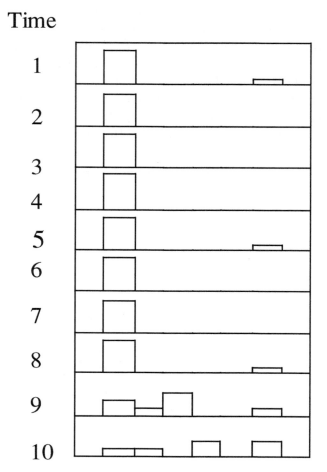

Figure 5.24. Time Lapsed Exponentially Weighted Pareto
Charts.

5.7   The Management and Planning Tools

All the techniques discussed so far, both those of exploratory character
and those directly applicable for quality improvement, are based on data
gathered from running processes and are for *on-line* step-wise optimization of these processes. By far the most important of these techniques
are control charts, but the others, i.e., simple run charts, schematic or
box plots, histograms, smoothing techniques, Pareto diagrams and cause
and effect or Ishikawa diagrams, are indispensable in some situations

too. When looking for correlations or nonlinear interdependencies between pairs of data, scatter diagrams or scattergrams which depict the pairs in a Cartesian plane should also be included into the set of such indispensable and simple means (in Chapter 6, we come directly to regression models for bivariate data and hence skip separate treatment of the scattergrams).

The seven techniques mentioned, run charts, control charts, schematic plots, histograms, Pareto and Ishikawa diagrams, and scattergrams, are often called *the seven SPC tools* (some authors prefer to use this term for the set of techniques with schematic plots replaced by data collection sheets; others, us included, consider the latter as tool number zero since data collection is an obvious and necessary, however nontrivial, prerequisite for any on-line analysis). To these, we have added in Section 5.6 the Bayesian Pareto model which, no doubt, is not so simple, but which is to be used when there is not enough data to begin scrutiny for opportunities for improvement with the standard Pareto diagram.

Within the Shewhart-Deming PDSA cycle, the seven SPC tools form the basic set of techniques for analysis of data coming from the cycle's Study stage. But what about the Plan stage, which is to be performed *off-line* and whose aim is to plan an innovation?

It is wise, when possible, to begin with correctly recognizing what is called the *true quality characteristic* or *true quality characteristics* if more than one are found. Such characteristics, which describe customers' quality requirements, are based on market research, on feedback from customers and on otherwise gathered information on the customers' needs, wants and expectations. As we already know, we can thus call them the *voice of the customer*, although they go beyond what a customer is able to formulate explicitly since customers do not have to be well aware of what innovation they will meet with delight, let alone they do not need to understand more distant aims of a system which provides product or service under consideration. As a rule, true quality characteristics are imprecise, as based on impressions and feelings rather than on technical specifications. They are the implicit multidimensional criterion function, with which we met at the beginning of Section 5.5 in the context of perfecting a car, such as car's safety, reliability, fuel efficiency, comfort of seating, etc. (in this example, only fuel efficiency can be considered well defined from the technical point of view).

It is the *substitute characteristics* which give technical flesh to the voice of the customer, as they describe technical requirements to be imposed on a product or service if true quality characteristics are to be met. That is, they are specifications and functionalities described in technical terms which, when fulfilled, make the voice of the processes conforming to the voice of the customer. Since providing substitute characteristics requires introduction of measures based on some well defined units, it paves the way for a structured and technically feasible step-wise improvement of processes.

At least three problems, however, need to be emphasized here. First, as a rule, several substitute characteristics should be considered surrogates for one true quality characteristic (e.g., the comfort of seating must evidently be described by more than one parameter). Second, neither of the substitutes is related to its true original in a mathematically rigorous way. Rather, the two are somehow correlated one with another. As a consequence, usually the choice of proper substitute characteristics, as well as their relation to the true characteristic, are established experimentally in a PDSA cycle. And third, the process of translating true quality characteristics into their surrogates is in fact a hierarchical one, most often with many levels. We saw this in a simple example of coming from the requirement for a car to be reliable and thus have possibly a long time between major repairs to that of maximizing the time until the first repair. This last requirement can already be considered a substitute and numerically defined characteristic. The cause and effect diagram helped us then to translate that requirement into the requirement to maximize time until first fault in, or in fact to improve, transmission. Of course, that was not the end of but rather a signal to begin the work on finding ways to better transmission, by properly stating aims and technical requirements for transmission modules and submodules and, finally, by turning to improving critical processes.

A structured method in which customer requirements are translated into appropriate technical requirements for each stage of product development and production, as it is put by the Big Three (Chrysler, Ford Motor and General Motors) in [4], is referred to as the *quality function deployment* (QFD). More specifically, in the document of the Big Three, QFD is decomposed into two dimensions: *quality deployment*, which consists in translating customer requirements into Product Design Requirements, and *function deployment*, which amounts to translating design requirements into appropriate Part, Process and Production Requirements. One can readily rephrase the above definition so that it fit the service industry context.

In the example of Section 5.5, the situation was rather simple. Hints gained from market research have led to a single most important substitute characteristic, and a clever use of the cause and effect diagram sufficed to deploy further stages of QFD. Essentially, one was faced with the need to improve existing processes. In general, e.g., when a true innovation or a breakthrough in design is sought, QFD requires more effort. In fact, already defining the true quality characteristics may be far from obvious.

In particular, it is a different story to think of improving one of the products or services delivered by a well established company with its rather stable market share and to undertake an effort to gain an edge in the market and substantially enlarge this very company's share in some way. It is also not easy to formulate a strategy for achieving a good and stable position in the competitive market by a newly established company which, in its relentless efforts to gain new customers, has been caught by too many challenges and has fallen into the state of fighting fires which spread everywhere in the company almost without a break (it is still another story that had the company started its operation from the outset with a sound program for continual improvement, it would have not ended in the state mentioned). Moreover, it is not always easy to see why a seemingly good product or service fails to bring customers, and profit, to a company. What is wrong with a French bakery and confectionery with a café, whose location is good for this type of product, quality of food is very good, prices are OK, no other offer of the same type can be found around and yet the owners complain that only a few customers come?

There are many ways of providing precise structure into planning a change in general, when the issue of what to accomplish needs to be worked out from scratch. We shall discuss briefly a planning model as described in [6]. In the model, the Plan stage of the PDSA cycle is decomposed into four phases – issue definition, analysis of action, organization of action and contingency analysis. The plans are developed using the so-called *seven management and planning tools*, also known as the *seven (Japanese) new tools*. These are: affinity diagram, relations diagram or interrelationship diagraph, tree or systematic diagram, matrix diagram, arrow diagram, process decision program chart (PDPC; sometimes another tool for contingency analysis is considered one of the seven tools) and matrix data analysis (in [6] a glyph is mentioned instead). Needless to say, the Plan stage requires work by multi-disciplined teams if the efforts are to be successful.

In planning phase I, devoted to defining the issue, the affinity diagram and relations diagram are of help. Clearly, at the very start of any project, there has to be clear and common understanding and agreement of what is to be accomplished. Common agreement is needed as to the issue statement. Given that, information about the situation has to be gathered and organized into logical groupings. This information is mostly if not only verbal but, of course, it is welcome in the form of numerical data whenever such are available.

Already the issue statement is not a trivial matter. It should facilitate thorough exposition of all the relevant issues and problems, not immediate answers, let alone proposals for action. First a detailed identification of problems is needed. It is noticed in [6] that good issue statements begin with phrases like: *factors that influence ...*, *elements of ...*, *what are the issues involved in ...?*, *what would be the characteristics of success of ...?*, *what makes ... effective?*, *what are the barriers/problems involved in ...?*

Having agreed on the issue statement, a team has to turn to generating and organizing information on that issue. It is the *affinity diagram* which helps in the matter. By open brainstorming team members come up with as many ideas as possible. The best way is to write them all on separate cards, to be referred to as the idea cards. In an organized manner, the idea cards are in turn grouped into logical clusters. This is an iterative process, with clusters sometimes broken up and recombined whenever appropriate. Within a cluster, subclusters can be built as well. Clusters consisting of just one idea are also a possibility. In turn, clusters are defined, that is, are given headings which encompass the meaning of the cluster (i.e., the common meaning of the idea cards in it). The headings are written down on the so-called header cards, so that the whole diagram could be posted on a wall or placed on a table in a neat and diagrammatic form. If, despite a long discussion, the heading for a cluster cannot be agreed upon, the team should move to another cluster. If, after returning to the cluster later, the team still cannot come up with a heading capturing the meaning of the cluster, the cluster has to be broken up and some logically consistent solution found. When ready, an affinity diagram may have the form like that sketched in Figure 5.25.

The affinity diagram, just as the rest of the seven new tools, can be applied to help provide a structured plan for solution of any problem faced by a company, whether it is a strict sense technical problem, that concerning a service segment or an organizational problem. The authors do not know of a company which, after having decided to implement the paradigm of continual improvement, did not turn to problems with internal cooperation within the company.

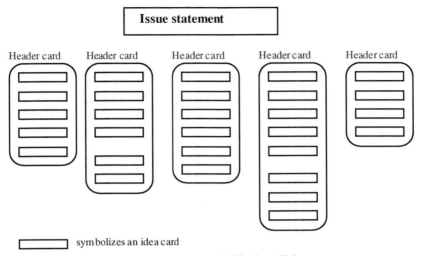

Figure 5.25. Affinity Diagram.

In one such company, a team has come up with an affinity diagram comprised of the following 5 clusters (in brackets, some or all problems which were included in a cluster are given): problems with superior - subordinate relations (lack of respect for a subordinate, hidden behind a mask of courtesy; distaste for meeting and discussion with subordinate; orders without explanation how to achieve the goal set; no cooperation with superiors; fear against asking a superior how to fulfill an order; frequent changes of superior's opinion on people and quality of work; frequent changes of orders and task assignments, seemingly without reason; subordinates' inclination to cheat; and others); problems with flow of information (superiors provide incomplete information; people learn of problems of their concern from peers, not from superiors; people receive incomplete instructions; information reaches rank-and-file with delay); poor organization of a company (a subordinate happens to receive contradictory orders from immediate and higher superiors; people responsible for one process impose their will on those working on other processes; lack of cooperation between departments); problems with peer-to-peer relations (peers treated as intruders; insensitivity to problems communicated by a peer working on the same or related process); personal flaws (blindness to one's own errors; inconsistency between words and deeds; distaste for cooperation; lack of creativity; insistence on repeating the same errors). Interestingly, the team proved unwilling to ascribe lack of creativity to poor relations between superiors and subordinates or some other flaw in the company as a whole and decided to include it in the last cluster.

The *relations diagram* or *interrelationship diagraph* has been developed to identify and describe cause and effect relationships between components of an issue or a problem. It does not have to but can be, and usually is, obtained from an affinity diagram. If it is, one begins by displaying the issue statement above affinity headings arranged in a circle, as depicted in Figure 5.26.

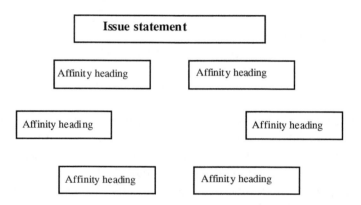

Figure 5.26. Towards Relations Diagram.

Each affinity heading describes some (clustered) idea (to be referred to as a *factor*). Now, one draws lines connecting related factors, and adds an arrow head to each line, thus indicating the direction from cause to effect. Once all such lines have been drawn, one includes at the bottom of each factor the number of arrows pointing away and pointing toward the factor (i.e., number away/number toward). This is the way to identify ideas which are the key cause factors (these are the factors with the most arrows pointing away from them) and those which are the key effect factors (i.e., the factors with the most arrows pointing toward them). In Figure 5.27, the key cause factors are those with 6 arrows pointing away and the key effect factor is the one with 6 arrows pointing toward it. As a rule, the key cause factors are marked by heavy lines around them.

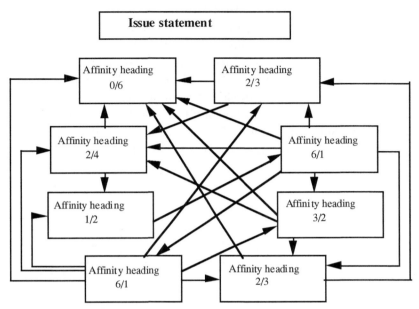

Figure 5.27. Relations Diagram.

We encourage the reader to return to our example with the French bakery and confectionery with a café and create a plausible affinity diagram for finding factors which influence the lack of enough customers. Clearly, one can think of clusters associated with service both in the shop and in the café, with the menu in the café, product variety in the shop, indoor and outdoor outlook, advertising, etc. Given the affinity diagram, one can turn to the relations diagram.

It is always a good idea to display the factors from left to right, with the key cause factors at the left end and the key effect factors at the right one. Of course, many variations of the above description of the affinity diagram can be used to advantage. For example, sometimes it may be recommended to begin with all idea cards from the affinity diagram, not only with the header cards. In any case, the obvious purpose of the relations diagram is to help plan priorities of actions.

This brings us to the second phase in planning, namely to the analysis of action or to answering the question what to do to accomplish an objective. The *tree* or *systematic diagram* is a tool to facilitate the task. If the affinity and relations diagrams were developed earlier, it is the latter which helps select the objective of immediate interest. The former can then be used to start constructing the tree diagram by referring to the ideas from the affinity cluster which has been chosen as the objective to deal with. The tree diagram is begun from the left and goes to the right, from general to specific. The initial (most general) objective is placed leftmost. Given an objective at a certain level of the hierarchy, one asks what actions to take to achieve this objective or what are this objective's subobjectives. In this way, the ideas displayed in the diagram go not only from general to specific but also from objectives to actions. The tree, as sketched in Figure 5.28, should be complete, that is, it should include all the action items needed to accomplish the main objective.

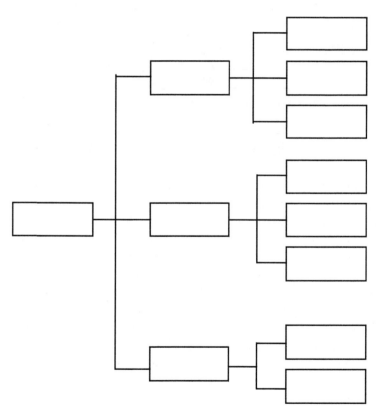

Figure 5.28. Tree Diagram.

In particular, if a rightmost item is an objective, not an action, the branch with this item is not complete — it has to be continued to the right, to answer how to achieve the objective in question. On the other hand, the tree should be pruned if some branches prove, upon final scrutiny, not necessary for achieving the main objective.

One should note that the difference between an objective and an action may not be clear from the outset. If, e.g., the rightmost item in a branch is "lessen the working load of truck drivers," while it sounds like an action, it is usually to be considered an objective – ways how to lessen the load mentioned are the actions which should be put at the end of the branch. It is good to consider actions as a means of achieving some objective. A means for achieving an objective, then, becomes an objective for the next level of the tree, as displayed in Figure 5.29, the process being continued for as long as found desirable.

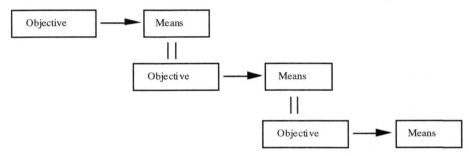

Figure 5.29. A Way to Develop Tree Diagram.

In turn, once it is agreed what to do, it is necessary to decide how to do what is to be done. Thus, phase III, i.e., organization of action, comes into play. In this context, one can mention the *matrix diagram* and *arrow diagram* as tools whose purpose is to help organize action. The matrix diagram has in fact much wider applicability. We shall not dwell on it, but the reader will readily realize that the matrix diagram can, for example, be used to relate true quality characteristics to substitute characteristics or to set priorities. If one aims at relating whats with hows, one can use the so-called L-matrix, given in Figure 5.30, where the symbols used (which can also be numerical scores) describe the strength of relationship between two factors involved. Whats and hows can be considered rather loosely. For instance, an L-matrix can be the responsibility matrix, when whats amount to tasks and hows to people who can have primary responsibility for a given task, secondary responsibility or who should only be kept informed on how the task is being performed (the given levels of responsibility can then be represented by the symbols in the figure).

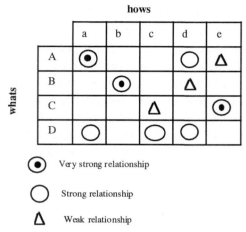

Figure 5.30. L-Matrix.

A more complex matrix can be used if one wants to see the relationship between two characteristics given a third one. For example, using the so-called T-matrix, one can relate simultaneously problems (factors) of our concern, countermeasures needed (whats) and action taken (hows), as sketched in Figure 5.31. In the figure, for illustration, only one combination of the three variates of interest is described. It follows for this combination that, if we are interested in factor or problem d, countermeasure 1 has a strong influence on this factor and, in turn, this countermeasure can very well be implemented by action E.

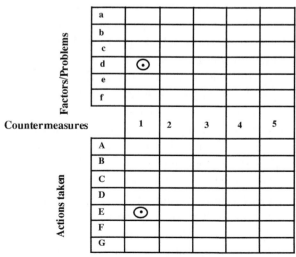

Figure 5.31. T-Matrix.

The *Arrow diagram* is an extremely efficient tool for providing a schedule for completing a project or process. In fact, it is a flowchart with explicitly displayed time to complete a task (Time in Figure 5.32), the earliest time the task can be started (EST in Figure 5.32) and the latest time the task can be started so that the whole project is finished on schedule (LST in Figure 5.32). If EST=LST for a given task, then there is no slack time for this task. Such tasks form a critical path (see thick line in the diagram).

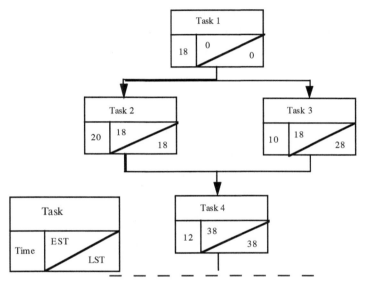

Figure 5.32. Arrow Diagram (Initial Part).

No project is immune to contingencies. Accordingly, carefully developed plans are needed either to prevent potential problems or to arrange for quick recovery from problems that do arise. The tools used most often to construct such plans are *process decision program chart* and *failure mode and effects analysis.* We shall skip their discussion and confine ourselves to referring the reader to [6] for a brief description of the former and to [4] for a comprehensive exposition of the latter (in the context of the automobile industry).

The last of the seven managerial and planning tools is the *matrix data analysis.* By that, one usually means a simple graphical tool which depicts bivariate data in a Cartesian plane. The data can, for example, be two dimensional quality characterizations of shops which are considered by our French bakery and confectionery as potential subcontractors to sell its product in some region. If we want to characterize potential sub-contractors by more than two characteristics, we can use, e.g., a *glyph*. There are several versions of glyphs, and we shall describe only one of them.

Let a circle of fixed (small) radius be given. Further, let each characteristic be represented by a ray emanating from the circle, the rays being equally spaced in the plane. Now, to each subcontractor and to each of its characteristics there corresponds some value of this characteristic which can be represented as a point on the corresponding ray. Joining up points representing a subcontractor, one obtains a polygon describing this subcontractor. For the sake of clarity, in Figure 5.33 polygons for only two subcontractors are displayed.

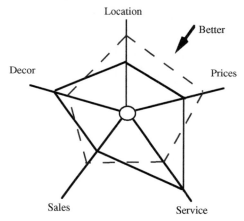

Figure 5.33. Glyph.

A glyph is thus a simple but efficient graphical means to compare objects described by multidimensional quality characteristics. If too many objects are to be displayed around one circle to retain clarity of the picture, each object can be assigned a separate circle (of the same radius) and the polygons obtained for the objects can be juxtaposed in some order. Let us note that the quality characteristics do not have to be numerical. If they are not, some numerical scores have to be assigned to possible "levels" of a characteristic. Whether characteristics are numerical or not, when interpreting a glyph one has to be careful about scales used for different characteristics. It is best when the scales can be made at least roughly equivalent. In our discussion of the seven new Japanese tools we have essentially followed [6], i.e., the way they are taught by the British experts whose approach is inspired by the theories of W. Edwards Deming. Needless to say, there are differences, however minor, in how different authors see and use these tools. For instance, the experts of the Japanese Union of Scientists and Engineers prefer a slightly different construction of the relation diagram. In any case, however brief, the given account of the seven managerial and planning tools already suffices to show the reader their great explanatory and analytic power. No wonder that they have become an indispensable set of means for planning purposes.

References

[1] Efron, B. (1979). "Bootstrap methods—another look at the jack-knife," *Annals of Statistics*, pp. 1-26.

[2] Martz, Harry F. and Waller, R. A.(1979). "A Bayesian zero-failures (BAZE) reliability analysis," *Journal of Quality Technology*, v.11., pp. 128-138.

[3] Martz, H. F. and Waller, R. A. (1982). *Bayesian Reliability Analysis*. New York: John Wiley & Sons.

[4] *QS-9000 Quality System Requirements*. (1995) Chrysler Corporation, Ford Motor Company, and General Motors. See in particular parts 7 and 4: *Advanced Product Quality & Control Plan* and *Potential Failure Mode Effects Analysis*.

[5] Simon, J.L.(1990). *Resampling Stats*. Arlington: Resampling Stats, Inc.

[6] *The Process Manager*. (1996) Process Management International Ltd.

[7] Thompson, J. R. (1985). "American quality control: what went wrong? What can we do to fix it?" *Proceedings of the 1985 Conference on Applied Analysis in Aerospace, Industry and Medical Sciences*, Chhikara, Raj, ed., Houston: University of Houston, pp. 247-255.

[8] Thompson, J.R. (1989). *Empirical Model Building*. New York: John Wiley & Sons, pp. 133-148.

[9] Thompson, J.R. and Walsh, R. (1996)"A Bayesian Pareto analysis for system optimization" in *Proceedings of the First Annual U.S. Army Conference on Applied Statistics*, pp. 71-83.

[10] Thompson, J.R. (1999). *Simulation: A Modeler's Approach.* New York: John Wiley & Sons, pp. 221-222.

[11] Tukey, J.W. (1977). *Exploratory Data Analysis.* Reading: Addison-Wesley.

[12] Waller, R. A., Johnson, M.M., Waterman, M.S. and Martz, H.F. (1977). "Gamma Prior Distribution Selection for Bayesian Analysis of Failure Rate and Reliability," in *Nuclear Systems Reliability Engineering and Risk Assessment,* Philadelphia: SIAM, pp. 584-606.

Problems

Remark: In Problems 5.1-5.3, a statistical software of the reader's choice has to be used. One of the aims of solving these problems is to gain a better understanding, and appreciation, of random phenomena. In particular, the problems show that statistical inference requires special caution when the size of a sample is small. Whenever a histogram is to be constructed, the reader is asked to use a default bar width first, and then to try using several other widths of his or her choice.

Problem 5.1. Generate two independent samples of size 100 each from the standard normal distribution.

a. Construct the schematic plots, normal probability plots and histograms of both samples.

b. Transform one of the original samples to a sample from $\mathcal{N}(1,1)$, and combine the sample obtained with the other original sample into one sample of size 200. Construct the schematic plot, a normal probability plot and a histogram for the last sample.

c. Transform one of the original samples to a sample from $\mathcal{N}(5,1)$, and combine the sample obtained with the other original sample into one sample of size 200. Construct the schematic plot, a normal probability plot and a histogram for the last sample.

d. Transform one of the original samples to a sample from $\mathcal{N}(0,.01)$, transform the other sample to a sample from $\mathcal{N}(1,.01)$ and combine the two samples obtained into one sample of size 200. Construct the schematic plot, a normal probability plot and a histogram for the last sample.

e. Compare the results for **a** through **d**. Repeat the whole experiment several times to find what is typical of the results and what is rather due to chance.

f. Repeat the whole experiment (**a** through **e**) starting with two samples of size 50 from the standard normal distribution.

g. Repeat the whole experiment (**a** through **e**) starting with two samples of size 25 from the standard normal distribution.

h. Construct the schematic plots, normal probability plots and histograms for the data set x_1 of Table 3.2, Table 3.3 and Table 4.3. Comment on the results obtained.

Problem 5.2. Consider the following sample of random variates drawn from an unknown probability distribution.

Lot	x	Lot	x
1	0.947	26	0.421
2	2.468	27	0.108
3	0.937	28	0.305
4	1.707	29	1.103
5	4.592	30	2.683
6	6.744	31	0.407
7	0.618	32	0.949
8	1.564	33	3.686
9	1.954	34	1.541
10	0.489	35	0.731
11	2.701	36	4.281
12	0.417	37	0.328
13	0.302	38	2.460
14	0.763	39	2.086
15	0.163	40	0.233
16	0.647	41	2.137
17	0.365	42	1.046
18	0.590	43	3.294
19	0.401	44	1.173
20	1.258	45	2.011
21	0.764	46	0.236
22	0.497	47	4.794
23	1.246	48	1.435
24	0.709	49	3.997
25	2.136	50	3.066

a. Use the schematic plot to verify that the data distribution is skewed. Try to symmetrize the distribution transforming the data to its square root, fourth root, natural logarithm and, possibly, using other transformations.

b. Use the normal probability plot to verify whether the data distribution can be readily transformed to an approximately normal distribution.

c. Given that the data comes from a lognormal distribution, is it possible to find the transformation that transforms the data to a normally distributed sample?

Problem 5.3. Consider the following sample of random variates drawn from an unknown probability distribution.

Lot	x	Lot	x
1	0.003	26	0.750
2	0.816	27	4.966
3	0.004	28	1.410
4	0.286	29	0.010
5	2.323	30	0.974
6	3.643	31	0.806
7	0.232	32	0.003
8	0.200	33	1.702
9	0.449	34	0.187
10	0.511	35	0.098
11	0.987	36	2.115
12	0.764	37	1.239
13	1.435	38	0.810
14	0.073	39	0.540
15	3.293	40	2.124
16	0.190	41	0.576
17	1.013	42	0.002
18	0.278	43	1.421
19	0.833	44	0.025
20	0.053	45	0.488
21	0.072	46	2.084
22	0.488	47	2.457
23	0.049	48	0.130
24	0.118	49	1.919
25	0.576	50	1.255

a. Use the schematic plot to verify that the data distribution is skewed. Try to symmetrize the distribution attempting several transformations of the data.

b. Use the normal probability plot to verify whether the data distribution can be readily transformed to an approximately normal distribution.

c. Given that the data comes from a χ^2 distribution, is it possible to find the transformation that transforms the data to a normally distributed sample?

Problem 5.4. Preliminary analysis of a criterion function of an unknown analytical form has to be performed. Function values can be observed, but each observation is corrupted by a random error. Below, the set of such "noisy" observations is given for the function's argument, x, varying from .1 to 3.

x	$f(x)$	x	$f(x)$
0.1	0.816	1.6	0.231
0.2	0.511	1.7	0.606
0.3	0.561	1.8	0.510
0.4	0.339	1.9	0.730
0.5	0.220	2.0	1.048
0.6	0.345	2.1	1.074
0.7	0.142	2.2	1.295
0.8	-0.108	2.3	1.646
0.9	0.287	2.4	1.874
1.0	-0.188	2.5	2.257
1.1	-0.110	2.6	2.576
1.2	-0.008	2.7	2.829
1.3	0.177	2.8	3.417
1.4	0.343	2.9	3.568
1.5	0.182	3.0	4.130

Use the 3R smooth algorithm to remove the jitters. Given that the data constitute noisy observations of the quadratic function $(x-1)^2$, does the smooth enable one to better approximate the true shape of the function?

Problem 5.5. In the following table, noisy observations of the quadratic function $(x-1)^2$ are given. For each x, the observation is in fact the sum of the function value and a normally distributed random variate with mean 0 and standard deviation .3.

Use the 3R smooth to remove the jitters. Does the smooth enable one to better approximate the function? Given that the noise in Problem 5.4 was normally distributed with mean 0 and standard deviation .1, compare the two results of using the smooth.

x	$f(x)$	x	$f(x)$
0.1	0.831	1.6	0.121
0.2	0.747	1.7	0.909
0.3	0.437	1.8	0.832
0.4	0.347	1.9	1.071
0.5	0.722	2.0	0.487
0.6	0.199	2.1	0.463
0.7	-0.038	2.2	1.304
0.8	-0.408	2.3	1.232
0.9	0.787	2.4	2.105
1.0	0.396	2.5	1.767
1.1	0.123	2.6	3.088
1.2	-0.170	2.7	2.676
1.3	-0.291	2.8	3.673
1.4	-0.093	2.9	3.617
1.5	0.307	3.0	3.690

Problem 5.6. Use bootstrap tests for the mean and standard deviation to determine which lots in the highly contaminated data set in Table 3.3 are out of control.

Problem 5.7. We have a system consisting of five independent modules which experience failures according to Poissonian flow. We incorporate our prior information about the distribution of θ in the form of Gamma distributions appropriately parameterized. Each of these has prior expectation of 4 and variance also 4 for modules 1−4. For module five, the prior expectation is 2 and the variance is 6. The failure rates are given, for the first eight time intervals by:

Table 5.6								
Module	z_{i1}	z_{i2}	z_{i3}	z_{i4}	z_{i5}	z_{i6}	z_{i7}	z_{i8}
1	2	1	2	24	15	2	2	0
2	0	1	2	0	1	2	0	1
3	1	0	0	1	1	1	2	1
4	0	0	0	2	1	1	1	0
5	4	1	5	6	7	5	6	8

Create time lapsed exponentially weighted Pareto charts for the first eight time intervals for each module.

Problem 5.8. Return to the problem of the French bakery and confectionery with a café from Section 5.7 and create a plausible affinity diagram for finding factors which influence the lack of enough customers. Given the diagram, complete the Plan stage of the PDSA cycle for finding ways out of the problem.

Chapter 6

Optimization Approaches

6.1 Introduction

Our activities to this point have largely been involved in the utilization of time indexed glitch information to find qualitative causes of excess variation and fix them. Our principal goal has been the reduction of variability. Our modes of intervention might be typically qualitative in nature, such as providing training for a joint replacement team or replacing faulty bearings. In other words, the use of control charts can be looked upon as a way to bring about conformity to the system as it was designed, always keeping an eye open to the possibility that improvements in the system over those suggested by the design are possible. And, speaking realistically, control charts are useful in improving systems which, if truth be told, were never fully designed, for which no complete flow chart exists. Our experiences in the United States and Poland indicate that such systems make up a large fraction of those in operation.

But what of "fine tuning"? That is, how shall we decide how to make small quantitative changes in the level of control variables in order to improve a process? And we recall that this improvement may be measured by an infinite number of possible criteria. These could include, say, a decrease in the deviations of the output product from specification. Or, we might use per unit cost of the output product. Or, we might use amount of product produced per unit time. And so on. Typically, we will be looking at several criteria simultaneously.

In the best of all possible worlds, optimization by changing continuous control variables would be well posed as a constrained optimization problem. For example, we might seek to maximize our total product output subject to the constraint that average squared deviation from product specification was below a specified amount and unit cost was kept below a certain level. Life is usually not so simple. A natural criterion function is generally not available. And it is frequently a mistake to attempt a strategy which pits quantity of output against quality. To view statistical process control as, somehow, a branch of "control theory," i.e., that field of engineering and applied mathematics which deals with well posed problems in constrained optimization, is a serious mistake, which can lead to disaster.

To utilize the "Control Theory" approach most effectively, we need to make a number of assumptions of "givens." These include but are not limited to:

1. A natural and well posed criterion function to be maximized (or minimized).

2. A natural set of constraints for various output variables.

3. A well posed mathematical model of the mechanism of the process.

Our experience indicates that it is seldom the case that any of these "givens" is actually available in other than a crude approximation form. Perhaps that is the reason that control theoretic approaches to SPC have usually proved to be so disappointing.

Returning again to the "best of all possible worlds" scenario, let us suppose that we have given to us a function of k variables subject to m constraints:

$$\text{Max } f(X_1, X_2, \ldots, X_k) \qquad (6.1)$$

subject to

$$g_i(X_1, X_2, \ldots, X_k) \leq 0 \; ; \; \text{i=1,2,\ldots,m} . \qquad (6.2)$$

In the rather unusual situation where we explicitly know f and the m constraints, there are a variety of algorithms for finding the value of $(X_1, X_2, \ldots, X_k)$ which maximizes f subject to the constraints (see, e.g., Thompson and Tapia [10], 253-286).

In the real world, we will usually not know the functional relationship between f and $(X_1, X_2, \ldots, X_k)$. In many situations, if we input specific values of the control variables $(X_1, X_2, \ldots, X_k)$, we will not be able to see f explicitly, but rather a value of f corrupted by experimental error. In other words, we will usually not know the function f and will only be able experimentally to observe pointwise evaluations of f corrupted by noise.

Let us then consider the model

$$Y(X_1, X_2, \ldots, X_k)(j) = f(X_1, X_2, \ldots, X_k) + \epsilon(X_1, \ldots, X_k)(j) \quad (6.3)$$

where $\epsilon(X_1, X_2, \ldots, X_k)(j)$ is the error at time j at a particular level of $(X_1, X_2, \ldots, X_k)$. We shall assume that the errors have zero averages and a variance which may be dependent on the level of the control variables $(X_1, X_2, \ldots, X_k)$. We shall assume that the lot samplings are sufficiently separated in time that an error at any time j is stochastically independent of an error at any other time i.

So, then, the situation is that we can construct an experiment, by setting the levels of $(X_1, X_2, \ldots, X_k)(j)$ to particular values and then observing a somewhat degraded version of $f(X_1, X_2, \ldots, X_k)$, namely $Y(X_1, X_2, \ldots, X_k)(j)$. We note here that we have avoided assuming as known the functional form of f. One possible f of interest would be the variability of the output process. This is a quantity we would surely wish to minimize.

The situation confronting the production staff would be, typically, that they believe they know roughly the value of $(X_1, X_2, \ldots, X_k)$ which gives good performance. The notion that the people in production will be ecstatic over anybody's notion that we should "fool around" with the process in order to achieve some improvement is naive. We recall that we observed that once a process was "in control," typically it was not a great deal of trouble to keep it in control. Changing the settings of key variables can be very risky. We have never experienced any production foreman who was keen on departing from the notion that "if it ain't broke, don't fix it."

If we knew a functional relationship between process variability and various variables whose levels we are free to control, then the optimization process would simply be numerical. Naturally, we would probably have a function sufficiently complex that we would require a computer to find numerically the value of the control variables which minimized the variance. That means, that to find the value of the function for a given value of the control variates, we would need to input the control variables into a computer. But in the real world we do not generally expect to know a good mathematical model of the variance. So, to find the variance for a given level of the control variables, we actually would need to use the process itself as our computer, namely we would have to run the process at the level of the control variables.

If the records of the process have been carefully recorded in user friendly fashion, then we may have some feel as to what levels of the control variables are likely to minimize the variability of the output process. There are always glitches in the control variables, times when their values are different from the standard. Perhaps at some of these glitches, it was noted that the variance appeared to be less. If such information is available, then it should be utilized in our search for minimum variance conditions.

6.2   A Simplex Algorithm for Optimization

In the following example, we shall deal with a continuous chemical reaction running in aqueous medium,

$$M + N \Rightarrow O .$$

For this reaction, the input compounds M and N are converted into product O. The conversion is, essentially, complete. But there are small traces of other, nonharmful compounds, which are by-products of the reaction, and these occur in very small amounts. This affects, somewhat, the output concentration of O (target value is 50 parts per thousand). The major factors in the reaction are the amounts of the input compounds introduced, and we shall not be dealing with these in this study. However, it has been suggested by some of the production staff that changes in the temperature and/or acidity (pH) of the solution might produce a less variable output concentration of O. Current operating conditions are 30 degrees Celsius and a pH of 7.2. We have observed, in the past, operating temperatures as low as 20 and as high as 40, and pH values as low as 5 and as high as 8.5. It appears that within these ranges the reaction runs essentially to completion and that satisfactory product is produced. But we would like to see what might be done to minimize lot to lot variation, by selecting "optimal" values for temperature and pH. How shall we proceed to find the optimal value of (T,pH) (if, indeed, there is one)?

The first approach we shall use is a bit in the time honored engineering approach of "flailing around" (called "random search" by some). But, rather than using random search, we shall employ the more orderly "simplex method" of Nelder and Mead [6] (not, of course, to be confused with the quite different simplex algorithm of linear programming). In order to proceed, we need to have collected lot output data for one more set of settings of the control variables vectors than is the dimension of the control variable. Here, the dimension of the control variable to be optimized is two (Temperature and pH). So we shall use three different settings and collect at each setting, say, ten lot measurements of output in parts per thousand. The (Temperature, pH) settings initially are (30, 7.2), (40, 6.5), and (35, 6.0). We note that, as it has turned out, the highest sample variance corresponds to (30, 7.2), so we shall designate this as the W(orst) point on the graph in Figure 6.1. The second highest variance was obtained in the lots with (Temperature, pH) set to (40, 6.5); so we shall call this Second Worst point, 2W. The lowest variance was obtained for (35, 6.0), which we designate in Figure 6.1 as the B(est) point.

Now the Nelder-Mead approach is to replace the W(orst) point, by another, superior set of conditions, which we find in an orderly fashion. So, next, let us find the midpoint of the line segment joining the B and 2W points and call this point the C(entroid) of all points except for the W(worst). We would like to go away from the W(orst) point to something better. So, let us proceed from W to C (with coordinates, here, of (37.5, 6.25)) and then continue the line to P, which is as far from C as C is from W. This point, P(rojection), with coordinates (45, 5.3) gives the conditions for our next experiment. In Table 6.1, we note that the variance has increased to a level which, not only is not less than that for B, but is actually worse than that for W and 2W. Accordingly, we move to point PP, the midpoint between points W and 2W (and the rest goes as is), carry out ten experiments, and, happily, find that the variance has dropped to a new low.

Shortly, we shall describe the Nelder-Mead "enlightened search" algorithm more generally, but here we only make a few observations about it. First of all, we note that the algorithm does not avail itself of the actual values of the variances at W, 2W, B, P and C, only whether the values there are greater or lower when compared internally. Thus, we are sacrificing some information in order to obtain robustness of the algorithm. Those familiar with Newton method approaches will note that we have denied ourselves the possibility of making the giant steps to the optimum associated with Newton methods. We do this, in part, because of the contamination of the underlying function value (the variance) by noise, and also because we do not know the functional relationship between the variance and the temperature and pH, so that derivative information is not available. Next, although we have chosen not to proceed with the algorithm further here, it is clear that, typically, many iterations will be required to arrive at a (local) minimum variance condition. The simplex algorithm is one of slow envelopment and collapsing toward an optimum. Recall that we are not evaluating a function on a computer. We are carrying out a real-world experiment. The Nelder-Mead approach is, frequently, a rough and ready method for making a satisfactory process better.

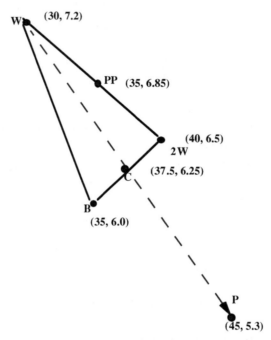

Figure 6.1. Nelder-Mead Experimental Optimization.

Table 6.1					
Points	W	2W	B	P	PP
1	46.6	48.1	49.5	52.8	50.6
2	52.5	49.1	51.2	45.8	48.4
3	47.2	47.4	48.3	51.5	50.7
4	52.9	48.4	49.5	54.1	50.8
5	51.8	52.3	52.2	46.0	49.1
6	52.0	51.6	51.8	46.4	51.1
7	50.9	48.2	49.4	54.3	50.7
8	49.5	51.9	52.2	50.4	48.5
9	49.3	50.8	48.1	51.5	48.9
10	46.9	52.2	47.7	47.2	51.3
$\bar{x}$	49.960	50.000	49.990	50.000	50.010
s^2	5.827	3.769	3.001	11.360	1.301

Let us note also that, obviously, the control variables are never allowed to assume all values from minus to plus infinity. Most often, they are confined to finite intervals. The requirement that the Nelder-Mead algorithm never leave an admissible set of values of the control variables can be satisfied in a very simple way. Whenever a function evaluation outside of the admissible set is needed, the function value at such a point is declared to be "worse" (e.g., smaller when function maximization is considered) than the function value at W.

Thirty-five years ago, an evolutionary method of operational improvement of various production criteria was proposed by Box and Draper [2] and given the name of Evolutionary Operations (EVOP). It was to be a means of continuing institutionalized optimization of processes within a plant. The production staff would simply expect, all the time, to take processes in control and jiggle them, so that improved strategies could be devised. Although the idea behind EVOP is quite intriguing, in retrospect it should not be surprising that EVOP is underutilized all over the world. Jiggles happen, and information obtained when they do occur should be carefully recorded and utilized. But production staffs are entitled to expect that most production time will be spent in an "in control" situation. To expect production personnel to function almost continuously "on the edge" is not particularly reasonable or desirable.

The kind of improvement we have been able to obtain in our brief example, utilizing five settings of (Temperature, pH), contains an essential flaw, which is not uncommon in textbook process optimization examples. Namely, the sample sizes are too small to give realistic recommendations for change in the standard operating values of temperature and acidity. If the observations are normally distributed, then for n lots, $(n-1)s^2/\sigma^2$ has the χ^2 distribution with $n-1$ degrees of freedom. Let us consider the 90% confidence interval for the true value of the variance at the present operating conditions of (30,7.2), $\sigma^2(30, 7.2)$. From the tabulated values of the χ^2 distribution, we can obtain the values of $\chi^2_{.05}$ and $\chi^2_{.95}$ in the chain inequality

$$\chi^2_{.95} \leq \frac{(n-1)s^2}{\sigma^2} \leq \chi^2_{.05} . \tag{6.4}$$

From the χ^2 tables, we then have

$$3.325 \leq \frac{(n-1)s^2}{\sigma^2} \leq 16.919 , \tag{6.5}$$

giving us that we are 90% certain that

$$3.100 \leq \sigma^2(30,7.2) \leq 15.772 . \tag{6.6}$$

The same inequality applied to the sample variance of the point (37.5, 6.25) gives us

$$.692 \leq \sigma^2(37.5,\ 6.25) \leq 3.52 . \tag{6.7}$$

In other words, the 90% confidence intervals for parts per thousand for the old conditions and the suggested improvement overlap. Without more confidence in the suggestions of our data, we are probably ill advised to change the conditions of temperature and acidity. Optimization of a well specified criterion function uncontaminated by noise is a different matter than optimization of noisy experimental pointwise evaluation.

Fortunately, in the kind of problem here, it is very frequently the case that sufficient data is available to overcome the problem of noise. Monitoring of output on a continuous basis is generally available. Moreover, we are well able to analyze very large data sets. In the example under consideration, experience shows that output records separated by one minute are stochastically independent. Thus, during less than one day, over 1,000 output data points are readily available. For the (30,7.2) setting, we show, in Figure 6.2, 1,000 successive output readings.

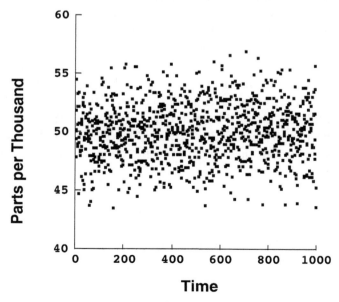

Figure 6.2. Output Under Existing Conditions.

In Figure 6.3, we show the data stream for each set of 1,000 observations under the five conditions considered earlier and recalled in Table 6.2.

Table 6.2					
Points	W	2W	B	P	C
Temperature	30	40	35	45	37.5
pH	7.2	6.5	6.0	5.3	6.25

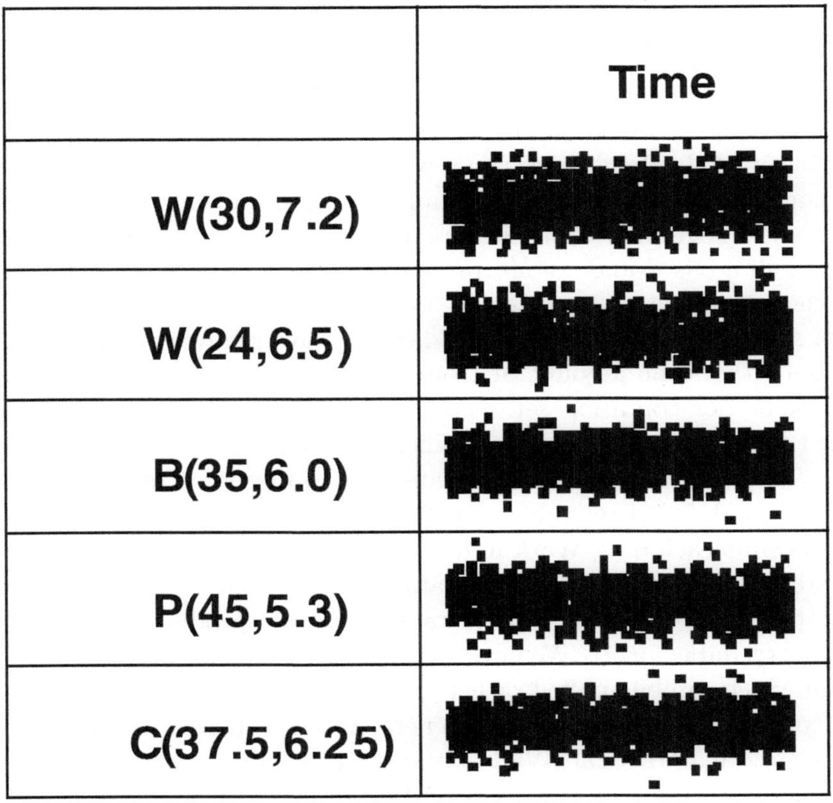

	Time
W(30,7.2)	
W(24,6.5)	
B(35,6.0)	
P(45,5.3)	
C(37.5,6.25)	

Figure 6.3. Output Under Varying Conditions.

We show summary results of our 5,000 readings in Table 6.3.

Table 6.3					
Points	W	2W	B	P	C
$\bar{x}$	49.943	49.963	49.972	49.944	50.067
s^2	5.826	4.109	3.054	12.236	1.345

For n large, the critical values of the χ^2 statistic can be readily obtained using the approximation

$$\chi^2 = \frac{(z + \sqrt{2n-3})^2}{2} \, , \qquad (6.8)$$

where z is the standard normal variate corresponding to the desired confidence level (e.g., for 90%, $z = \pm 1.645$). Thus, for the W point, we have 90% certainty that σ^2 is between 5.42 and 6.28. For the C point, we have 90% certainty that σ^2 is between 1.25 and 1.45. Hence, our recommendation to change the operating conditions of temperature and acidity to 37.5 and 6.25, respectively can be made with some confidence.

In the real world where noise and empiricism replace determinism and exact knowledge of the dynamics of the production system, we must be ever mindful of the dangers we run if we recommend changes based upon what amounts to weak evidence of improvement. Nothing so much undermines the credibility of innovators as a recommendation which turns out to be ill advised and counterproductive. In the example considered here where a modest improvement was obtained as the result of five days of sampling (1,000 points each day) we have really caused production about the same amount of trouble as if we had asked for the 10 observations per condition set. The very fact that the conditions are changed five times will require significant skill in effecting the adjustments. Time will be required for transients to die out after conditions have been modified. Probably, we will wish to make the changes during the same shift each day. Work may be required to port the data out of the system (in many factories, the data stream is continuously recorded on a paper roll, but no arrangements are made to produce the data in computer readable form). We now show a version of the Nelder-Mead algorithm which can be applied rather generally. Although we develop the argument for two control variables, generalization to higher dimensions is rather clear. In a practical EVOP situation, it is unlikely that more than four control variables will be optimized in a given study. Consequently, the rather slow convergence properties of the algorithm for dimensions greater than six are not generally a problem. Although the algorithm is presented here for the two dimensional situation, it may be used generally, if one remembers that the "C(entroid)" point is always taken to be the centroid of all points other than the W(orst). The parameter γ_E is frequently taken to be 1.0, and γ_C to be 0.5. There must be one more point than the dimensionality of the problem, and the initial points must be selected such that none is a linear combination of the others.

A Simplex Algorithm
Expansion Mode
Let $P = C + \gamma_E (C - W)$
If $Y(P) < Y(B)$, then
[a] Let $PP = P + \gamma_E(C - W)$
If $Y(PP) < Y(P)$, then
[c] Replace W with PP as new vertex
Else
[b] Accept P as new vertex
End If
Else
[b] If $Y(P) < Y(2W)$, then accept P as new vertex
Else
Contraction Mode
If $Y(W) < Y(P)$, then
[a*] $PP = C + \gamma_C(W - B)$
If $Y(PP) < Y(W)$, then
[b*] Replace W with PP as new vertex
Else
[c*] **Total Contraction**
Replace W with $\frac{W+B}{2}$ and 2W with $\frac{2W+B}{2}$
End If
Else
Contraction Mode
If $Y(2W) < Y(P)$, then
[aa] $PP = C + \gamma_C(P-B)$
If $Y(PP) < Y(P)$, then
[bb] Replace W with PP as new vertex
Else
[cc] **Total Contraction**
Replace W with $\frac{W+B}{2}$ and 2W with $\frac{2W+B}{2}$
End If
Else
Replace W with P
End If
End If

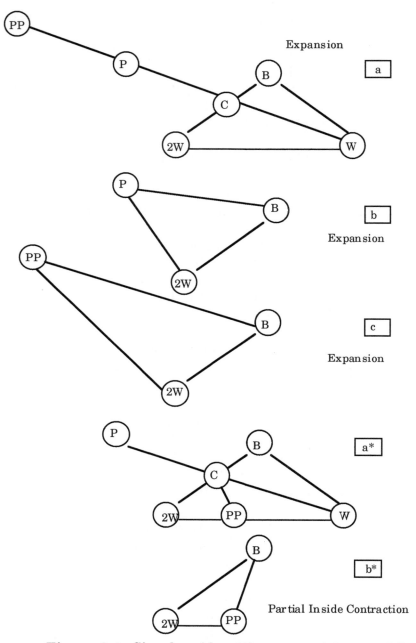

Figure 6.4. Simplex Algorithm. $\gamma_E = 1.0$, $\gamma_C = 0.5$.

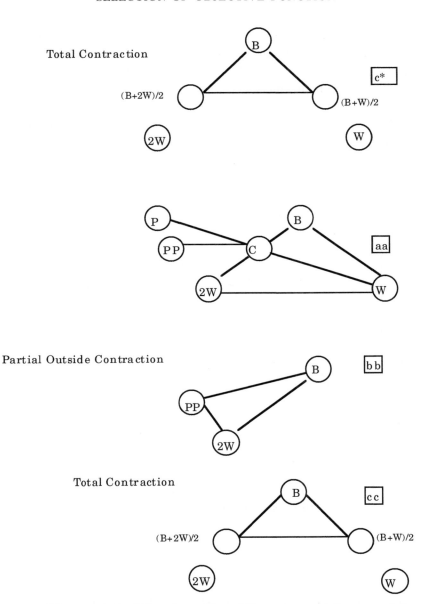

Figure 6.5. Simplex Algorithm. $\gamma_E = 1.0$, $\gamma_C = 0.5$.

6.3 Selection of Objective Function

As a practical matter the selection of the function to be maximized or minimized (frequently called the "objective" function) is seldom very objective. There are the relatively rare situations where we want to maximize some explicit dollar profit function. Even these are seldom completely objective. For example, let us suppose that we wish to maximize profit per item produced. Because of "economies of scale" (i.e., within limits, the per unit cost is generally lower as we increase the volume of items produced), the profit per item profile is frequently as shown in Figure 6.6. Past a limit, the situation usually deteriorates as the volume of goods we produce becomes sufficient to depress the profit margin or we get to the capacity of a factory and further volume would cause large, perhaps rather unpredictable costs.

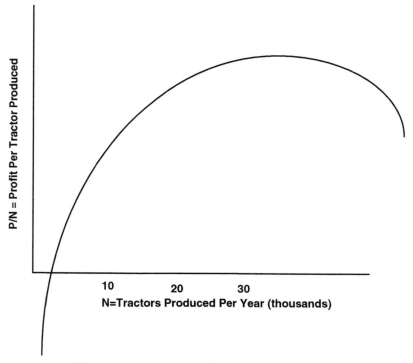

Figure 6.6. Profitability Profile.

If we know the curve in Figure 6.6, then we might decide to use the production level which gives maximum per unit profit. More likely we might choose to maximize total profit:

$$\text{Total Profit} = N\frac{P}{N}. \tag{6.9}$$

As a practical matter, we will not know the P/N curve, and even if we did know it, it would be changing over time. That is the situation in most economic scenarios. If we expand capacity in a rising demand period, what shall we do with excess capacity when the market recedes? The point here is that planning is very hard. If it were not so, then everyone who could carry out a numerical optimization would be rich.

In statistical process control, we are frequently locked into a product, a plant design, etc. No one should expect us, using the same production process, to "upgrade" our automobile production to airplane production. (Although both SAAB and Mitsubishi produce both cars and airplanes, they do so in processes of production specific to the one or the other.)

In the petro-chemical industry (where much of the EVOP philosophy was developed), hardware is not necessarily devised to produce particular products in particular quantities. So, for example, a 100,000 barrel/day petroleum distillation column can produce a wide variety of products from gasoline, to jet fuel, to asphalt, in varying quantities depending on the composition of the input stream and the demands of the market. In such situations, we frequently have optimization problems in which we seek to maximize, say, the quantity of the jet fuel stream, subject to constraints in terms, say, of the quantity and characteristics of all the output streams. In such a case, we probably will not know explicitly the functional form of the objective function or those of the constraints. But we will be in a situation where there is a direct profit related objective function to be maximized.

In most SPC situations, such as the manufacturing of a particular car, our task will be to produce a definite product composed of 10,000 definite parts, with specifications designed years earlier. In such situations, our objective functions will not be directly related to profit, but will rather be related to variability of output and conformance to design specifications.

Let us consider some commonly used objective functions used in the case where we are seeking to minimize departures from conformity to design specifications. The first to be considered is simply the sample variance, as considered in our example of a continuous chemical reaction:

$$s^2 = \frac{1}{n-1} \sum_{i=1}^{n} (x_i - \bar{x})^2. \tag{6.10}$$

We may find it useful to use the sample mean square error from the target μ_0:

$$\text{MSE} = s^2 + (\bar{x} - \mu_0)^2. \tag{6.11}$$

The second term, the square of the *bias*, is generally something which we can adjust essentially to zero by shifting operating conditions. As a practical matter, the hard part in lowering the mean square error is lowering the variance. Hence, there is little to choose between s^2 and the MSE as an objective function.

Sometimes, the argument is made that variance and MSE are not naturally scaled. If there is such a problem, then it can frequently be resolved by dividing the sample standard deviation by the sample mean to give the *coefficient of variation*:

$$CV = \frac{s}{\bar{x}}. \tag{6.12}$$

As a practical matter, the value of $\bar{x}$ will generally change but little as we seek to minimize departures from the target μ_0. Hence, there will be little change whether we use as objective function the sample variance or the square of the CV.

Among other objective functions in common use by the SPC professional, we have the "signal to noise" ratios of the Japanese quality control leader Genichi Taguchi [7].

One of these deals with "smaller is better" situations:

$$\frac{S}{N} = -10\log_{10}(\frac{\sum_{i=1}^{n}(x_i - \mu_0)^2}{n}). \tag{6.13}$$

Another deals with "larger is better" situations, e.g.:

$$\frac{S}{N} = -10\log_{10}(\frac{\sum_{i=1}^{n}\frac{1}{(x_i-\mu_0)^2}}{n}). \tag{6.14}$$

A third objective function of Taguchi is essentially the logarithm of the reciprocal of the square of the coefficient of variation. It is recommended by its creator for "nominal/target is best" situations:

$$\frac{S}{N} = 10\log_{10}\frac{\bar{x}^2}{s^2}. \tag{6.15}$$

Some take great care about selection of the appropriate objective function. There has been interest about using, as an alternative to the sample variance or its square root, the sample standard deviation, the mean absolute deviation:

$$MAD = \frac{1}{n}\sum_{i=1}^{n}|x_i - \bar{x}|. \tag{6.16}$$

The MAD objective function has the advantage of not giving a large deviation squared, and, hence perhaps unduly emphasized, importance. In Figure 6.7, we compare Taguchi's S/N ratio $-10\log(\text{MSE})$ with the MSE for various functional forms of the MSE. Someone who is trying to minimize the MSE reaches a point of diminishing returns, where there is very little difference in the criterion function as we move θ closer to zero. There is no such diminishing returns phenomenon with the Taguchi S/N ratio (which, of course, is to be maximized rather than, as is the case with MSE, minimized). This function goes to infinity for both MSE functions shown here as θ goes to zero. Such an approach appears consistent with a philosophy which seeks to bring the MSE all the way to zero.

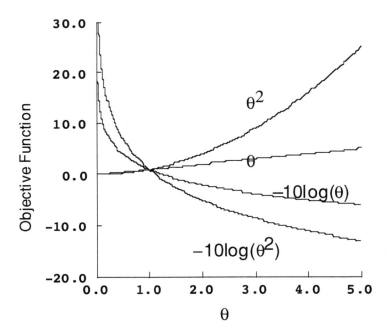

Figure 6.7. A Variety of Objective Functions.

Naturally, it could be argued that it is senseless to attempt unlimited lowering of the MSE. Suppose, for example, there is a "wall," i.e., a boundary below which the error cannot be lessened without fundamental change in design. In Figure 6.8, we consider such situations, using MSE models of the form

$$MSE(\theta) = a + \theta^2.$$

Here, of course, the constant a serves the function of a "wall." The Taguchi S/N ratio, as we note, is not completely insensitive to the phenomenon of diminishing returns, though it is much less so than using simply the MSE.

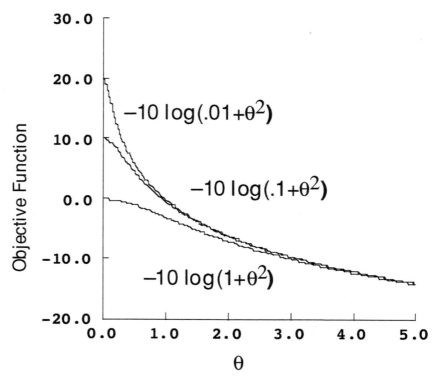

Figure 6.8. A Variety of Objective Functions.

6.4　Motivation for Linear Models

The simplex algorithm for optimization is essentially model free. One proceeds sequentially by experimentation to obtain (generally noisy) pointwise evaluations of the function to be optimized. There are advantages to such procedures, but they frequently require a large number of experiments carried out sequentially over time. Most experimental design is carried out according to a different, more batch oriented, approach. Namely, we carry out experiments in an orderly procedure over a predetermined grid and fit a simple function to it.

In Figure 6.9, we have 20 points to which we have fitted the quadratic

$$Y = 1 - 2X + X^2. \tag{6.17}$$

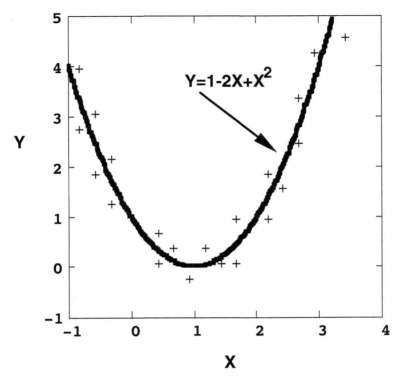

Figure 6.9. A Quadratic Fit to Data.

We can now use the simple optimization approach of taking the derivative of Y with respect to X and setting it equal to zero. This gives us

$$Y' = -2 + 2X = 0 \qquad (6.18)$$

with solution $X = 1$.

Some of the advantages of fitting a quadratic to a set of data and then minimizing the fitted quadratic (or maximizing as the situation dictates) are clear. First of all, the procedure is "holistic." Unlike the simplex approach in which each point was taken as a "stand-alone," the fitting procedure uses all the points to find one summary curve. Generally, this hanging together of the data tends to lessen the effect of noise. Next, if the fitting model is simple, say linear or quadratic, then the optimization algorithms we can use may be relatively efficient. And, of course, by knowing beforehand precisely how many experiments we are going to make, we can tell the production people that we will be inconveniencing them for so long and no longer, as opposed to the situation with the more open-ended simplex strategy. There is no doubt that the "fooling around" strategies, of which the simplex is among the most satisfactory, account for most planned experimentation in industrial systems. But fitted quadratic based approaches are, in general, more appropriate.

We recall from calculus Taylor's formula:

$$f(X) = f(a) + (X-a)f'(a) + \frac{(X-a)^2}{2!}f''(a) + \ldots +$$
$$\frac{(X-a)^{k-1}}{(k-1)!}f^{(k-1)}(a) + R_k \qquad (6.19)$$

where f has continuous derivatives through the kth over the interval $a \leq X \leq b$ and

$$R_k = \int_a^X \frac{(X-t)^{k-1}}{(k-1)!}f^{(k)}(t)dt. \qquad (6.20)$$

The remainder formula R_k is generally unobtainable. However, if f has derivatives of all orders, then an infinite series version of Taylor's formula, one requiring no remainder term is available:

$$f(X) = \sum_{j=0}^{\infty} \frac{f^{(j)}(a)(X-a)^j}{j!} \qquad (6.21)$$

provided, for example, that $|f^{(j)}(X)| \leq M^j$ for all X, $a \leq X \leq b$, all j and some constant M. Recall that the use of Taylor's formula formally requires that we know the precise form of $f(X)$. However, in the case where we do not know $f(X)$ but do have the observations of $f(X)$ for a number of values of X, we might write Taylor's formula as

$$f(X) \approx \sum_{j=0}^{k} A_j X^j. \qquad (6.22)$$

If we have n observations of $f(x_i); i = 1, 2, \ldots, n$, then we might find the values $\hat{A}_j; j = 1, 2, \ldots, n$ which minimize

$$S(\hat{A}_1, \hat{A}_2, \ldots, \hat{A}_k) = \sum_{i=1}^{n}[f(x_i) - \sum_{j=0}^{k} \hat{A}_j x_i^j]^2. \qquad (6.23)$$

Taylor's formula gives many a sense of overconfidence. (This overconfidence has greatly lessened the usefulness of whole fields of research, for example, econometrics.) They feel that it really is not so important whether they actually understand the underlying process. If only enough terms are included in the Taylor's approximation, they believe that empirical data can somehow be used to obtain an excellent fit to reality.

Unfortunately, this sense of confidence is misplaced whenever, as is almost always the case, the observations have a component of noise. To note the ambiguities encountered in practice, we show in Figure 6.9, over the X range of 0 to 3, graphs of

$$X = X, \tag{6.24}$$

$$Y = X + .2\sin(20x) \tag{6.25}$$

and

$$W = X + .2Z \tag{6.26}$$

where Z is a normal random variable with mean 0 and variance 1. Most observers presented with such data would declare that the underlying mechanism for each was, simply, a straight line, and that the Y and W data had simply been contaminated by noise. In general, we will probably be well advised to attempt to use a relatively simple approximating function rather than to entertain notions of a "rich" model. Naturally, if we understand the mechanism of the process so that we can replace some of our ad hocery by a concrete model, we will be well advised to do so. However, in a statistical process control setting, our main criterion function will generally be a measure of the variability of an output variable. Even in cases where we have a fair idea of the mechanism of the process under consideration, the mechanism of the variation will not be perceived very well. Thus, as a practical matter, we will generally be choosing approximating models for their simplicity, stability and ease of manipulation. Generally speaking, these models will be polynomials of degree one or two.

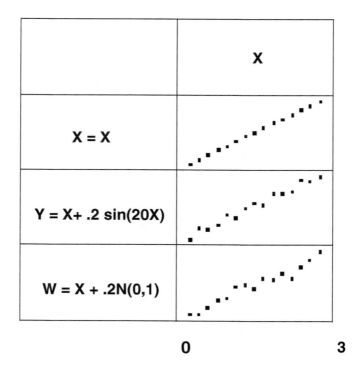

Figure 6.10. Deterministic and Stochastic Functions.

Let us consider the simple exponential function and its Taylor's series representation about $a = 0$.

$$e^X = 1 + X + \frac{X^2}{2!} + \frac{X^3}{3!} + \dots \, . \tag{6.27}$$

In Figure 6.11, we note linear and quadratic approximations to e^x over the interval $0 \le X \le .50$. We note that over the range considered probably the linear and certainly the quadratic approximations will be satisfactory for most purposes. The cubic is indistinguishable from e^x at the level of resolution of Figure 6.11. We should observe that e^x is an "exploding" function when compared with any polynomial. But still, over a modest range, the low order polynomial approximations work rather well.

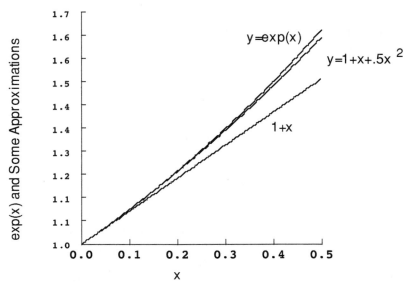

Figure 6.11. Approximations to e^X.

Note the relatively disastrous quality of the three polynomial approximations to e^x in Figure 6.12. Our strategy should be to choose an approximation interval which is sufficiently tight so that our approximation will be of reasonable quality.

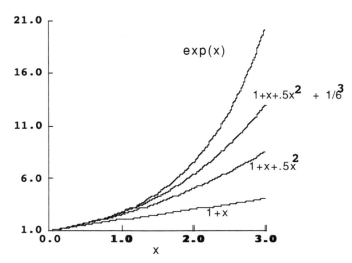

Figure 6.12. Approximations to e^X.

Sometimes, it is clear that a function is growing much faster (or slower) in x than a polynomial of degree one or two. For example we may have a data base like that shown in the X, Y columns of Table 6.4 and graphed in Figure 6.13.

Table 6.4				
X	Y	$\sqrt{Y}$	$Y^{.25}$	$\ln(Y)$
0.10	0.6420	0.8013	0.8951	-0.4431
0.34	1.2820	1.1323	1.0641	0.2484
0.46	1.1672	1.0804	1.0394	0.1546
0.51	1.2751	1.1292	1.0626	0.2431
0.73	1.9933	1.4118	1.1882	0.6898
0.91	1.9022	1.3792	1.1744	0.6430
1.06	2.9460	1.7164	1.3101	1.0805
1.21	3.2087	1.7913	1.3384	1.1659
1.43	4.2880	2.0707	1.4390	1.4558
1.59	5.5137	2.3481	1.5324	1.7072
1.86	6.9380	2.6340	1.6230	1.9370
2.01	9.1154	3.0192	1.7376	2.2100
2.16	11.0467	3.3237	1.8231	2.4021
2.33	13.4119	3.6622	1.9137	2.5961
2.44	15.2298	3.9025	1.9755	2.7233
2.71	20.9428	4.5763	2.1392	3.0418
2.88	25.7533	5.0748	2.2527	3.2486
2.95	27.4802	5.2422	2.2896	3.3135
3.21	37.0459	6.0865	2.4671	3.6122
3.51	53.4271	7.3094	2.7036	3.9783
3.71	68.5668	8.2805	2.8776	4.2278
3.99	96.0248	9.7992	3.1304	4.5646

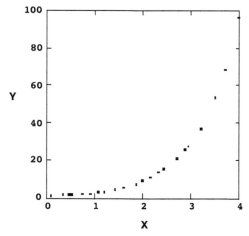

Figure 6.13. Rapidly Growing Function.

Let us consider one version of the "transformational ladder."

$$\exp(e^Y)$$
$$e^Y$$
$$Y^4$$
$$Y^2$$
$$Y$$
$$\sqrt{Y}$$
$$Y^{.25}$$
$$\ln(Y)$$
$$\ln(\ln(Y))$$

Functions which are growing faster than linear are transformed by transformations below Y in the ladder. The only function which is readily identified by the eye is the straight line. We will, accordingly, continue going down the ladder until the transformed data is approximately linear.

As a first try, we show a plot of $\sqrt{Y}$ versus X in Figure 6.14. The transformed curve is still growing much faster than a straight line.

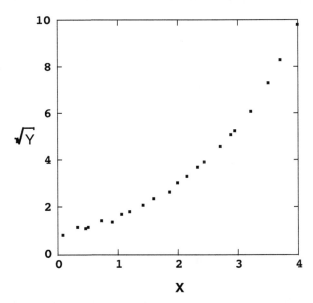

Figure 6.14. Square Root Transformation.

Continuing to the next rung in the ladder, we plot the fourth root of Y versus X in Figure 6.15. We note that the curve still is distinctly concave upward.

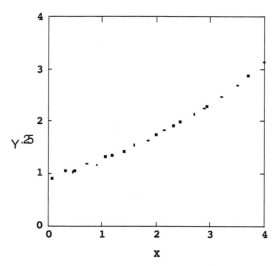

Figure 6.15. Fourth Root Transformation.

Continuing down the ladder to the logarithmic transformation in Figure 6.16, we note that something very close to a straight line has been achieved.

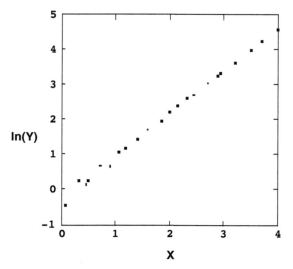

Figure 6.16. Logarithmic Transformation.

Fitting a straight line (by least squares), that is, essentially by (6.23), gives the approximation:

$$\ln(Y) \approx -.34 + 1.24X. \tag{6.28}$$

Exponentiating both sides, we have

$$Y \approx .71 e^{1.24X}.$$ (6.29)

This might suggest that for the data set considered, we might consider in our further work fitting the curve:

$$Y = A_0 + A_1 Z + A_2 Z^2$$ (6.30)

where

$$Z = e^{1.24X}.$$ (6.31)

Returning to the fitting of the straight line to $\ln(y_i)$, we recall that, due to its simplicity and nonnegativity, least squares was selected by K.F. Gauss as a natural criterion function whose minimization frequently gives ready estimates of parameters. In the present context, suppose we have n observations of a (generally noisy) variable, $\{w_i\}$, to which we wish to fit a linear function of x:

$$w_i = a + bx_i + \epsilon_i.$$ (6.32)

Here, ϵ_i denotes the miss between w_i and the approximating straight line. (In the immediate example, here, we recall that $w_i = \ln y_i$.) So we shall find a and b to minimize

$$S(a,b) = \sum_{i=1}^{n} [w_i - (a + bx_i)]^2.$$ (6.33)

We recall that a necessary condition for a maximum or minimum of S is

$$\frac{\partial S}{\partial a} = 0$$ (6.34)

$$\frac{\partial S}{\partial b} = 0.$$ (6.35)

This gives:

$$\hat{b} = \frac{\sum_{i=1}^{n} x_i w_i - n \bar{w} \bar{x}}{\sum_{i=1}^{n} x_i^2 - n \bar{x}^2}$$ (6.36)

and

$$\hat{a} = \bar{w} - \hat{b}\bar{x}. \tag{6.37}$$

For the solution to the necessary conditions give a minimum, it is sufficient that

$$D = (\frac{\partial^2 S}{\partial a \partial b})^2 - \frac{\partial^2 S}{\partial a^2}\frac{\partial^2 S}{\partial b^2} > 0. \tag{6.38}$$

As we see here

$$D = (2n\bar{x})^2 - (2n)(-2\sum_{i=1}^{n} x_i^2) > 0. \tag{6.39}$$

For the particular example at hand, using the data in Table 6.4, we obtain the fit $\hat{a} = -.34$ and $\hat{b} = 1.23$.

6.5 Multivariate Extensions

Let us proceed to problems where there are a number of potentially causal ("independent") variables involved in a function f. Then, assuming we have continuous partial derivatives of all orders, we obtain the Taylor's series representation of f expanding about the vector $(a_1, a_2, \ldots, a_p)$

$$\begin{aligned} f(X_1, X_2, \ldots, X_p) &= f(a_1, a_2, \ldots, a_p) + \sum_{j=1}^{p}(X_j - a_j)\frac{\partial f}{\partial X_j}|_{a_j} \quad (6.40) \\ &+ \sum_{j=1}^{p}\sum_{i=1}^{p}(X_j - a_j)(X_i - a_i)\frac{\partial^2 f}{2\partial X_j \partial X_i}|_{a_j, a_i} + \ldots \, . \end{aligned}$$

Writing out the first few terms of the Taylor's expansion in increasing powers, we have

$$\begin{aligned} f(X_1, X_2, \ldots, X_p) &= A_0 + \sum_{j=1}^{p} A_j X_j + \sum_{j=1}^{p} A_{jj} X_j^2 + \sum_{j=1}^{p}\sum_{i=j+1}^{p} A_{ji} X_j X_i \\ &+ \sum_{j=1}^{p}\sum_{i=j}^{p}\sum_{h=i}^{p} A_{jih} X_j X_i X_h + \ldots \, . \quad (6.41) \end{aligned}$$

It is very rare that we will have occasion to use a polynomial of degree higher than three. Two will generally suffice, and very frequently a polynomial of only the first degree will do nicely. We recall that we are laboring under several major disadvantages. First of all, we expect that our observations will be degraded by noise. And, of course, if we are using a polynomial model in ad hoc fashion, that implies that we do not actually know the true model. If we had no noise and an infinite data base, then the Taylor's expansion of appropriately high degree might well be appropriate to consider. As it is, putting too many terms in the expansion will lead to instability in the sense that if we changed one of the data points, we might obtain very different coefficients for our fitting model.

We recall that we are not asking miracles of our polynomial fit. We will simply use it for guidance as to how we might gingerly change conditions to lower variance, or to improve purity of product, etc. We will not be enthusiastic in using the fitted polynomial for extrapolations well outside the range of the data base. The particular quadratic, say, fit we obtain will be used simply for operating within the narrow ranges where we believe it to be a satisfactory fit.

Having said all this, we must now address the issue of fitting a polynomial to a data set. We shall generally follow orthodox statistical practise, commenting from time to time where the standard assumptions are false to the level where we must be very careful if we are to stay out of trouble.

6.6   Least Squares

We consider the standard "linear statistical model"

$$Y = \sum_{j=0}^{k} \theta_j X_j + \epsilon \, ,$$

where θ_j, $j = 0, 1, \ldots, k$, are unknown coefficients, X_j, $j = 0, 1, \ldots, k$ can be considered as causal variables and ϵ as an "error term."

First of all, we note that the X_j can be almost anything. For example, we could let (and generally do) $X_0 = 1$. Then, we could allow X_2 to equal X_1^2. And, we could allow $X_3 = \exp(\sin(X_1))$. The term "linear" applies to the θ_j coefficients rather than to the power of the variables.

Next, we note that ϵ is supposed to serve many convenient functions. First of all, any departures of truth from our fitting model are subsumed in the ϵ process, which has always zero expectation and constant variance. These assumptions are generally overly optimistic, and we need to be aware that they are.

We want estimates for the θ_j coefficients, given, say, n observations of the random variable Y. Actually, therefore, the linear statistical model assumes the form

$$Y_i = \sum_{j=0}^{k} \theta_j X_{ij} + \epsilon_i \; . \tag{6.42}$$

Here i goes from 1 to n, and Y_i denotes the ith observation, corresponding to the ith vector of variables $(X_{i1}, X_{i2}, \ldots, X_{ik})$. The error term ϵ_i ensemble has the miraculous property that, for all i,

$$
\begin{aligned}
E(\epsilon_i) &= 0 \; , & (6.43) \\
E(\epsilon_i^2) &= \sigma^2 \; , & (6.44) \\
E(\epsilon_i \epsilon_l) &= 0, \text{ if } i \neq l \; . & (6.45)
\end{aligned}
$$

Again we observe that the X_i can be virtually anything. We will generally let $X_0 = 1$. Then, we could allow X_2 to equal X_1^9. We could set $X_3 = \exp(\exp(X_1))$. The term "linear" applies to the θ_j coefficients.

Generally speaking, we shall not concern ourselves much here with addressing the question of dealing with those situations where model inadequacy becomes confounded with error due simply to noise. If we use the model carefully without attempting to make giant steps outside the region of reasonable approximation, frequently the massive amounts of empiricism, which are an essential part of linear modeling, do us no great harm.

Next, we obtain estimates for the θ_j, which we shall denote by $\hat{\theta}_j$, by minimizing the sum of squared deviations of fit from reality,

$$
\begin{aligned}
\text{Min}_{\theta_0, \ldots, \theta_k} S(\theta_0, \theta_1, \ldots, \theta_k) &= \text{Min}_{\theta_0, \ldots, \theta_k} \sum_{i=1}^{n} \hat{\epsilon}_i^2 & (6.46) \\
&= \text{Min}_{\theta_0, \ldots, \theta_k} \sum_{i=1}^{n} [y_i - \sum_{j=0}^{k} \theta_j x_{ij}]^2 \; .
\end{aligned}
$$

Necessary conditions for existence of a minimum are

$$\frac{\partial S}{\partial \theta_j} = -2 \sum_{i=1}^{n} x_{ij}[y_i - \sum_{j=0}^{k} \theta_j x_{ij}] = 0; \; j = 0, 1, \ldots, k \; . \tag{6.47}$$

Summarizing in matrix notation, we have

$$\frac{\partial S}{\partial \Theta} = 2\mathbf{X}'(\mathbf{y} - \mathbf{X}\Theta) = 0, \tag{6.48}$$

where

$$\mathbf{y} = \begin{bmatrix} y_1 \\ y_2 \\ \vdots \\ y_n \end{bmatrix}; \tag{6.49}$$

$$\Theta = \begin{bmatrix} \theta_0 \\ \theta_2 \\ \vdots \\ \theta_k \end{bmatrix}; \tag{6.50}$$

and

$$\mathbf{X} = \begin{bmatrix} x_{10} & x_{11} & \cdots & x_{1k} \\ x_{20} & x_{21} & \cdots & x_{2k} \\ \vdots & \vdots & & \vdots \\ x_{n0} & x_{n1} & \cdots & x_{nk} \end{bmatrix}. \tag{6.51}$$

Then we have

$$\mathbf{X}'\mathbf{y} = \mathbf{X}'\mathbf{X}\Theta. \tag{6.52}$$

Assuming $\mathbf{X}'\mathbf{X}$ is invertible, we then have

$$\hat{\Theta} = (\mathbf{X}'\mathbf{X})^{-1}\mathbf{X}'\mathbf{y}. \tag{6.53}$$

Now we should observe that so far we have not utilized the prescribed conditions of the $\{\epsilon\}$ process. We could, formally, use the least squares criterion absent these. And, in general, the least squares procedure frequently works satisfactorily when one or more of the conditions is not satisfied. We show below a very powerful consequence of least squares estimation in conjunction with the indicated conditions on the error process $\{\epsilon\}$.

Rewriting these conditions in matrix form, we can write the summarized *dispersion matrix* as

$$\mathbf{V}(\epsilon) = E(\epsilon\epsilon') = \sigma^2 \mathbf{I}, \tag{6.54}$$

where $\mathbf{I}$ is the n by n identity matrix. Then, substituting the least squares estimate for Θ, we have

$$\hat{\boldsymbol{\Theta}} = (\mathbf{X}'\mathbf{X})^{-1}\mathbf{X}'(\mathbf{X}\boldsymbol{\Theta} + \epsilon)$$
$$= \boldsymbol{\Theta} + (\mathbf{X}'\mathbf{X})^{-1}\mathbf{X}'\epsilon. \tag{6.55}$$

Then, we easily see by substitution that the dispersion matrix of $\hat{\boldsymbol{\Theta}}$ is given by

$$\begin{aligned}
\mathbf{V}(\hat{\boldsymbol{\Theta}}) &= E[(\hat{\boldsymbol{\Theta}} - \boldsymbol{\Theta})(\hat{\boldsymbol{\Theta}} - \boldsymbol{\Theta})'] \\
&= E[\{(\mathbf{X}'\mathbf{X})^{-1}\mathbf{X}'\epsilon\}\{(\mathbf{X}'\mathbf{X})^{-1}\mathbf{X}'\epsilon\}'] \\
&= (\mathbf{X}'\mathbf{X})^{-1}\mathbf{X}'E(\epsilon\epsilon')\mathbf{X}(\mathbf{X}'\mathbf{X})^{-1} \\
&= \sigma^2(\mathbf{X}'\mathbf{X})^{-1}. \tag{6.56}
\end{aligned}$$

The last step in (6.56) (based upon assumptions in (6.44) and (6.45)) has important implications in the construction of experimental designs. For example, if we wished the estimates for the components of $\boldsymbol{\Theta}$ to be independent of each other, we could simply see to it that the off diagonal components of $\mathbf{X}'\mathbf{X}$ were all zero. That is, we could construct design $\mathbf{X}$ matrices with *orthogonal* columns.

As an example of such a design matrix, let us suppose we wish to obtain least squares estimates appropriate for the model

$$Y = \theta_0 + \theta_1 X_1 + \theta_2 X_2 + \epsilon. \tag{6.57}$$

Here, we will create X_0, a variable which is always equal to 1. Now, if we wish to design an experiment of size 4, we could consider using the design matrix

$$\mathbf{X} = \begin{bmatrix} 1 & 2 & 5 \\ 1 & -2 & 5 \\ 1 & 2 & -5 \\ 1 & -2 & -5 \end{bmatrix}. \tag{6.58}$$

This gives

$$\mathbf{X}'\mathbf{X} = \begin{bmatrix} 4 & 0 & 0 \\ 0 & 16 & 0 \\ 0 & 0 & 100 \end{bmatrix}. \tag{6.59}$$

Thus, for such a design, again assuming the rather strong conditions in (6.44) and (6.45) hold, we would have

$$\mathbf{V}(\hat{\boldsymbol{\Theta}}) = \sigma^2 \begin{bmatrix} \frac{1}{4} & 0 & 0 \\ 0 & \frac{1}{16} & 0 \\ 0 & 0 & \frac{1}{100} \end{bmatrix}. \tag{6.60}$$

Thus, we would be assured that there was no correlation between the estimates for each of the θ_j.

Next, let us suppose that we had obtained the least squares estimates for the θ_j. Suppose then that we wished to obtain estimates for some linear combination of the θ_j, say $-.1\theta_0 + 27\theta_1 + 1002\theta_2$. In the best of all possible worlds we might hope that the optimum estimator would be given by $-.1\hat{\theta}_0 + 27\hat{\theta}_1 + 1002\hat{\theta}_2$. We show below, following Kendall and Stuart's explication ([5], v. II, 79-80) of Plackett's simplification of the proof of the Gauss-Markov Theorem, that we are able to come close to the "best of all possible worlds" result.

Let us suppose that we have a matrix of dimension $k+1$ by r, say, C, and we wish to estimate r linear combinations of the unknown parameters, say

$$\mathbf{c} = \mathbf{C\Theta}. \tag{6.61}$$

Suppose, moreover, that we wish to restrict ourselves to estimators which are linear combinations of the n by 1 observation vector $\mathbf{y}$, say

$$\mathbf{t} = \mathbf{Ty}. \tag{6.62}$$

We shall employ Markov's restriction of unbiasedness, namely that

$$E(\mathbf{Ty}) = \mathbf{c} = \mathbf{C\Theta}. \tag{6.63}$$

And our criterion shall be to minimize the diagonal elements of the dispersion matrix of $\mathbf{t}$, namely

$$\mathbf{V(t)} = E[(\mathbf{t} - \mathbf{C\Theta})(\mathbf{t} - \mathbf{C\Theta})']. \tag{6.64}$$

Now, the condition of unbiasedness gives us

$$E[\mathbf{Ty}] = E[\mathbf{T}(\mathbf{X\Theta} + \epsilon)] = \mathbf{C\Theta}. \tag{6.65}$$

So then, we have

$$\mathbf{TX} = \mathbf{C}. \tag{6.66}$$

Thus, the dispersion matrix of $\mathbf{t}$ becomes

$$\mathbf{V(t)} = E(\mathbf{T}\epsilon\epsilon'\mathbf{T}') = \sigma^2\mathbf{TT}'. \tag{6.67}$$

Next, we observe Plackett's clever decomposition of $\mathbf{TT}'$, namely

$$\begin{aligned} \mathbf{TT}' &= [\mathbf{C}(\mathbf{X'X})^{-1}\mathbf{X}'][\mathbf{C}(\mathbf{X'X})^{-1}\mathbf{X}']' \\ &+ [\mathbf{T} - \mathbf{C}(\mathbf{X'X})^{-1}\mathbf{X}'][\mathbf{T} - \mathbf{C}(\mathbf{X'X})^{-1}\mathbf{X}']' \end{aligned} \tag{6.68}$$

as may easily be verified by multiplication of the right hand side. Now, a matrix of the form $\mathbf{MM}'$ has only non-negative elements on the main diagonal. Thus, both of the terms on the right hand side of (6.68) can only add to the magnitude of each of the terms on the diagonal of $\mathbf{TT}'$. The best that we can hope for is to minimize the amount they add to the dispersion. But we have no control over anything except our choice of $\mathbf{T}$, so there is no way we can do anything about the first term on the right hand side of (6.68). But we have a means of reducing the second term to $\mathbf{0}$, namely, by letting

$$\mathbf{T} = \mathbf{C}(\mathbf{X}'\mathbf{X})^{-1}\mathbf{X}'. \tag{6.69}$$

So, finally, we have demonstrated that once the least squares estimates for the θ's have been found, we can readily obtain estimators for any set of linear functions of the θ's, say $\mathbf{C\Theta}$, by using $\mathbf{C\hat{\Theta}}$.

Let us turn now to the situation where the dispersion matrix of the errors is given by $\sigma^2 \mathbf{V}$, where $\mathbf{V}$ is completely general. Then it can be shown that the unbiased minimum variance (Gauss-Markov) estimator of $\mathbf{C\Theta}$ is given by

$$\mathbf{t} = \mathbf{C}(\mathbf{X}'\mathbf{V}^{-1}\mathbf{X})^{-1}\mathbf{X}'\mathbf{V}^{-1}\mathbf{y}, \tag{6.70}$$

and

$$\mathbf{V(t)} = \sigma^2 \mathbf{C}(\mathbf{X}'\mathbf{V}^{-1}\mathbf{X})^{-1}\mathbf{C}'. \tag{6.71}$$

6.7   Model "Enrichment"

One might suppose that if we managed to fit a linear model to a set of operations data, and if, on the basis of that model, it appeared that we should modify current control conditions to a different level of one or more variables, we would be well advised to exercise caution in recommending a change from the current level. The reasons are fairly obvious. Under the current conditions, the production staff has most likely achieved a certain degree of control, and the process probably is, in most senses, satisfactory. We would not have carried out the analysis had we not been willing to make recommendations using it. But we cannot do so on the basis of shaky information.

Even to carry out a scheduled experimental study on a production process is generally regarded as an intrusive inconvenience by some production staffs. Thirty years ago, one of us, as a young chemical engineer (Thompson), was carrying out a factorial design on operating conditions for a large petroleum distillation column to determine the feasibility of a change in operating conditions in order to increase by 50% the volume of the medium weight "jet fuel" sidestream, without significant specification degradation to the other fractions of output. To get the unit to steady state after changing conditions required around 36 hours of continuous attention by the instigator (the production staff not being inclined to assist with such foolishness). Steady state having been achieved, the instigator returned briefly to his office for a short nap. One hour later, a frantic call was received from the production staff. The pressure at the top of the column had opened the safety valves, and a fine spray was being diffused from the great height of the column over large sections of Baton Rouge, Louisiana. The instigator drove furiously to the column, pouring over in a fatigued mind what thermodynamic anomalies possibly could have caused this multimillion dollar catastrophe. Anxiously, he gazed unsuccessfully through a dirty windscreen and the hot July haze to see signs of the ecological disaster. Stopping the car and jogging furiously to the operations room, he was greeted with loud guffaws of laughter, from the members of the operations staff, who were having a bit of not so good-natured fun with the smart aleck, who was fixing something that was not "broke." Naturally, from the standpoint of the company, changing market conditions had made jet fuel very profitable relative to other products—hence the study. But from the standpoint of the production staff, changes in a perfectly satisfactory production paradigm were being made, and they were not overjoyed. From their standpoint, little upside good was likely to come from the whole business. And their little prank indicated that it was just possible that something awful could occur on the downside.

It is most important that, before we go to the stage of designing experiments to change "in control" processes, we should establish a thorough familiarity of the more basic notions of statistical process control, such as control charting, with the entire production staff. There is nothing more terrifying to a staff, most of whom have little idea of how to bring a system into control, than purposefully taking an in control system out of control.

Assuming that we have a level of confidence on the part of all concerned which enables us to carry out experiments, it is even more important that we not recommend ineffective, let alone disastrous, changes. If we discover from our ad hoc model that we may be able to lower the variability of the output by, say 50%, then such an improvement is probably well worth having. But it is a bad idea to recommend change unless we are confident of our ground. An important part of gaining this confidence is making sure our model is sufficiently supported by the data to recommend a change in operating conditions.

6.8   Testing for Model "Enrichment"

We start, following Kendall and Stuart [5], consideration of two alternative hypotheses:

$$H_0 : y_i = \epsilon_i \tag{6.72}$$

or

$$H_1 : y_i = x_{i0}\theta_0 + x_{i1}\theta_1 + \ldots + x_{ip}\theta_p + \epsilon_i. \tag{6.73}$$

In both cases, we assume the ϵ_j are independent and identically distributed $\mathcal{N}(0, \sigma^2)$ random variables.

Clearly, there may well be a potential gain in process enhancement if we know that H_1 is true rather than H_0. On the other hand, opting for H_1 when H_0 is true may well cause us to move the process out of control to no good purpose.

Clearly, we need to be able to analyze data to distinguish between strong evidence that H_1 is true as opposed to a mere statistical fluke. We will now develop a testing procedure for this purpose.

First of all, on the basis of a sample $\{y_i, X_i\}_{i=1}^{n}$, we shall obtain the least squares estimates of $\mathbf{\Theta}$, assuming H_1 to be true. Then we shall observe the differences between each y_j and its least squares estimate:

$$\begin{aligned} \mathbf{y} - \mathbf{X}\hat{\mathbf{\Theta}} &= [\mathbf{X}\mathbf{\Theta} + \epsilon] - \mathbf{X}[(\mathbf{X}'\mathbf{X})^{-1}\mathbf{X}'\mathbf{y}] \\ &= [\mathbf{X}\mathbf{\Theta} + \epsilon] - \mathbf{X}[(\mathbf{X}'\mathbf{X})^{-1}\mathbf{X}'(\mathbf{X}\mathbf{\Theta} + \epsilon)]. \end{aligned} \tag{6.74}$$

Canceling the terms in $\mathbf{\Theta}$ we have

$$\mathbf{y} - \mathbf{X}\hat{\mathbf{\Theta}} = [\mathbf{I_n} - \mathbf{X}(\mathbf{X}'\mathbf{X})^{-1}\mathbf{X}']\epsilon. \tag{6.75}$$

Here, $\mathbf{I_n}$ is the n by n identity matrix.

Next, multiplying the term in brackets by its transpose, we have

$$\begin{aligned} [\mathbf{I_n} - \mathbf{X}(\mathbf{X}'\mathbf{X})^{-1}\mathbf{X}']'[\mathbf{I_n} - \mathbf{X}(\mathbf{X}'\mathbf{X})^{-1}\mathbf{X}'] &= \\ \mathbf{I_n} - 2\mathbf{X}(\mathbf{X}'\mathbf{X})^{-1}\mathbf{X}' + \mathbf{X}(\mathbf{X}'\mathbf{X})^{-1}\mathbf{X}' &= [\mathbf{I_n} - \mathbf{X}(\mathbf{X}'\mathbf{X})^{-1}\mathbf{X}']. \end{aligned} \tag{6.76}$$

Thus,

$$\mathbf{B} = [\mathbf{I_n} - \mathbf{X}(\mathbf{X}'\mathbf{X})^{-1}\mathbf{X}'] \tag{6.77}$$

is idempotent, and, from Appendix A, we know that its rank is equal to its trace. Recall that we have assumed $\mathbf{X'X}$ is invertible, and hence it has rank $p + 1$. Now,

$$
\begin{aligned}
tr[\mathbf{I_n} - \mathbf{X}(\mathbf{X'X})^{-1}\mathbf{X'}] &= [tr(\mathbf{I_n})] - tr[(\mathbf{X'X})^{-1}\mathbf{X'X})] \quad (6.78) \\
&= n - (p + 1).
\end{aligned}
$$

Then, we have

$$
\begin{aligned}
(\mathbf{y} - \mathbf{X}\hat{\boldsymbol{\Theta}})'(\mathbf{y} - \mathbf{X}\hat{\boldsymbol{\Theta}}) &= \epsilon'\mathbf{B}\epsilon \quad (6.79) \\
&= \sum_{i=1}^{n} b_{ii}\epsilon_i^2 + \sum_{i \neq j} b_{ij}\epsilon_i\epsilon_j. \quad (6.80)
\end{aligned}
$$

By the i.i.d. assumption of the ϵ_j we then have

$$
E[(\mathbf{y} - \mathbf{X}\hat{\boldsymbol{\Theta}})'(\mathbf{y} - \mathbf{X}\hat{\boldsymbol{\Theta}})] = \sigma^2 tr(\mathbf{B}) = \sigma^2(n - p - 1). \quad (6.81)
$$

Thus, an unbiased estimator of σ^2 is given by

$$
s^2 = \frac{1}{n - (p + 1)}(\mathbf{y} - \mathbf{X}\hat{\boldsymbol{\Theta}})'(\mathbf{y} - \mathbf{X}\hat{\boldsymbol{\Theta}}). \quad (6.82)
$$

At this point, we bring normality into the argument, for

$$
\frac{n - (p + 1)}{\sigma^2} s^2 = \frac{\epsilon'\mathbf{B}\epsilon}{\sigma^2} \quad (6.83)
$$

is an idempotent quadratic form in n independent $\mathcal{N}(0, 1)$ variates. Hence, from Appendix B, we have that $(n - p - 1)s^2/\sigma^2$ is a χ^2 variate with $(n - p - 1)$ degrees of freedom *regardless of the true value of* $\boldsymbol{\Theta}$.

Next, we note that

$$
\mathbf{y'y} = (\mathbf{y} - \mathbf{X}\hat{\boldsymbol{\Theta}})'(\mathbf{y} - \mathbf{X}\hat{\boldsymbol{\Theta}}) + (\mathbf{X}\hat{\boldsymbol{\Theta}})'(\mathbf{X}\hat{\boldsymbol{\Theta}}). \quad (6.84)
$$

The first term on the right, when divided by σ^2, is $\chi^2(n-p-1)$ regardless of $\boldsymbol{\Theta}$. Let us investigate $(\mathbf{X}\hat{\boldsymbol{\Theta}})'(\mathbf{X}\hat{\boldsymbol{\Theta}})$.

$$
\begin{aligned}
(\mathbf{X}\hat{\boldsymbol{\Theta}})'(\mathbf{X}\hat{\boldsymbol{\Theta}}) &= \hat{\boldsymbol{\Theta}}'\mathbf{X'X}\hat{\boldsymbol{\Theta}} \quad (6.85) \\
&= [(\mathbf{X'X})^{-1}\mathbf{X'y}]'\mathbf{X'X}(\mathbf{X'X})^{-1}\mathbf{X'y} \\
&= \mathbf{y'X}(\mathbf{X'X})^{-1}\mathbf{X'y} \\
&= (\boldsymbol{\Theta}'\mathbf{X'} + \epsilon')\mathbf{X}(\mathbf{X'X})^{-1}\mathbf{X'}(\mathbf{X}\boldsymbol{\Theta} + \epsilon).
\end{aligned}
$$

Now if $\boldsymbol{\Theta} = \mathbf{0}$, then we have

$$(\mathbf{X}\hat{\boldsymbol{\Theta}})'(\mathbf{X}\hat{\boldsymbol{\Theta}}) = \epsilon'\mathbf{X}(\mathbf{X}'\mathbf{X})\mathbf{X}'\epsilon. \qquad (6.86)$$

Now $\mathbf{X}(\mathbf{X}'\mathbf{X})^{-1}\mathbf{X}'$ is idempotent, since

$$\begin{aligned}
[\mathbf{X}(\mathbf{X}'\mathbf{X})^{-1}\mathbf{X}']'\mathbf{X}(\mathbf{X}'\mathbf{X})^{-1}\mathbf{X}' &= \mathbf{X}(\mathbf{X}'\mathbf{X})^{-1}\mathbf{X}'\mathbf{X}(\mathbf{X}'\mathbf{X})^{-1}\mathbf{X}' \\
&= \mathbf{X}(\mathbf{X}'\mathbf{X})^{-1}\mathbf{X}'. \qquad (6.87)
\end{aligned}$$

Thus its rank is given by its trace

$$tr[\mathbf{X}(\mathbf{X}'\mathbf{X})^{-1}\mathbf{X}'] = tr[(\mathbf{X}'\mathbf{X})^{-1}\mathbf{X}\mathbf{X}'] = p + 1, \qquad (6.88)$$

since $(\mathbf{X}'\mathbf{X})^{-1}$ is assumed to be invertible. Moreover

$$\frac{\epsilon'\mathbf{X}(\mathbf{X}'\mathbf{X})^{-1}\mathbf{X}'\epsilon}{\sigma^2} \text{ is } \chi^2(p+1), \qquad (6.89)$$

since it is an idempotent quadratic form in standard normal variates.

But then, since the ranks $n - p - 1$ and $p + 1$ add to n, the rank of the left hand side, we know from Cochran's Theorem (see Appendix B) that the residual sum of squares $(\mathbf{y} - \mathbf{X}\hat{\boldsymbol{\Theta}})'(\mathbf{y} - \mathbf{X}\hat{\boldsymbol{\Theta}})$ and $\epsilon'\mathbf{X}(\mathbf{X}'\mathbf{X})^{-1}\mathbf{X}'\epsilon$ are independently distributed. Indeed, if the null hypothesis is true, all the terms of (6.84) are quadratic forms in ϵ. Now we have a means of testing H_0, for if H_0 be true, then

$$\frac{(\mathbf{X}\hat{\boldsymbol{\Theta}})'(\mathbf{X}\hat{\boldsymbol{\Theta}})/(p+1)}{(\mathbf{y} - \mathbf{X}\hat{\boldsymbol{\Theta}})'(\mathbf{y} - \mathbf{X}\hat{\boldsymbol{\Theta}})/(n-p-1)} \text{ is distributed as } F_{p+1,n-(p+1)},$$
$$(6.90)$$

since we then have the quotient of two χ^2 variables with $p + 1$ and $n - (p + 1)$ degrees of freedom, divided by their respective degrees of freedom.

We have seen that whatever be $\boldsymbol{\Theta}$,

$$E[(\mathbf{y} - \mathbf{X}\hat{\boldsymbol{\Theta}})'(\mathbf{y} - \mathbf{X}\hat{\boldsymbol{\Theta}})] = (n - p - 1)\sigma^2. \qquad (6.91)$$

But

$$\begin{aligned}
E[(\mathbf{X}\hat{\boldsymbol{\Theta}})'(\mathbf{X}\hat{\boldsymbol{\Theta}})] &= E[\epsilon'\mathbf{X}(\mathbf{X}'\mathbf{X})^{-1}\mathbf{X}'\epsilon] + \boldsymbol{\Theta}'\mathbf{X}'\mathbf{X}\boldsymbol{\Theta} \\
&= (p+1)\sigma^2 + (\mathbf{X}\boldsymbol{\Theta})'(\mathbf{X}\boldsymbol{\Theta}). \qquad (6.92)
\end{aligned}$$

Thus, if H_0 be false and H_1 be true, for a fixed significance level α, we should expect that

$$\frac{(\mathbf{X}\hat{\boldsymbol{\Theta}})'(\mathbf{X}\hat{\boldsymbol{\Theta}})/(p+1)}{(\mathbf{y} - \mathbf{X}\hat{\boldsymbol{\Theta}})'(\mathbf{y} - \mathbf{X}\hat{\boldsymbol{\Theta}})/(n-p-1)} > F_{p+1,n-p-1}(\alpha). \qquad (6.93)$$

Next, let us suppose we have been using, with some success, a linear model

$$y_i = x_{i0}\theta_0 + x_{i1}\theta_1 + \ldots + x_{ip_1}\theta_{p_1} + \epsilon_i, \qquad (6.94)$$

which we wish to consider enhancing to the model

$$y_i = x_{i0}\theta_0 + x_{i1}\theta_1 + \ldots + x_{ip_1}\theta_{p_1} + x_{i,p_1+1}\theta_{p_1+1} + \ldots + x_{i,p_1+p_2}\theta_{p_1+p_2} + \epsilon_i. \qquad (6.95)$$

Once again we are making the customary assumption that the ϵs are independent and identically distributed as $\mathcal{N}(0, \sigma^2)$. We shall write this in shorthand notation for n experiments

$$\mathbf{y} = \mathbf{Z_1}\boldsymbol{\Theta_1} + \mathbf{Z_2}\boldsymbol{\Theta_2} + \epsilon. \qquad (6.96)$$

That is,

$$\mathbf{X} = [\mathbf{Z_1} \ \mathbf{0}] + [\mathbf{0} \ \mathbf{Z_2}], \qquad (6.97)$$

where

$$\mathbf{X} = \begin{bmatrix} x_{1,0} & x_{1,1} & \cdots & x_{1,p_1} & x_{1,p_1+1} & \cdots & x_{1,p_1+p_2} \\ x_{2,0} & x_{2,1} & \cdots & x_{2,p_1} & x_{2,p_1+1} & \cdots & x_{2,p_1+p_2} \\ \vdots & \vdots & \vdots & \vdots & \vdots & \vdots & \vdots \\ x_{n,0} & x_{n,1} & \cdots & x_{n,p_1} & x_{n,p_1+1} & \cdots & x_{n,p_1+p_2} \end{bmatrix}. \qquad (6.98)$$

So $\mathbf{Z_1}$ is the $(n, p_1 + 1)$ matrix consisting of the first $p_1 + 1$ columns on the right hand side, and $\mathbf{Z_2}$ is the next p_2 columns.

The null hypothesis which we wish to test here is

$$\mathrm{H_0} : \theta_{p_1+1} = \theta_{p_1+2} = \ldots = \theta_{p_1+p_2} = 0. \qquad (6.99)$$

We shall first consider the case where $\mathbf{Z_1}$ and $\mathbf{Z_2}$ are orthogonal, i.e.,

$$\mathbf{Z_2}'\mathbf{Z_1} = \mathbf{0}_{p_2,p_1+1}. \qquad (6.100)$$

Then, for the enhanced model, the least squares estimates for $\boldsymbol{\Theta}$ are given by

$$\begin{aligned} \hat{\boldsymbol{\Theta}} &= (\mathbf{X}'\mathbf{X})^{-1}\mathbf{X}'\mathbf{y} & (6.101) \\ &= (\mathbf{Z}_1'\mathbf{Z}_1)^{-1}\mathbf{Z}'_1\mathbf{y} + (\mathbf{Z}_2'\mathbf{Z}_2)^{-1}\mathbf{Z}'_2\mathbf{y}. & (6.102) \end{aligned}$$

Recalling the argument in (6.79), we have

$$\begin{aligned} [(\mathbf{y} - \mathbf{X}\hat{\boldsymbol{\Theta}})'(\mathbf{y} - \mathbf{X}\hat{\boldsymbol{\Theta}})] &= \epsilon'[\mathbf{I_n} - \mathbf{Z}_1(\mathbf{Z}_1'\mathbf{Z}_1)^{-1}\mathbf{Z}'_1 - \mathbf{Z}_2(\mathbf{Z}_2'\mathbf{Z}_2)^{-1}\mathbf{Z}_2']\epsilon \\ &= \epsilon'[\mathbf{I_n} - \mathbf{Z}_1(\mathbf{Z}_1'\mathbf{Z}_1)^{-1}\mathbf{Z}'_1]\epsilon - \epsilon'\mathbf{Z}_2(\mathbf{Z}_2'\mathbf{Z}_2)^{-1}\mathbf{Z}_2'\epsilon. \end{aligned}$$

$$(6.103)$$

Rearranging terms, we have

$$\epsilon'[\mathbf{I_n} - \mathbf{Z}_1(\mathbf{Z}_1'\mathbf{Z}_1)^{-1}\mathbf{Z}'_1]\epsilon = [(\mathbf{y} - \mathbf{X}\hat{\Theta})'(\mathbf{y} - \mathbf{X}\hat{\Theta})] + \epsilon'\mathbf{Z}_2(\mathbf{Z}_2'\mathbf{Z}_2)^{-1}\mathbf{Z}_2'\epsilon. \tag{6.104}$$

But from the argument leading to (6.83), we know that the term on the left hand side is χ^2 with $n - (p_1 + 1)$ degrees of freedom. The first term on the right, regardless of the value of Θ_1 or that of Θ_2, we have shown, again in the argument leading to (6.83), is χ^2 with $n - (p_1 + p_2 + 1)$ degrees of freedom. Finally, by the argument leading to (6.89) the second term on the right hand side is χ^2 with p_2 degrees of freedom. So, again by Cochran's Theorem, since the degrees of freedom of the two quadratic forms on the right hand side add to that of the quadratic form on the left, they are stochastically independent. Thus, if $\Theta_2 = \mathbf{0}$, then

$$\frac{(\mathbf{Z_2}\hat{\Theta}_2)'(\mathbf{Z_2}\hat{\Theta}_2)/(p_2)}{(\mathbf{y} - \mathbf{X}\hat{\Theta})'(\mathbf{y} - \mathbf{X}\hat{\Theta})/(n - p_1 - p_2 - 1)} \tag{6.105}$$

is distributed as $F_{p_2, n-(p_1+p_2+1)}$, since the null hypothesis implies that

$$\epsilon'\mathbf{Z}_2(\mathbf{Z}_2'\mathbf{Z}_2)^{-1}\mathbf{Z}_2'\epsilon = (\mathbf{Z_2}\hat{\Theta})'(\mathbf{Z_2}\hat{\Theta}).$$

Next, let us consider the situation where $\mathbf{Z_2}$ is not necessarily orthogonal to $\mathbf{Z_1}$. Consider

$$\begin{align}
\mathbf{y}_{.1} &= \mathbf{y} - \mathbf{Z_1}\hat{\Theta}_1 \tag{6.106} \\
&= \mathbf{y} - \mathbf{Z_1}[(\mathbf{Z_1}'\mathbf{Z_1})^{-1}\mathbf{Z}'_1\mathbf{y}]. \tag{6.107}
\end{align}$$

Multiplying by $\mathbf{Z_1}'$, we have

$$\begin{align}
\mathbf{Z_1}'\mathbf{y}_{.1} &= \mathbf{Z_1}'\mathbf{y} - \mathbf{Z_1}'\mathbf{Z_1}[(\mathbf{Z_1}'\mathbf{Z_1})^{-1}\mathbf{Z_1}'\mathbf{y}] \tag{6.108} \\
&= \mathbf{0}. \tag{6.109}
\end{align}$$

Thus, we have established that $\mathbf{Z}_1$ is orthogonal to $\mathbf{y}_{.1}$.

Now, let us "regress" $\mathbf{Z}_2$ on $\mathbf{Z}_1$ to obtain

$$\begin{align}
\hat{\mathbf{Z}}_2 &= (\mathbf{Z_1}'\mathbf{Z_1})^{-1}\mathbf{Z_1}'\mathbf{Z_2}\mathbf{Z_1} \tag{6.110} \\
&= \mathbf{A}\mathbf{Z}_1, \tag{6.111}
\end{align}$$

where

$$\mathbf{A} = (\mathbf{Z_1}'\mathbf{Z_1})^{-1}\mathbf{Z_1}'\mathbf{Z_2}. \tag{6.112}$$

Next, let

$$\mathbf{Z}_{2.1} = [\mathbf{I} - \mathbf{Z}_1(\mathbf{Z}_1'\mathbf{Z}_1)^{-1}\mathbf{Z}_1']\mathbf{Z}_2 \tag{6.113}$$
$$= \mathbf{Z}_2 - \mathbf{Z}_1\mathbf{A}. \tag{6.114}$$

We note that $\mathbf{Z}_1$ is orthogonal to $Z_{2.1}$, for

$$\mathbf{Z}_1'\mathbf{Z}_{2.1} = \mathbf{Z}_1'\mathbf{Z}_2 - (\mathbf{Z}_1'\mathbf{Z}_1)(\mathbf{Z}_1'\mathbf{Z}_1)^{-1}\mathbf{Z}_1'\mathbf{Z}_2 \tag{6.115}$$
$$= \mathbf{0}. \tag{6.116}$$

So, we have, in a sense, reduced the more general situation to the orthogonal case if we consider the decomposition:

$$\mathbf{y} = \mathbf{Z}_1\mathbf{\Theta}_1 + \mathbf{Z}_2\mathbf{\Theta}_2 + \epsilon \tag{6.117}$$
$$= \mathbf{Z}_1(\mathbf{\Theta}_1 + \mathbf{A}\mathbf{\Theta}_2) + (\mathbf{Z}_2 - \mathbf{Z}_1\mathbf{A})\mathbf{\Theta}_2 + \epsilon \tag{6.118}$$
$$= \mathbf{Z}_1\mathbf{\Theta}^* + \mathbf{Z}_{2.1}\mathbf{\Theta}_2 + \epsilon, \tag{6.119}$$

where

$$\mathbf{\Theta}^* = \mathbf{\Theta}_1 + \mathbf{A}\mathbf{\Theta}_2. \tag{6.120}$$

As in (6.104), we have

$$\epsilon'[\mathbf{I_n} - \mathbf{Z}_1(\mathbf{Z}_1'\mathbf{Z}_1)^{-1}\mathbf{Z}'_1]\epsilon = [(\mathbf{y} - \mathbf{X}\hat{\mathbf{\Theta}})'(\mathbf{y} - \mathbf{X}\hat{\mathbf{\Theta}})] \tag{6.121}$$
$$+ \epsilon'\mathbf{Z}_{2.1}(\mathbf{Z}'_{2.1}\mathbf{Z}_{2.1})^{-1}\mathbf{Z}_{2.1}'\epsilon.$$

Now, by the argument in (6.104), the left hand side of (6.121) is χ^2 with $n - p_1 + 1$ degrees of freedom and the first term on the right hand side of (6.122) is χ^2 with $n - p_1 - p_2 - 1$ degrees of freedom regardless of $\mathbf{\Theta}_2$. Now if $\mathbf{\Theta}_2 = 0$, then

$$(\mathbf{Z}_{2.1}\hat{\mathbf{\Theta}}_2)'(\mathbf{Z}_{2.1}\hat{\mathbf{\Theta}}_2) = \epsilon'\mathbf{Z}_{2.1}(\mathbf{Z}'_{2.1}\mathbf{Z}_{2.1})^{-1}\mathbf{Z}_{2.1}'\epsilon, \tag{6.122}$$

so that the second term on the right hand side (6.122) is χ^2 with p_2 degrees of freedom. Since the degrees of freedom of the quadratic forms on the two sides add to $n - p_1 + 1$, by Cochran's Theorem, we have that the two on the right are independent of each other. Thus, if $\mathbf{\Theta}_2 = 0$, then

$$\frac{(\mathbf{Z}_{2.1}\hat{\mathbf{\Theta}}_2)'(\mathbf{Z}_{2.1}\hat{\mathbf{\Theta}}_2)/(p_2)}{(\mathbf{y} - \mathbf{X}\hat{\mathbf{\Theta}})'(\mathbf{y} - \mathbf{X}\hat{\mathbf{\Theta}})/(n - p_1 - p_2 - 1)} \text{ is distributed as } F_{p_2, n-(p_1+p_2+1)}. \tag{6.123}$$

6.9 2^p Factorial Designs

Let us consider designing an experiment for estimating the regression coefficients in the model:

$$y_i = \beta_0 + \beta_1 X_{i1} + \beta_2 X_{i2} + \epsilon_i \qquad (6.124)$$

where ϵ_i is $\mathcal{N}(0, \sigma^2)$.

We will assume that the X_1 and X_2 variables, or *factors*, have been recoded via a simple linear transformation. Suppose that the original variables are W_1 and W_2 and we wish to sample each at two levels. Then we let

$$X_1 = 2\frac{W_1 - W_{1\ \text{lower}}}{W_{1\ \text{upper}} - W_{1\ \text{lower}}} - 1 \qquad (6.125)$$

$$X_2 = 2\frac{W_2 - W_{2\ \text{lower}}}{W_{2\ \text{upper}} - W_{2\ \text{lower}}} - 1. \qquad (6.126)$$

A convenient two factor design with each recoded factor at two possible levels 1 and -1 has the form

Table 6.5				
Experiment Number = i	X_0	X_{i1}	X_{i2}	y_i
1	1	1	1	y_1
2	1	-1	1	y_2
3	1	1	-1	y_3
4	1	-1	-1	y_4

The given design is orthogonal in that the columns of the design matrix $\mathbf{X}$, $\mathbf{X}_0$, $\mathbf{X}_1$ and $\mathbf{X}_2$, are mutually orthogonal. We note that the design consists of 2^2 different experimental points.

Then, the least squares estimator for $(\beta_0,\ \beta_1,\ \beta_2)'$ is given by

$$\left[\begin{pmatrix} 1 & 1 & 1 & 1 \\ 1 & -1 & 1 & -1 \\ 1 & 1 & -1 & -1 \end{pmatrix} \begin{pmatrix} 1 & 1 & 1 \\ 1 & -1 & 1 \\ 1 & 1 & -1 \\ 1 & -1 & -1 \end{pmatrix} \right]^{-1} \begin{pmatrix} 1 & 1 & 1 & 1 \\ 1 & -1 & 1 & -1 \\ 1 & 1 & -1 & -1 \end{pmatrix} \begin{pmatrix} y_1 \\ y_2 \\ y_3 \\ y_4 \end{pmatrix}$$

$$= \begin{bmatrix} 4 & 0 & 0 \\ 0 & 4 & 0 \\ 0 & 0 & 4 \end{bmatrix}^{-1} \begin{pmatrix} 1 & 1 & 1 & 1 \\ 1 & -1 & 1 & -1 \\ 1 & 1 & -1 & -1 \end{pmatrix} \begin{pmatrix} y_1 \\ y_2 \\ y_3 \\ y_4 \end{pmatrix}$$

$$= \frac{1}{4} I \begin{pmatrix} 1 & 1 & 1 & 1 \\ 1 & -1 & 1 & -1 \\ 1 & 1 & -1 & -1 \end{pmatrix} \begin{pmatrix} y_1 \\ y_2 \\ y_3 \\ y_4 \end{pmatrix} \qquad (6.127)$$

$$= \begin{pmatrix} \frac{y_1+y_2+y_3+y_4}{4} \\ \frac{y_1+y_3-y_2-y_4}{4} \\ \frac{y_1+y_2-y_3-y_4}{4} \end{pmatrix}.$$

The geometrical motivation for the estimator is clear. $\hat{\beta}_0$ is simply the average of all the observed y_j. $\hat{\beta}_1$ is the difference of the y_j levels at $X_1 = 1$ and $X_1 = -1$. $\hat{\beta}_2$ is the difference of the y_j levels at $X_2 = 1$ and $X_2 = -1$.

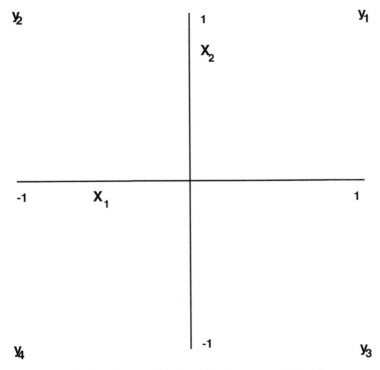

Figure 6.17. Two Factor Orthogonal Design.

In a more general setting, if the linear model has p factors,

$$y_i = \beta_0 + \beta_1 X_{i1} + \beta_2 X_{i2} + \ldots + \beta_p X_{ip} + \epsilon_i,$$

the orthogonal 2^p factorial design is constructed in the following way:
- Each column has length 2^p.

- The zeroth column, corresponding to factor X_0, consists of ones.

- In the first column, corresponding to factor X_1, the signs alternate in groups of 2^0, that is, they alternate each time.

- In the second column, corresponding to factor X_2, the signs alternate in groups of 2^1, that is, they alternate in pairs.

- In the third column, corresponding to factor X_3, the signs alternate in groups of 2^2, that is, they alternate in groups of four.

- In general, in the kth column, corresponding to factor X_k, $k \geq 1$, the signs alternate in groups of 2^{k-1}.

Again, we have assumed here that the factors have been recoded and can take values ± 1 only.

Now, let us return to the two factor model. Typically, we might wish to test whether the levels of the variables X_1 and X_2 really affect y. So we could argue that the value of y fluctuates about the constant $\hat{\beta}_0$, which is typically not equal to zero, but any appearance of effect by X_1 and X_2 is simply due to statistical fluke. So then, the model presently under use is:

$$y_i = \beta_0 + \epsilon_i . \tag{6.128}$$

The new model, which may be based on illusion rather than on reality, is

$$y_i = \beta_0 + \beta_1 X_{i1} + \beta_2 X_{i2} + \epsilon_i \tag{6.129}$$

where ϵ_i is $\mathcal{N}(0, \sigma^2)$ for both models. Another way of expressing the above is by stating the null hypothesis

$$H_0 : \beta_1 = \beta_2 = 0. \tag{6.130}$$

We can now employ equation (6.105), where

$$Z_1 = \begin{pmatrix} 1 \\ 1 \\ 1 \\ 1 \end{pmatrix} \tag{6.131}$$

and

$$Z_2 = \begin{pmatrix} 1 & 1 \\ -1 & 1 \\ 1 & -1 \\ -1 & -1 \end{pmatrix} \tag{6.132}$$

and

$$\hat{\Theta}_2 = \begin{pmatrix} \hat{\beta}_1 \\ \hat{\beta}_2 \end{pmatrix} . \tag{6.133}$$

So, then, if H_0 be true

$$\frac{(\mathbf{Z_2\hat{\Theta}_2})'(\mathbf{Z_2\hat{\Theta}_2})/(p_2)}{(\mathbf{y} - \mathbf{X\hat{\Theta}})'(\mathbf{y} - \mathbf{X\hat{\Theta}})/(n - p_1 - p_2 - 1)} =$$

$$\frac{(4\hat{\beta}_1^2 + 4\hat{\beta}_2^2)/2}{\sum_{i=1}^{n}(y_j - \hat{\beta}_0 - \hat{\beta}_1 X_{i1} - \hat{\beta}_2 X_{i2})^2} \qquad (6.134)$$

is distributed as $F_{2,1}$.

Below, we show the results of carrying out such an experiment.

Table 6.6. Two Factor Experiment

Experiment Number = i	X_0	X_{i1}	X_{i2}	y_i
1	1	1	1	7.01984
2	1	-1	1	14.08370
3	1	1	-1.	18.70730
4	1	-1	-1	27.73476

The fitted model is given by

$$\hat{y} = 16.886 - 4.023X_1 - 6.334X_2. \qquad (6.135)$$

In (6.134), the numerator and denominator are given by 112.621 and .064, respectively. This gives an F statistic of 116.836, which is significant at the .065 level. That means, formally, that there is a chance of only .065 that such a large value would have been expected if the null hypothesis were in fact true.

In the example considered, y represents the variation in the output of a process, so it is desirable to minimize it. We note that to achieve this, our model indicates it to be best to increase X_1 and X_2 in the *steepest descent* direction indicated by the estimated linear coefficients. Thus we might profitably march away from the origin according to the proportions:

$$X_1 = \frac{4.023}{6.334}X_2. \qquad (6.136)$$

If we wished to find the logical point on the circle inscribed in the current design (which has unit radius), we could try $(X_1, X_2) = (.536, .844)$. Or, we might go to the steepest descent point on the circle with radius equal to two, namely, $(X_1, X_2) = (.758, 1.193)$.

There are many ways we might proceed. But in many production systems the question of whether X_1 and X_2 are of relevance to the output has been answered affirmatively beforehand. And in such cases, the conditions may be reasonably close to optimality already. The use of a model which includes only linear terms in X_1 and X_2 always pushes us to the borders of the design region, and beyond. It might be argued that the use of a model containing quadratic terms in X_1 and X_2 is appropriate, since the quadratic model may admit of a "bottom of the trough" solution.

6.10   Some Rotatable Quadratic Designs

Let us consider the more complicated model

$$y_i = \beta_0 + \beta_1 X_{i1} + \beta_2 X_{i2} + \beta_{11} X_{i1}^2 + \beta_{22} X_{i2}^2 + \beta_{12} X_{i1} X_{i2} + \epsilon_i. \quad (6.137)$$

If we tried to use the experiment already designed to estimate the six coefficients of the quadratic model, we know, intuitively, that we would be in trouble, since we have only four experimental points. Nevertheless, let us consider the resulting design table.

i	X_0	X_{i1}	X_{i2}	X_{i1}^2	X_{i2}^2	$X_{i1}X_{i2}$	y_i
1	1	1	1	1	1	1	7.01984
2	1	-1	1	1	1	-1	14.08370
3	1	1	-1	1	1	-1	18.70730
4	1	-1	-1	1	1	1	27.73476

Table 6.7. Two Factor Experiment

We note that the X_{i1}^2 and X_{i2}^2 levels are precisely those of the X_0 vector. There is thus no hope for using the existing design to estimate the coefficients of these two terms. With $X_{i1}X_{i2}$ we have a better chance, though we would then have no possibility of estimating the variance via (see (6.82))

$$\hat{\sigma}^2 = \frac{[(\mathbf{y} - \mathbf{X}\hat{\beta})'(\mathbf{y} - \mathbf{X}\hat{\beta})]}{\text{sample size - number of } \beta \text{s estimated}}. \quad (6.138)$$

Clearly, the two level design is insufficient for estimating the coefficients of the quadratic model. We would like to effect modifications to the four level model which would enable estimation of the coefficients of the quadratic model, while retaining the useful property of orthogonality.

Returning to the simple model in (6.124), we recall another useful property of this model. Namely, if we look at the variance of a predictor $\hat{y}$ at some new (X_1, X_2) value, we have

$$Var(\hat{y}) = Var(\hat{\beta}_0) + Var(\hat{\beta}_1)X_1^2 + Var(\hat{\beta}_2)X_2^2. \qquad (6.139)$$

This simple form is due to the orthogonality of the design matrix, $\mathbf{X_D}$, which insured the independence of the estimated coefficients, for we recall that

$$\mathbf{V}(\hat{\beta}_0, \hat{\beta}_1, \hat{\beta}_2) = (\mathbf{X_D}'\mathbf{X_D})^{-1}\sigma^2 \qquad (6.140)$$

$$= \frac{1}{4}I\sigma^2. \qquad (6.141)$$

We note, then, that the variance of $\hat{y}$ increases as (X_1, X_2) moves away from the origin in such a way that it is constant on (hyper)spheres about the origin. Such a property is called *rotatability* by Box and Hunter, and it is, for the case where only the linear terms are included in the model, a natural consequence of the type of orthogonal design employed here. Rotatability is desirable, inasmuch as is generally the case that the center of a design is the present perceived optimum value of $\mathbf{X}$, and generally we will choose experimental points which we believe may produce approximately the same order of effect on the response variable y. Thus, prior to the experiment, we would like for our confidence in the result to be higher and higher as we approach the prior best guess for the optimum, namely the origin. If we should relax too much approximate rotatability of design, we might encounter the bizarre spectacle of patches of low variability of y away from the origin, paid for by increased variability near the origin. One might suppose that rotatability can be most nearly achieved by placing design points in the $\mathbf{X}$ space on concentric (hyper)spheres about the origin. And, indeed, this is the strategy proposed in 1957 by Box and Hunter [3],[4]. They have suggested creating a quadratic design which builds upon the original $(\pm 1, \pm 1, \ldots, \pm 1)$ design points of the orthogonal design for estimating the coefficients of a model of first degree. In other words, it would be desirable to create a design which might utilize the data from a ± 1 model. Naturally, such points, in p dimensions, lie on the surface of a (hyper)sphere of radius

$$r = \sqrt{1^2 + 1^2 + \ldots 1^2} = \sqrt{p}. \qquad (6.142)$$

There are 2^p such points. Next, we place several points at the origin, thus on the surface of a hypersphere of radius zero. And, finally, we place "star" points on a (hyper)sphere of radius α. These points are placed such that all the **X** coordinates except one are zero. Thus, they are obtained by going from the origin through each of the faces of the $(\pm 1, \pm 1, \ldots, \pm 1)$ hypercube a distance of α, i.e., these points are of the form $(\pm \alpha, 0, \ldots, 0), (0, \pm \alpha, \ldots, 0), (0, 0, \ldots, \pm \alpha)$. There are $2p$ such points.

Let us consider such a design in Figure 6.18.

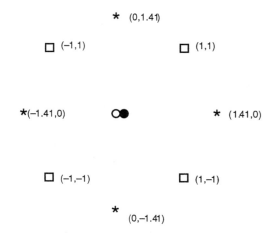

Figure 6.18. Rotatable Two Factor Design.

We examine the enhanced experimental design with two center points and four star points in Table 6.8

\multicolumn{8}{c}{Table 6.8. Two Factor Experiment}
i	X_0	X_{i1}	X_{i2}	X_{i1}^2	X_{i2}^2	$X_{i1}X_{i2}$	y_i
1	1	1	1	1	1	1	7.01984
2	1	-1	1	1	1	-1	14.08370
3	1	1	-1	1	1	-1	18.70730
4	1	-1	-1	1	1	1	27.73476
5	1	0	0	0	0	0	15.01984
6	1	0	0	0	0	0	14.08370
7	1	$\sqrt{2}$	0	2	0	0	11.05540
8	1	$-\sqrt{2}$	0	2	0	0	23.36286
9	1	0	$\sqrt{2}$	0	2	0	8.56584
10	1	0	$-\sqrt{2}$	0	2	0	27.28023

The quadratic fitted by least squares is given by

$$\hat{y} = 14.556 - 4.193X_1 - 6.485X_2 + 1.159X_1^2 + 1.518X_2^2 + .490X_1X_2.$$
$$(6.143)$$

We now attempt to see if we have gained anything by including the quadratic terms, i.e., we wish to test the null hypothesis

$$\beta_{11} = \beta_{22} = \beta_{12} = 0.$$

We have to employ now (6.123) with

$$Z_2 = \begin{pmatrix} 1 & 1 & 1 \\ 1 & 1 & -1 \\ 1 & 1 & -1 \\ 1 & 1 & 1 \\ 0 & 0 & 0 \\ 0 & 0 & 0 \\ 2 & 0 & 0 \\ 2 & 0 & 0 \\ 0 & 2 & 0 \\ 0 & 2 & 0 \end{pmatrix} \qquad (6.144)$$

and

$$\hat{\Theta}_2 = \begin{pmatrix} \hat{\beta}_{11} \\ \hat{\beta}_{22} \\ \hat{\beta}_{12} \end{pmatrix}. \qquad (6.145)$$

Then, we have

$$\frac{(\mathbf{Z}_{2.1}\hat{\Theta}_2)'(\mathbf{Z}_{2.1}\hat{\Theta}_2)}{3} = 4.491. \qquad (6.146)$$

For the residual from the full quadratic model, we have

$$\frac{1}{10-6} \sum_{i=1}^{10} (y_i - \hat{\beta}_0 - \hat{\beta}_1 X_{i1} - \hat{\beta}_2 X_{i2} - \hat{\beta}_{11} X_{i1}^2 - \hat{\beta}_{22} X_{i2}^2 - \hat{\beta}_{12} X_{i1} X_{i2})^2$$
$$(6.147)$$
$$= .435.$$

The resulting ratio of these two estimates (under the null hypothesis) for σ^2 is given by $4.491/.435 = 10.324$. Looking at the $F_{3,4}$ tables, we find this value to be significant at the .025 level. Thus, we most likely will wish to use the full quadratic model. From (6.143), we may seek an optimum by solving the necessary conditions

$$\frac{\partial \hat{y}}{\partial X_1} = \frac{\partial \hat{y}}{\partial X_2} = 0. \qquad (6.148)$$

This gives, as the supposed minimizer of $E(\hat{y})$, $X_1 = 1.405$ and $X_2 = 1.909$. In point of fact, the data in this two by two factorial experiment were all generated from the model

$$y = 2 + (X_1 - 2)^2 + (X_2 - 3)^2 + \epsilon, \qquad (6.149)$$

where ϵ is $\mathcal{N}(0,1)$. In fact, substituting the experimentally determined minimum conditions for y in (6.149), we note that we achieve an average y of 3.544, an improvement well worth having over that at the conditions of $X_1 = 0$ and $X_2 = 0$, namely, $E(y) = 15$.

For the design at hand, the variance of an estimate at a new (X_1, X_2) value is given by:

$$\mathrm{Var}(\hat{y}) = \sigma^2 \begin{pmatrix} 1 & X_1 & X_2 & X_1^2 & X_2^2 & X_1X_2 \end{pmatrix} \times \qquad (6.150)$$

$$\left[\begin{pmatrix} 1 & 1 & 1 & 1 & 1 & 1 & 1 & 1 & 1 & 1 \\ 1 & -1 & 1 & -1 & 0 & 0 & \sqrt{2} & -\sqrt{2} & 0 & 0 \\ 1 & 1 & -1 & -1 & 0 & 0 & 0 & 0 & \sqrt{2} & \sqrt{2} \\ 1 & 1 & 1 & 1 & 0 & 0 & 2 & 2 & 0 & 0 \\ 1 & 1 & 1 & 1 & 0 & 0 & 0 & 0 & 2 & 2 \\ 1 & -1 & -1 & 1 & 0 & 0 & 0 & 0 & 0 & 0 \end{pmatrix} \begin{pmatrix} 1 & 1 & 1 & 1 & 1 & 1 \\ 1 & -1 & 1 & 1 & 1 & -1 \\ 1 & 1 & -1 & 1 & 1 & -1 \\ 1 & -1 & -1 & 1 & 1 & 1 \\ 1 & 0 & 0 & 0 & 0 & 0 \\ 1 & 0 & 0 & 0 & 0 & 0 \\ 1 & \sqrt{2} & 0 & 2 & 0 & 0 \\ 1 & -\sqrt{2} & 0 & 2 & 0 & 0 \\ 1 & 0 & \sqrt{2} & 0 & 2 & 0 \\ 1 & 0 & -\sqrt{2} & 0 & 2 & 0 \end{pmatrix} \right]^{-1}$$

$$\times \begin{pmatrix} 1 & X_1 & X_2 & X_1^2 & X_2^2 & X_1X_2 \end{pmatrix}'.$$

Indeed,

$$\hat{y} = \hat{\Theta}' \begin{pmatrix} 1 \\ X_1 \\ X_2 \\ X_1^2 \\ X_2^2 \\ X_1X_2 \end{pmatrix}$$

and the required result follows easily from the fact that

$$\mathrm{Var}(\hat{y}) = E(\hat{y}^2) - [E(\hat{y})]^2.$$

After a little algebra, we have that

$$Var(\hat{y}) = \sigma^2[.5 + \rho^2(-.5 + .2188\rho^2)] \qquad (6.151)$$

where

$$\rho^2 = X_1^2 + X_2^2. \qquad (6.152)$$

We show a plot of $Var(\hat{y})/(\sigma^2)$ from (6.151) in Figure 6.19.

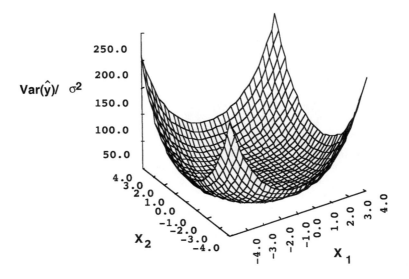

$Var(\hat{y})/ \; \sigma^2$

Figure 6.19. Variance Profile of Rotatable Design.

A seemingly reasonable alternative to the two factor rotatable design considered above is the two factor design with each X at three levels. We show such a design in Table 6.9.

Table 6.9. 3^2 Experimental Design					
i	X_0	X_{i1}	X_{i2}	X_{i1}^2	X_{i2}^2	$X_{i1}X_{i2}$
1	1	-1	-1	1	1	1
2	1	0	-1	0	1	0
3	1	1	-1	1	1	-1
4	1	-1	0	1	0	0
5	1	0	0	0	0	0
6	1	1	0	1	0	0
7	1	-1	1	1	1	-1
8	1	0	1	0	1	0
9	1	1	1	1	1	1

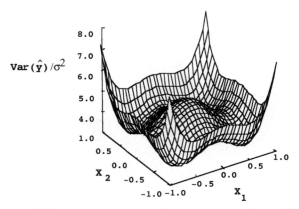

Figure 6.20. Variance Profile of 3^2 Orthogonal Design.

In general, the rotatable designs of Box, Hunter and Draper are very useful when one believes himself to be sufficiently close to the optimal value of the **X** value that it makes sense to fit a quadratic model. For dimensionality p, we start with the simple orthogonal factorial design having 2^p points at the vertices of the (hyper)cube $(\pm 1, \pm 1, \ldots, \pm 1)$. Then we add $2p$ star points at $(\pm \alpha, 0, \ldots, 0)$ $(0, \pm \alpha, 0, \ldots, 0)$, ..., $(0, 0, \ldots, 0, \pm \alpha)$. Then, we generally add two points (for, say, $p \leq 5$, more for larger dimensionality) at the origin. A sufficient condition for rotatability of the design, i.e., that, as above, $Var(\hat{y})$ is a function only of

$$\rho^2 = X_1^2 + X_2^2 + \ldots + X_p^2, \qquad (6.153)$$

can be shown to be [4] that

$$\alpha = (2^p)^{.25}. \qquad (6.154)$$

In Table 6.10 we show rotatable designs for dimensions 2,3,4, 5 and 6.

Table 6.10. Some Rotatable Designs				
Dimension	Num. Cube Points	Num. Center Points	Num. Star Points	α
2	4	2	4	$\sqrt{2}$
3	8	2	6	$2^{.75}$
4	16	2	8	2
5	32	2	10	$2^{1.25}$
6	64	2	12	$2^{1.5}$

6.11 Saturated Designs

Let us return to the two factor experiment in Table 6.6.

Table 6.11. Saturated Three Factor Design					
Experiment Number = i	X_0	X_{i1}	X_{i2}	$X_{i1}X_{i2} = X_{i3}$	y_i
1	1	1	1	1	y_1
2	1	-1	1	-1	y_2
3	1	1	-1	-1	y_3
4	1	-1	-1	1	y_4

We note that the $X_{i1}X_{i2}$ column is orthogonal to the $\mathbf{X}_1$ and $\mathbf{X}_2$ columns. If we assume that there is no interaction effect between $\mathbf{X}_1$ and $\mathbf{X}_2$, then we "saturate" the 2^2 design by confounding $X_{i1}X_{i2}$ with a third variable X_3, using the design indicated. Let us extend this notion to the saturation of the 2^3 design in Table 6.12.

Table 6.12. Saturated Seven Factor Design							
X_{i1}	X_{i2}	X_{i3}	$X_{i1}X_{i2} = X_{i4}$	$X_{i1}X_{i3} = X_{i5}$	$X_{i2}X_{i3} = X_{i6}$	$X_{i1}X_{i2}X_{i3} = X_{i7}$
1	1	1	1	1	1	1
-1	1	1	-1	-1	1	-1
1	-1	1	-1	1	-1	1
-1	-1	1	1	-1	-1	1
1	1	-1	1	-1	-1	-1
-1	1	-1	-1	1	-1	1
1	-1	-1	-1	-1	1	1
-1	-1	-1	1	1	1	-1

In the design indicated, we note that each of the columns is orthogonal to the others, so our least squares estimation procedure is unusually simple. Let us consider the simple linear model

$$y_i = \beta_0 + \sum_{j=1}^{7} \beta_j X_{ij} + \epsilon_i. \tag{6.155}$$

We note, for example, that we will have

$$
\begin{pmatrix} \hat{\beta}_0 \\ \hat{\beta}_1 \\ \hat{\beta}_2 \\ \hat{\beta}_3 \\ \hat{\beta}_4 \\ \hat{\beta}_5 \\ \hat{\beta}_6 \\ \hat{\beta}_7 \end{pmatrix}
= \frac{1}{8}
\begin{pmatrix}
1 & 1 & 1 & 1 & 1 & 1 & 1 & 1 \\
1 & -1 & 1 & -1 & 1 & -1 & 1 & -1 \\
1 & 1 & -1 & -1 & 1 & 1 & -1 & -1 \\
1 & 1 & 1 & 1 & -1 & -1 & -1 & -1 \\
1 & -1 & -1 & 1 & 1 & -1 & -1 & 1 \\
1 & -1 & 1 & -1 & -1 & 1 & -1 & 1 \\
1 & 1 & -1 & -1 & -1 & -1 & 1 & 1 \\
1 & -1 & 1 & 1 & -1 & 1 & 1 & -1
\end{pmatrix}
\begin{pmatrix} y_1 \\ y_2 \\ y_3 \\ y_4 \\ y_5 \\ y_6 \\ y_7 \\ y_8 \end{pmatrix}.
\tag{6.156}
$$

Generally speaking, one would seldom wish to be quite so ambitious as to carry out such a massive saturation. In particular, we would have zero degrees of freedom for the purposes of estimating the variance. However, some feel that such designs are useful in "fishing expeditions," situations where we have vague feelings that some variables might just possibly be useful in minimizing, say, the production variability. It is clear how one could construct designs with less than full saturation. For example, we might decide to confound only one variable, say X_4 with one of the interaction terms, for example, with the $X_1X_2X_3$ term. All the least squares computations for orthogonal designs can be employed, and then one would have three degrees of freedom remaining for the estimation of the underlying variance.

6.12   A Simulation Based Approach

There are many situations where we have reasonable comprehension, at the microaxiom level, but have little justification for using an ad hoc Taylor's expansion type model to describe what is going on at the macro aggregate level. Attempts to utilize such ad hocery, to assume that if one throws in enough terms all will be well, has rendered the utility of some fields, such as econometrics, of marginal utility. There is generally a very great problem in getting from the micro axioms to the macro aggregate model ("closed form") which is its consequence. Happily, rapid computing gives us hope in many cases of estimating the parameters of the micro model from aggregate data without the necessity of explicitly computing, say, a closed form likelihood function. The SIMEST algorithm is explicated more fully elsewhere (e.g., [8], [9], [10]). Here we simply give an indication of its potential in a statistical process control setting. The reader may find it useful to review the Poisson process section in the Appendix.

Let us consider the following axiomitization of a plausible mechanism by which system errors are generated and corrected.

(1) Following the standard assumptions of an homogeneous Poisson process, the probability that a system error not caused by a prior system error will appear in the infinitesimal interval $[t, t + \Delta t]$ is given by

$$P_1(\text{system error in } [t, t + \Delta t]) = \lambda \Delta t.$$

Thus, the cumulative distribution function of time to occurrence of a new system error is given by

$$F_1(t) = 1 - e^{-\lambda t}.$$

(2) At its origin, the "effect" of the system error is 1. As time progresses, it grows exponentially, i.e., s time units after the error is created, its effect is given by

$$E = e^{\alpha s}.$$

(3) The probability that a system error, not discovered previously, is caught and eliminated in the time interval $[s, s + \Delta s]$ is proportional to the "effect" of the system error, i.e.,

$$P_2(\text{ system error caught in } [s, s + \Delta s]) = \gamma e^{\alpha s} \Delta s.$$

After a little integration and algebra, we see that the cdf of time after its generation until the detection of a system error is given by

$$\begin{aligned} F_2(s_D) = 1 - P[\text{no detection by } s_D] &= 1 - \exp(-\int_0^{s_D} \gamma e^{\alpha \tau} d\tau) \\ &= 1 - \exp(-\frac{\gamma}{\alpha}[e^{\alpha s_D} - 1]). \end{aligned}$$

We note that the time origin of s_D is the instant where the new system error came into being.

(4) The probability that an existing system error will itself generate a new system error somewhere else in the system during the interval $[s, s + \Delta s]$ is proportional to the effect of the system error, i.e.,

$$P_3(\text{new secondary system error generated in } [s, s + \Delta s]) = \phi e^{\alpha s} \Delta s.$$

Hence, the cdf of time until creation of a new (secondary) system error is given by

$$F_3(s_{S,1}) = 1 - \exp(-\frac{\phi}{\alpha}[e^{\alpha s_{S,1}} - 1]).$$

(5) Once generated, each system error has the same underlying kinetic characteristics in terms of effect, spread and detection as any other system error.

Now, from the real aggregate world, all that we observe is the discovery and correction times of system errors. In order to take these times, say $\{T_1, T_2, \ldots, T_M\}$, and employ them for the estimation of the characterizing parameters, α, λ, γ, ϕ, in a "closed form" setting would require the computation of the likelihood function or some such surrogate. Experience [8] indicates that this kind of exercise is exceptionally time consuming. Rapid computing gives us a means of bypassing this step.

Let us take a time interval in a simulation which is equal to that of the observed data train. Assuming a value for the vector parameter (α, λ, γ, ϕ), we begin to generate "pseudo" system errors. We note that the task for accomplishing this is relatively easy, since, for any random variable say, X, with increasing (in X) cdf $F(X)$, the random variable $F(X)$ is uniform on the unit interval. To see this, we note that if we look at the cdf G of $F(X)$, we have

$$G(y) = P[F(X) \le y] = P[X \le F^{-1}(y)] = F(F^{-1}(y))$$
$$= y.$$

Thus, we can simulate the time of a system error by generating a random uniform variate u and then solving for t in

$$u = F_1(t) = 1 - e^{-\lambda t}. \tag{6.157}$$

Using this t as the starting point for the time at which there is a risk of the generation of "secondary" system errors, we can generate these using the relationship that if u is a uniform random observation over the unit interval, then we can solve for $s_{S,1}$ using

$$u = F_3(s_{S,1}) = 1 - \exp(-\frac{\phi}{\alpha}[e^{\alpha s_{S,1}} - 1]). \tag{6.158}$$

We then generate the time of the discovery of the first primary system error by generating the uniform random variable u and then solving for s_D from

$$u = F_2(s_D) = 1 - \exp(-\frac{\gamma}{\alpha}[e^{\alpha s_D} - 1]). \tag{6.159}$$

In the event that

$$s_D < s_{S,1} \tag{6.160}$$

then we will not have to worry that the primary system error has generated a secondary system error. But if

$$s_D \ge s_{S,1} \tag{6.161}$$

then a secondary system error will have been generated, and we will have to go through a similar exercise to see what additional errors it might have caused, and so on. (Indeed, in such a case, we will have to generate tertiary system errors which may have been generated by the secondary ones.) For the first of these, clearly we have

$$u = F_2(s_{S,2}) = 1 - \exp(-\frac{\gamma}{\alpha}[e^{\alpha s_{S,2}} - e^{\alpha s_{S,1}}]). \tag{6.162}$$

A complete flowcharting of this sort of simulation is straightforward though nontrivial. For further details in dealing with such a simulation, we refer the reader to [8], [9], [10]. But, after we have computed, say 10,000, simulations, we note the average number of pseudo system errors discovered up to time T_1, say n_1, the average number of pseudo system errors after T_1 and before T_2, say n_2, etc. Then a natural criterion function for the appropriateness of the parameters assumed might be a goodness of fit type of function, such as

$$\chi^2(\alpha, \lambda, \gamma, \phi) = \sum_{i=1}^{M+1} (n_i - 1)^2. \qquad (6.163)$$

Utilizing a standard optimization routine such as the Nelder-Mead algorithm described in Section 6.2, we then have a straightforward means of moving through the parameter space until we have good concordance between our real world data and that generated assuming a value of (α, λ, γ, ϕ).

References

[1] Adams, B.M. and Woodall, W.H. "An analysis of Taguchi's on-line process control model under a random-walk model," *Technometrics, 31*, pp. 401-413.

[2] Box, G.E.P. and Draper, N.R. (1969). *Evolutionary Operation*. New York: John Wiley & Sons.

[3] Box, G.E.P. and Draper, N.R. (1989). *Empirical Model-Building and Response Surfaces*. New York: John Wiley & Sons.

[4] Box, G.E.P. and Hunter, J.S. (1957). "Multifactor experimental designs for exploring response surfaces," *Annals of Statistics, 28*, pp. 195-241.

[5] Kendall, M.G. and Stuart, A.(1958). *The Advanced Theory of Statistics, I & II*. New York: Hafner.

[6] Nelder, J.A. and Mead, R. (1965). "A simplex method for function minimization," *Computational Journal, 7*, pp. 308-313.

[7] Roy, R.(1990). *A Primer on the Taguchi Method*. New York: Van Nostrand Reinhold.

[8] Thompson, J.R., Atkinson, E.N. and Brown, B.W. (1987). "SIMEST: An algorithm for simulation-based estimation of parameters characterizing a stochastic process," *Cancer Modeling*, Thompson, J.R. and Brown, B.W., eds., New York: Marcel Dekker, pp. 387-415.

[9] Thompson, J.R. (1989). *Empirical Model Building.* New York: John Wiley & Sons.

[10] Thompson, J.R. and Tapia, R.A. (1990). *Nonparametric Function Estimation, Modeling and Simulation.* Philadelphia: Society for Industrial and Applied Mathematics.

Problems

Remark: In Problems 6.1-6.3, extrema of functions of one variable are to be found. Such problems can also be solved using the Nelder-Mead simplex algorithm. When compared with problems with two control variables, each step of the algorithm is simplified in that the Best and the Second Worst points coincide. Whatever the dimension of an optimization problem, successful implementation of the simplex algorithm requires that a stopping rule be incorporated into the algorithm, so that it terminates in a finite time. The reader is asked to use stopping rules of his or her own choice.

Problem 6.1. Consider the function

$$f(x) = .1(x^4 - 20x^2 + 5x)$$

over the interval $x \in [-5, 5]$.

   a. Apply the simplex algorithm to find a minimum of the function (in the given interval). Use two points of your choice from the function domain as the starting points of the procedure. After stopping the algorithm, repeat the search several times, using another starting point. Plot the function using a computer graphics software and compare your results with those given by the plot.

   b. Repeat **a** with a noise corrupting the readings of the function values: Whenever a function evaluation is needed, add to the function value a random variable from $\mathcal{N}(0, .0025)$, generated by a computer's random number generator. Random errors corrupting the readings should form a set of independent random variables. Use the same starting points as in **a**.

c. Repeat **b** with a random noise from $\mathcal{N}(0, .01)$.
d. Repeat **b** with a random noise from $\mathcal{N}(0, .25)$.
e. Repeat **b** with a random noise from $\mathcal{N}(0, 1)$.

Problem 6.2. Consider the function

$$f(x) = \frac{\sin(4x)}{x}$$

over the interval $x \in [-1, 6]$.

a. Apply the simplex algorithm to find a maximum of the function (in the given interval). Use two points of your choice from the function domain as the starting points of the procedure. After stopping the algorithm, repeat the search several times, using other starting points. Plot the function using a computer graphics software and compare your results with those given by the plot.

b. Repeat **a** with a noise corrupting the readings of the function values: Whenever a function evaluation is needed, add to the function value a random variable from $\mathcal{N}(0, .0025)$, generated by a computer's random number generator. Random errors corrupting the readings should form a set of independent random variables. Use the same starting points as in **a**.

c. Repeat **b** with a random noise from $\mathcal{N}(0, .01)$.
d. Repeat **b** with a random noise from $\mathcal{N}(0, .25)$.
e. Repeat **b** with a random noise from $\mathcal{N}(0, 1)$.

Problem 6.3. Consider the function

$$f(x) = x\sin(x)$$

over the interval $x \in [0, 12]$.

a. Apply the simplex algorithm to find a minimum of the function (in the given interval). Use two points of your choice from the function domain as the starting points of the procedure. After stopping the algorithm, repeat the search several times, using other starting points. Plot the function using a computer graphics software and compare your results with those given by the plot.

b. Repeat **a** with a noise corrupting the readings of the function values: Whenever a function evaluation is needed, add to the function value a random variable from $\mathcal{N}(0, .0025)$, generated by a computer's random number generator. Random errors corrupting the readings should form a set of independent random variables. Use the same starting points as in **a**.

c. Repeat **b** with a random noise from $\mathcal{N}(0, .01)$.

d. Repeat **b** with a random noise from $\mathcal{N}(0, .25)$.

e. Repeat **b** with a random noise from $\mathcal{N}(0, 1)$.

f. Repeat **a** to **e** replacing minimization function by maximization.

Problem 6.4. Consider the function

$$f(x_1, x_2) = x_1 \sin(x_1) + x_2 \sin(x_2)$$

over the square $x_1 \in [0, 12]$, $x_2 \in [0, 12]$.

a. Apply the simplex algorithm to find a minimum of the function (in the given square). Use three points of your choice from the function domain as the starting points of the procedure. After stopping the algorithm, repeat the search several times, using other starting points. Plot the function using a computer graphics software and compare your results with those given by the plot.

b. Repeat **a** with a noise corrupting the readings of the function values: Whenever a function evaluation is needed, add to the function value a random variable from $\mathcal{N}(0, .0025)$, generated by a computer's random number generator. Random errors corrupting the readings should form a set of independent random variables. Use the same starting points as in **a**.

c. Repeat **b** with a random noise from $\mathcal{N}(0, .01)$.

d. Repeat **b** with a random noise from $\mathcal{N}(0, .25)$.

e. Repeat **b** with a random noise from $\mathcal{N}(0, 1)$.

Problem 6.5. Consider the function

$$f(x_1, x_2) = .5(x_1^2 + x_2^2)$$

over the square $x_1 \in [-4, 4]$, $x_2 \in [-4, 4]$.

a. Apply the simplex algorithm to find a minimum of the function (in the given square). Use three points of your choice from the function domain as the starting points of the procedure. After stopping the algorithm, repeat the search several times, using other starting points. Plot the function using a computer graphics software and compare your results with those given by the plot.

b. Repeat **a** with a noise corrupting the readings of the function values: Whenever a function evaluation is needed, add to the function value a random variable from $\mathcal{N}(0, .0025)$, generated by a computer's random number generator. Random errors corrupting the readings should form a set of independent random variables. Use the same starting points as in **a**.

c. Repeat **b** with a random noise from $\mathcal{N}(0, .01)$.

d. Repeat **b** with a random noise from $\mathcal{N}(0, .25)$.

e. Repeat **b** with a random noise from $\mathcal{N}(0, 1)$.

Problem 6.6. Consider the function

$$f(x_1, x_2) = \frac{2\sin(2\sqrt{x_1^2 + x_2^2})}{\sqrt{x_1^2 + x_2^2}}$$

over the square $x_1 \in [-5, 5]$, $x_2 \in [-5, 5]$ (the function has been borrowed from a manual of the SYSTAT statistical package).

a. Apply the simplex algorithm to find a minimum of the function (in the given square). Use three points of your choice from the function domain as the starting points of the procedure. After stopping the algorithm, repeat the search several times, using other starting points. Plot the function using a computer graphics software and compare your results with those given by the plot.

b. Repeat **a** with a noise corrupting the readings of the function values: Whenever a function evaluation is needed, add to the function value a random variable from $\mathcal{N}(0, .0025)$, generated by a computer's random number generator. Random errors corrupting the readings should form a set of independent random variables. Use the same starting points as in **a**.

c. Repeat **b** with a random noise from $\mathcal{N}(0, .01)$.

d. Repeat **b** with a random noise from $\mathcal{N}(0, .25)$.

e. Repeat **b** with a random noise from $\mathcal{N}(0, 1)$.

Problem 6.7. Consider the following data set.

X	Y	X	Y
1.0	1.5	2.6	1.262
1.1	1.022	2.7	1.277
1.2	1.050	2.8	1.295
1.3	1.055	2.9	1.306
1.4	1.079	3.0	1.323
1.5	1.120	3.1	1.344
1.6	1.113	3.2	1.342
1.7	1.184	3.3	1.352
1.8	1.160	3.4	1.354
1.9	1.174	3.5	1.369
2.0	1.174	3.6	1.383
2.1	1.198	3.7	1.382
2.2	1.218	3.8	1.391
2.3	1.218	3.9	1.393
2.4	1.250	4.0	1.405
2.5	1.258		

Use the transformational ladder to find the transformation that (approximately) linearizes the relationship between the independent variable X and dependent variable Y.

Problem 6.8. Consider the following data set.

X	Y	X	Y
1.0	0.320	2.6	45.652
1.1	0.587	2.7	53.058
1.2	1.279	2.8	62.228
1.3	2.775	2.9	70.982
1.4	2.805	3.0	80.754
1.5	3.709	3.1	94.021
1.6	8.808	3.2	104.442
1.7	9.917	3.3	119.730
1.8	10.548	3.4	135.101
1.9	13.620	3.5	150.323
2.0	15.847	3.6	166.241
2.1	19.652	3.7	184.782
2.2	23.211	3.8	208.577
2.3	28.933	3.9	230.051
2.4	33.151	4.0	256.714
2.5	39.522		

Use the transformational ladder to find the transformation that (approximately) linearizes the relationship between the independent variable X and dependent variable Y.

Problem 6.9. Perform the following experiment. Tabulate the function $Y = \ln(X)$ for $X = 3.1, 3.2, 3.3, \ldots, 5.8, 5.9, 6.0$.

a. Add normal noise of mean 0 and standard deviation .01 to the function readings, that is, use a computer's random number generator to generate a sequence of thirty normal variates from $\mathcal{N}(0, (.01)^2)$ and add these variates to successive readings of the function. Using the transformational ladder, attempt to find the transformation that (approximately) linearizes the relationship between X and Y.

b. Repeat **a** with normal noise of mean 0 and standard deviation .02.

c. Repeat **a** with normal noise of mean 0 and standard deviation .05.

d. Repeat **a** with normal noise of mean 0 and standard deviation .1. Comment on your results.

Problem 6.10. Perform the following experiment. Tabulate the function $Y = \exp(X)$ for $X = 2.1, 2.2, 2.3, \ldots, 4.8, 4.9, 5.0$.

a. Add normal noise of mean 0 and standard deviation .01 to the function readings, that is, use a computer's random number generator to generate a sequence of thirty normal variates from $\mathcal{N}(0, (.01)^2)$ and add these variates to successive readings of the function. Using the transformational ladder, attempt to find the transformation that (approximately) linearizes the relationship between X and Y.

b. Repeat **a** with normal noise of mean 0 and standard deviation .02.

c. Repeat **a** with normal noise of mean 0 and standard deviation .05.

d. Repeat **a** with normal noise of mean 0 and standard deviation .1. Comment on your results. Compare your conclusions with those from Problem 6.9.

Problem 6.11. It is believed that the variability of a production process is a quadratic function of two decision variables, X_1 and X_2. The variables have been linearly transformed in such a way that the perceived optimum lies at $X_1 = X_2 = 0$. For the given rotatable quadratic design, the following function readings have been obtained.

j	X_0	X_{1j}	X_{2j}	X_{1j}^2	X_{2j}^2	$X_{1j}X_{2j}$	y_j
1	1	1	1	1	1	1	3.365
2	1	-1	1	1	1	-1	3.578
3	1	1	-1	1	1	-1	12.095
4	1	-1	-1	1	1	1	11.602
5	1	0	0	0	0	0	5.699
6	1	0	0	0	0	0	5.320
7	1	1.414	0	2	0	0	9.434
8	1	-1.414	0	2	0	0	8.145
9	1	0	1.414	0	2	0	0.098
10	1	0	-1.414	0	2	0	12.431

Estimate the coefficients of the quadratic model in a neighborhood of the origin. Test the null hypothesis that the term with X_1X_2 is negligible. What should be the next step (or steps) of minimizing the process variability?

Problem 6.12. It is believed that the variability of a production process is a quadratic function of three decision variables, X_1, X_2 and X_3. The variables have been linearly transformed in such a way that the perceived optimum lies at $X_1 = X_2 = X_3 = 0$. For the given rotatable quadratic design, the following function readings have been obtained.

j	X_0	X_{1j}	X_{2j}	X_{3j}	y_j
1	1	1	1	1	4.912
2	1	-1	1	1	4.979
3	1	1	-1	1	13.786
4	1	-1	-1	1	13.065
5	1	1	1	-1	4.786
6	1	-1	1	-1	4.253
7	1	1	-1	-1	14.295
8	1	-1	-1	-1	13.659
9	1	0	0	0	6.188
10	1	0	0	0	5.650
11	1	1.682	0	0	11.023
12	1	-1.682	0	0	11.236
13	1	0	1.682	0	2.196
14	1	0	-1.682	0	15.159
15	1	0	0	1.682	6.699
16	1	0	0	1.682	6.320

Estimate the coefficients of the quadratic model in a neighborhood of the origin (note that the given design implies the values of X_{1j}^2, X_{2j}^2, ..., $X_{2j}X_{3j}$). Test the null hypothesis that the terms with X_1X_2, X_1X_3

and X_2X_3 are all negligible. Test the negligibility of each of the terms mentioned separately. Test the null hypothesis that the term with X_1^2 is negligible. Test the null hypothesis that the term with X_2^2 is negligible. Test the null hypothesis that the term with X_3^2 is negligible. Estimate the variance of the error terms. What should be the next step (or steps) of minimizing the process variability?

Problem 6.13. Perform the following experiment. Suppose the problem consists in finding the minimum of the quadratic function

$$f(x) = 3 + (x_1 - 3)^2 + (x_2 - 2)^2 + 2(x_3^2 - 4)^2$$

when function evaluations are subject to normally distributed random errors with mean 0 and variance 1. Simulate the process of searching for the minimum, starting from a neighborhood of the origin and:

a. estimating the coefficients of the linear approximation of the function then making one step in the steepest descent direction, and, finally, approximating the function by a quadratic;

b. implementing the Nelder-Mead simplex algorithm;

c. comparing the two approaches.

Chapter 7

Multivariate Approaches

7.1 Introduction

In very many cases, multivariate time indexed data are available for analysis. Yet it is rather standard for investigators to carry out their searches for Pareto glitches by control charting the data base one dimension at a time. A little modeling of the SPC situation gives us an indication that we are dealing with a rather special multivariate scenario. Multivariate tests can easily be constructed which give us the potential of substantial improvement over the one factor at a time approach.

We recall that the density function of a variable $\mathbf{x}$ of dimension p from a multivariate normal distribution is given by

$$f(\mathbf{x}) = |2\pi\boldsymbol{\Sigma}|^{-1/2} \exp\{-\frac{1}{2}(\mathbf{x} - \boldsymbol{\mu})'\boldsymbol{\Sigma}^{-1}(\mathbf{x} - \boldsymbol{\mu})\}, \qquad (7.1)$$

where $\boldsymbol{\mu}$ is a constant vector and $\boldsymbol{\Sigma}$ is a constant positive definite matrix. By analogy with the use of control charts to find a change in the distribution of the output and/or the input of a module, we can describe a likely scenario of a process going out of control as the mean suddenly changes from, say, $\boldsymbol{\mu}_0$ to some other value. One approach to discovering the point in time where a glitch occurred would be to find lots where $\boldsymbol{\mu} \neq \boldsymbol{\mu}_0$.

Let us note that dealing with multivariate data often requires that the data be standardized prior to an analysis, i.e., that they be mean centered and rescaled to have components with unit variance. In particular, standardization is usually recommended if observation components are measured in different units.

7.2   Likelihood Ratio Tests for Location

First of all, let us consider the situation where we believe that the shift in the mean will not be accompanied by a significant shift in the variance. In this case, if we have a long history of data from a process in control, we can (see Appendix) transform the data so that the covariance matrix is diagonal, i.e.,

$$f(\mathbf{x}; \boldsymbol{\mu}) = \prod_{i=1}^{p}(2\pi\sigma_{ii})^{-1/2}\exp\{-\frac{1}{2}(x_i - \mu_i)\sigma_{ii}^{-1}(x_i - \mu_i)\}. \qquad (7.2)$$

We will assume that the variables have been shifted so that in the "in control" situation

$$\boldsymbol{\mu} = \mathbf{0}. \qquad (7.3)$$

Here, the likelihood ratio statistic is given by

$$\lambda(\mathbf{x}_1, \mathbf{x}_2, \ldots, \mathbf{x}_n) = \frac{\prod_{i=1}^{n} f(\mathbf{x}; \bar{\mathbf{x}})}{\prod_{i=1}^{n} f(\mathbf{x}; \mathbf{0})}. \qquad (7.4)$$

But then, after a little algebra, we are left with the test statistic

$$Q_1 = \sum_{j=1}^{p}(\frac{\bar{x}_j}{\sigma_j/\sqrt{n}})^2. \qquad (7.5)$$

If, indeed, the process is under control (i.e., $\boldsymbol{\mu} = \mathbf{0}$), then we are left with the sum of squares of p independent normal random variables with means zero and unit variances and the statistic Q_1 is distributed as χ^2 with p degrees of freedom.

In the situation where we believe that the variance structure may change when the mean changes, we can estimate the covariance matrix $\boldsymbol{\Sigma}$ in the obvious fashion (using the maximum likelihood estimator) via

$$\mathbf{S} = \frac{1}{n-1}\sum_{i=1}^{n}(\mathbf{x_i} - \bar{\mathbf{x}})(\mathbf{x_i} - \bar{\mathbf{x}})'. \qquad (7.6)$$

We wish to obtain a likelihood ratio test to test the null hypothesis

$$\mathrm{H}_0\colon \boldsymbol{\mu} = \boldsymbol{\mu}_0. \qquad (7.7)$$

Going through a fair amount of straightforward algebra, we obtain the likelihood ratio statistic

$$T^2 = n(\bar{\mathbf{x}} - \boldsymbol{\mu}_0)'\mathbf{S}^{-1}(\bar{\mathbf{x}} - \boldsymbol{\mu}_0). \tag{7.8}$$

Essentially, the likelihood ratio test (Hotelling's T^2) is based on comparing the spread of a lot of size n about its sample mean with the spread of the same lot about $\boldsymbol{\mu}_0$. Further algebra demonstrates that we can obtain a probability of rejecting the null hypothesis when it is truly equal to α, by using as the critical region those values of T^2 which satisfy

$$T^2 > \frac{p(n-1)}{n-p} F_{p,n-p}(\alpha), \tag{7.9}$$

where $F_{p,n-p}(\alpha)$ is the upper (100α)th percentile of the $F_{p,n-p}$ distribution.

We have assumed that a process goes "out of control" when the normal distribution of the output changes its vector mean from $\boldsymbol{\mu}_0$ to some other value, say $\boldsymbol{\mu}$. Let us define the dispersion matrix (covariance of the set of estimates $\hat{\boldsymbol{\mu}}$),

$$\mathbf{V}_{(p \times p)} = \begin{bmatrix} \mathrm{Var}(\hat{\mu}_1) & \mathrm{Cov}(\hat{\mu}_1, \hat{\mu}_2) & \cdots & \mathrm{Cov}(\hat{\mu}_1, \hat{\mu}_p) \\ \mathrm{Cov}(\hat{\mu}_1, \hat{\mu}_2) & \mathrm{Var}(\hat{\mu}_2) & \cdots & \mathrm{Cov}(\hat{\mu}_2, \hat{\mu}_p) \\ \vdots & \vdots & & \vdots \\ \mathrm{Cov}(\hat{\mu}_1, \hat{\mu}_p) & \mathrm{Cov}(\hat{\mu}_2, \hat{\mu}_p) & \cdots & \mathrm{Var}(\hat{\mu}_p) \end{bmatrix}. \tag{7.10}$$

Here, naturally,

$$\hat{\mu}_j = \bar{x}_j = \frac{1}{n}\sum_{i=1}^{n} x_{ij} \tag{7.11}$$

for each j.

It is of interest to investigate the power (probability of rejection of the null hypothesis) of the Hotelling T^2 test. This can be done as a function of the noncentrality

$$\lambda = (\boldsymbol{\mu} - \boldsymbol{\mu}_0)'\mathbf{V}_{(p \times p)}^{-1}(\boldsymbol{\mu} - \boldsymbol{\mu}_0). \tag{7.12}$$

For n very large we can use an asymptotic (in n) χ^2 approximation for the power of the Hotelling T^2 test:

$$P(\lambda) = \int_{(\frac{p+\lambda}{p+2\lambda})\chi_\alpha^2(p)}^{\infty} d\chi^2(p + \frac{\lambda^2}{p+2\lambda}), \tag{7.13}$$

where $d\chi^2(p)$ is the differential of the cumulative distribution function of the central χ^2 distribution with p degrees of freedom and $\chi^2_\alpha(p)$ its $100(1 - \alpha)$ per cent point. We demonstrate below power curves for the Hotelling T^2 test for various sample sizes, n, and various noncentralities λ. The dimensionalities are one through five in Figures 7.1 through 7.5 and ten in Figure 7.6.

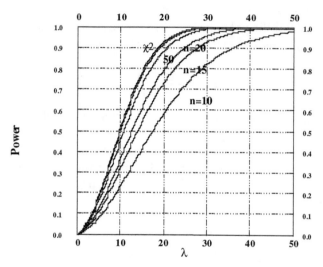

Figure 7.1. Power Curves for One Dimensional Data.

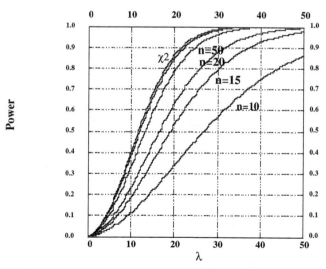

Figure 7.2. Power Curves for Two Dimensional Data.

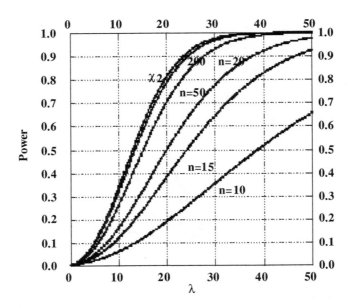

Figure 7.3. Power Curves for Three Dimensional Data.

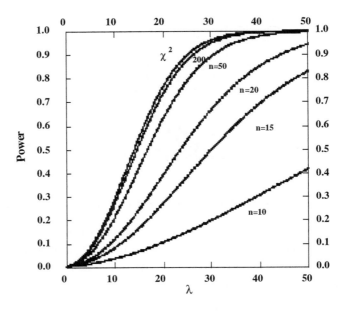

Figure 7.4. Power Curves for Four Dimensional Data.

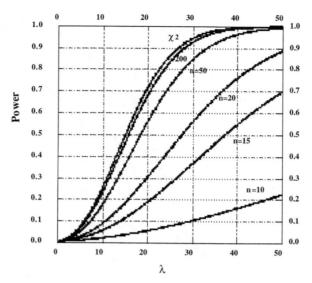

Figure 7.5. Power Curves for Five Dimensional Data.

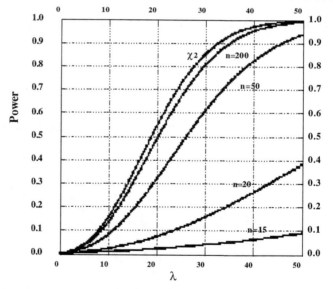

Figure 7.6. Power Curves for Ten Dimensional Data.

Let us consider, next, the situation where we wish to compare the performance of the multidimensional Hotelling T^2 test with α = probability of Type I error equal to .002 with the one dimension at a time test. We consider, by way of comparison, a situation where

$$H_0 : \lambda = (\boldsymbol{\mu} - \boldsymbol{\mu}_0)' \mathbf{V}^{-1}_{(p \times p)} (\boldsymbol{\mu} - \boldsymbol{\mu}_0) = 0. \qquad (7.14)$$

We consider the alternative

$$H_1 : \lambda > 0. \tag{7.15}$$

We are assuming here that the noncentrality is equal in each dimension. Suppose, for example, that we wish to compare the one dimension at a time test with Hotelling's T^2 in the case of p dimensions. For noncentrality of 0, we wish to have probability of rejection equal to .002. This means that for each one dimensional test, we design a test with $\alpha = 1 - .998^p$.

In Figure 7.7, we show the ratio of noncentrality per dimension for the battery of one dimensional tests required to give the same power as that of the multivariate likelihood test given in (7.5). The efficiency is plotted versus the noncentrality per dimension used in the Hotelling T^2 test.

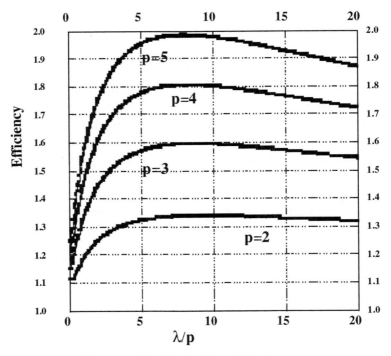

Figure 7.7. Relative Efficiency.

We note that the efficiencies in this case exceed 100%. Let us consider a particular example where the efficiencies of the multivariate test are even more impressive. Suppose that

$$H_0 : \boldsymbol{\mu} = \mathbf{0}. \tag{7.16}$$

We shall assume a covariance matrix of

$$\boldsymbol{\Sigma}_{(5\times5)} = \begin{bmatrix} 1 & .8 & .8 & .8 & .8 \\ .8 & 1 & .8 & .8 & .8 \\ .8 & .8 & 1 & .8 & .8 \\ .8 & .8 & .8 & 1 & .8 \\ .8 & ..8 & .8 & .8 & 1 \end{bmatrix}. \tag{7.17}$$

We are assuming here an α level of .002 for both the battery of five one dimensional tests and the Hotelling T^2 five dimensional test. For equal slippage in each dimension, we show the ratio of the sample size required to obtain the power of .9 with the battery of one dimensional tests with that required to obtain a power of .9 with the Hotelling test. The "slippage per dimension" abscissa is the absolute value per dimension of each of the means. In this case, we have assumed that the "slippages" in the first two dimensions are positive; the last three are negative. The irregularities in Figure 7.8 are due to the granularity effects of discrete sample sizes.

Let us now consider explicitly the situation when we are given several, say N, lots of size n of p-dimensional data. In order to verify whether the lots are "in control," we can then use (7.8) repeatedly. Whenever the value of the Hotelling's T^2 statistic is greater than the right hand side of (7.9), the lot is deemed to be "out of control." Thus, the right hand side of (7.9) is in fact the Upper Control Limit for the in-control values of the lots' T^2 statistics. The Lower Control Limit is simply equal to zero and is of no interest, since the T^2 statistic is necessarily non-negative. Charting the lots' T^2 values against the UCL gives us a multidimensional counterpart of the standard $\bar{X}$ chart for univariate data.

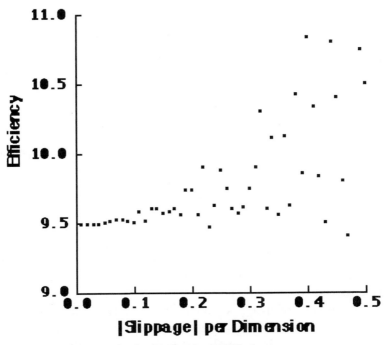

Figure 7.8. Relative Efficiency.

If the lot size is small, as is frequently the case, the sample dispersion matrix $\mathbf{S}$ can sometimes fail to reveal the true correlation structure of data under scrutiny. A way out of the problem consists in constructing a "pooled" sample dispersion matrix, the same for all lots and equal to the average of N lot sample dispersion matrices.

Let us consider this last proposal in greater detail. Let $\mathbf{S}_j$ denote the sample dispersion matrix of the jth lot and let

$$\bar{\mathbf{S}} = \frac{1}{N} \sum_{j=1}^{N} \mathbf{S}_j.$$

In order to be in full analogy with the one dimensional case, let us also replace $\boldsymbol{\mu}_0$ by its sample counterpart,

$$\bar{\bar{\mathbf{x}}} = \frac{1}{N} \sum_{j=1}^{N} \bar{\mathbf{x}}_j,$$

where $\bar{\mathbf{x}}_j$ is the sample mean for the jth lot. Now the T^2-like statistic for the jth lot assumes the form

$$T_j^2 = n(\bar{\mathbf{x}}_j - \bar{\bar{\mathbf{x}}})' \bar{\mathbf{S}}^{-1} (\bar{\mathbf{x}}_j - \bar{\bar{\mathbf{x}}}), \qquad (7.18)$$

where $j = 1, 2, \ldots, N$. It can be shown (see Alt, Goode and Wadsworth [5]) that

$$\frac{nN - N - p + 1}{p(n-1)(N-1)} T_j^2$$

has F distribution with p and $nN - N - p + 1$ degrees of freedom. Thus, we consider the jth lot to be out of control if

$$T_j^2 > \frac{p(n-1)(N-1)}{nN - N - p + 1} F_{p,nN-N-p+1}(\alpha), \qquad (7.19)$$

where $F_{p,nN-N-p+1}(\alpha)$ is the upper (100α)th percentile of the F distribution with p and $nN - N - p + 1$ degrees of freedom. In other words, the right hand side of (7.19) is the UCL for the T_j^2 statistics. Usually, α is set equal to .002.

As in the case of using $\bar{X}$ charts for detecting Pareto glitches among lots of one dimensional data, the above analysis enables one to examine future multivariate data on the basis of the past. Suppose N_1 lots have already been examined, N of them being "in control." Suppose also that assignable causes for $N_1 - N$ glitches have been found and removed. Thus, we can recompute the sample mean $\bar{\bar{\mathbf{x}}}$ and the pooled dispersion matrix $\mathbf{S}$ for the remaining N lots. Now, the T^2-like statistic can be constructed for a new lot with the sample mean $\bar{\mathbf{x}}_*$,

$$T_*^2 = n(\bar{\mathbf{x}}_* - \bar{\bar{\mathbf{x}}})' \bar{\mathbf{S}}^{-1} (\bar{\mathbf{x}}_* - \bar{\bar{\mathbf{x}}}). \qquad (7.20)$$

Alt, Goode and Wadsworth have shown that

$$\frac{nN - N - p + 1}{p(n-1)(N+1)} T_*^2$$

has the F distribution with p and $nN - N - p + 1$ degrees of freedom. Thus, we can consider the new lot to be out of control if

$$T_*^2 > \frac{p(n-1)(N+1)}{nN - N - p + 1} F_{p,nN-N-p+1}(\alpha). \qquad (7.21)$$

We note that the UCL's provided by (7.19) and (7.21) are slightly different. This difference is due to the fact that in the latter case the new lot is not used when computing the sample mean $\bar{\bar{\mathbf{x}}}$ and the pooled dispersion matrix $\mathbf{S}$. If N is sufficiently large, both limits are practically the same. Similarly, we can replace the sample mean $\bar{\bar{\mathbf{x}}}$ by $\boldsymbol{\mu}_0$ in (7.18) and (7.20), and still use the UCL's given by (7.19) and (7.21), provided N is not too small.

The above T^2 tests are, of course, for location or the process mean. The first of them, given by (7.19), is for past observations, while that given by (7.21) is for future observations. Both tests are for p-dimensional observations grouped into lots of size greater than 1 (actually, of size $n > 1$ in our considerations). Given individual (i.e., ungrouped) past observations, it is an easy exercise to construct a Hotelling's T^2 chart for future observations, but developing such a chart for past individual data requires more effort. The latter chart has been provided by Wierda, and both are described in Wierda [21]; see also a discussion by Ryan [17], Section 9.5.

We shall conclude this Section with an illustration why a T^2 test behaves differently from a corresponding battery of one dimensional tests. The reason is in fact rather obvious and can be stated briefly: it is only the former test which takes into account possible correlations between vector components. For the sake of illustration, we shall confine ourselves to the case $p = 2$.

However, before we proceed with the illustration, let us pause for a moment on the problem of proper choice of (common) significance level for one dimensional tests. If the two vector components were stochastically independent, one would require that the two one dimensional tests have significance level β, where

$$1 - \alpha = (1 - \beta)^2.$$

Clearly, given that α is small, we would then have

$$1 - \alpha \approx 1 - 2\beta$$

and hence we could use

$$\beta = \alpha/2.$$

Interestingly, it follows from the celebrated Bonferroni inequality that we should use $\beta = \alpha/2$ also when the vector components are not independent. Namely, in its full generality, the Bonferroni inequality states that if we are given p events A_i, $i = 1, 2, \ldots, p$, and if

$$P(A_i) = 1 - \alpha/p,$$

then

$$P(A_1, A_2, \ldots, A_p) \geq 1 - \alpha.$$

Thus, if we construct p (2 in our case) one dimensional tests, each of significance level α/p, then the significance level for the whole battery of p tests applied simultaneously is not greater than α, whatever the stochastic relationships between the p variables.

Returning to our illustration of how a T^2 test differs from a corresponding battery of one dimensional tests, let us assume that we are given a sample of 25 bivariate normally distributed observations with known covariance matrix

$$\boldsymbol{\Sigma} = \begin{bmatrix} 1 & 0.7 \\ 0.7 & 1 \end{bmatrix}. \tag{7.22}$$

Assume that we want to test hypothesis (7.3) at the $\alpha = 0.002$ significance level. Since the covariance matrix is known, in lieu of Hotelling's (7.8), we can use the χ^2 test with statistic

$$25\bar{\mathbf{x}}'\boldsymbol{\Sigma}^{-1}\bar{\mathbf{x}}$$

distributed under H_0 as χ^2 with p (in our case 2) degrees of freedom. Accordingly, we get an elliptical acceptance set with boundary

$$25\bar{\mathbf{x}}'\boldsymbol{\Sigma}^{-1}\bar{\mathbf{x}} = 12.429. \tag{7.23}$$

On the other hand, the two one dimensional tests (each at the 0.001 level) lead to the square acceptance set with vertices $(\pm 0.658, \pm 0.658)$, since $3.29/5 = 0.658$.

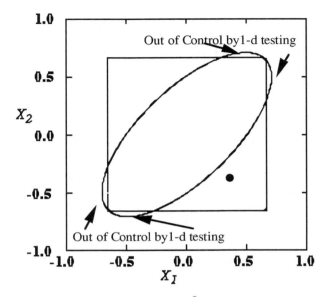

Figure 7.9. Acceptance Sets for the T^2 and Simultaneous One Dimensional Tests.

Both sets are given in Figure 7.9. It is clear that due to a rather strong positive dependence between the two variates of an observation (correlation is equal to 0.7) the two dimensional test will ring an alarm if one variate is large while the other is small (see observation $(0.4, -0.4)$ in Figure 7.9). At the same time, each of the variates may well stay within the one dimensional limits, since the one dimensional tests are by definition incapable of taking into account the association between the variates comprising an observation. The two dimensional test and the corresponding battery of one dimensional tests become close to one another when the variates involved are independent and both have the same variance (or both are standardized), since then an ellipse (ellipsoid for $p > 2$) becomes a circle (sphere or hypersphere for $p > 2$).

Remark: In our exposition of multivariate tests, and hence of control charts, which are all based on normality assumption for observations, we skip tests and charts for process variability or dispersion. Their use is rather limited and they have rather apparent drawbacks. For their discussion see, e.g., Alt [3] and [4] and Ryan [17]. Let us mention, however, that charts for future observations either draw from known results for likelihood ratio tests for the dispersion matrix Σ or are based on tests for the determinant $|\mathbf{S}|$ of $\mathbf{S}$, the sample dispersion matrix. Charts for past observations are usually based on the determinant mentioned.

7.3   Compound and Projection Tests

Most SPC professionals analyze their time indexed data one dimension at a time. We propose here a compound test which begins by looking at all the one dimensional control charts. As an example, let us suppose p = 5. Then, if we use the customary α level of .002, we will declare that the null hypothesis appears to be false and that a lot appears to be out of control if $\bar{x}_j$ falls outside the interval $(\mu_{0,j} \pm 3\sigma_j/\sqrt{n})$ for any j between 1 and 5. In the event that we wish to use an estimate of variance based on the lot variability rather than that of the record of previous "in control" lots, we will reject the null hypothesis if, for any j, $\bar{x}_j$ falls outside the interval $(\mu_{0,j} \pm \mathbf{t}_{.998,n-1}s_j/\sqrt{n})$. Here, we recall that $\boldsymbol{\mu}_0$ is the vector of "in-control" means. Let us suppose that all five dimensional sample means have outputs stochastically independent of each other. Then, the actual α level of the five one dimensional tests is

$$\alpha = 1 - (1 - .002)^5 = .01. \tag{7.24}$$

In the proposed test, we next construct all two dimensional T^2 tests of nominal α level .002. There are 10 such tests. Then, we construct all three dimensional T^2 tests. There are 10 such tests. This is followed by all four dimensional T^2 tests. There are 5 such tests. Finally, we construct the 1 five dimensional T^2 test. Altogether, we will construct $2^5 - 1$ tests. If all the tests were independent, we would expect an actual α level of

$$\alpha = 1 - (1 - .002)^{31} = .06. \tag{7.25}$$

A false alarm level of 6 out of 100 lots would generally be rather high for practical quality control work. Fortunately, however, the tests are not independent. The failure of a lot on the two dimensional $\bar{x}_1, \bar{x}_2$ test is more likely also to fail the $\bar{x}_1$ test than if the two dimensional test had been passed.

In Table 7.1 we show simulation based estimates (50,000 simulated tests per table entry) of the actual α levels of the proposed compound test for a nominal α level of .002 for each test performed. The second column represents the actual α level if one uses only the one dimensional tests, with stochastic independence between the dimensions.

				Table 7.1				
p	1-d Tests	$n=5$	$n=10$	$n=15$	$n=20$	$n=50$	$n=100$	$n=200$
2	0.00400	0.00434	0.00466	0.00494	0.00508	0.00512	0.00528	0.00546
3	0.00599	.00756	0.00722	0.00720	0.00720	0.00704	0.00826	0.00802
4	0.00798	0.01126	0.01072	0.01098	0.01098	0.01108	0.01136	0.01050
5	0.00996	0.01536	0.01582	0.01552	0.01544	0.01706	0.01670	0.01728

We observe that the increase in Type I error by using all the T^2 tests as opposed to only the one dimensional tests is rather modest. For example, for the three dimensional case, using the two and three dimensional tests in addition to the one dimensional tests increases the actual α only around a third. And we know that the compound test always has greater power than using only the one dimensional tests. As a practical matter, the general default to the purely one dimensional tests may be due simply to the increased computational complexity associated with higher dimensional T^2 tests. But present generation hand held calculators easily admit of programming the more informative compound test.

Indeed, the compound test is extremely useful from the practical point of view. It not only happens that a multidimensional test detects a true Pareto glitch while the one dimensional tests fail to do so, but also the opposite is possible. For example, let us consider two dimensional data that are strongly positively dependent and have correlation close to 1. If, for a few lots, rather large values of one variate are associated with rather small values of the other, the two dimensional test may ring an alarm, since the T^2-like tests are sensitive to departures from the correlation structure implied by **S**. If, however, these values are not extremely large, one dimensional tests may fail to ring the alarm. On the other hand, if the two variates "vary together," the two dimensional test may not ring an alarm when some of these values happen to be unusually large (cf. Figure 7.9). It is the one dimensional test which is more sensitive to this sort of a glitch.

Using the compound test to advantage is a very good idea as long as the dimensionality of data is not too large. Clearly, with dimensionality increasing, explanatory power of the approach diminishes very quickly. It is here where statistical projection methods, most notably principal component analysis (PCA), should come in (for an excellent chapter-long introduction to PCA see Krzanowski [12] and for thorough book-long expositions see Jolliffe [10] and Jackson [8]).

In order to briefly introduce the reader to PCA concepts, let us return to our illustrative example in Section 7.2 and assume that we are given observations from bivariate normal distribution with mean zero and covariance matrix (7.22). Ellipses like the one given by (7.23) and Figure 7.9 are called the contours of the distribution (note that the normal density is constant on these ellipses). In general, for a p-variate normally distributed random vector $\mathbf{x}$ with mean $\boldsymbol{\mu}$ and covariance matrix $\boldsymbol{\Sigma}$, the contours assume the form of an ellipsoid

$$(\mathbf{x} - \boldsymbol{\mu})'\boldsymbol{\Sigma}^{-1}(\mathbf{x} - \boldsymbol{\mu}) = c^2, \tag{7.26}$$

where c is a positive constant. Let $\boldsymbol{\Gamma}\boldsymbol{\Lambda}\boldsymbol{\Gamma}'$ be the spectral decomposition of $\boldsymbol{\Sigma}$, i.e., $\boldsymbol{\Sigma} = \boldsymbol{\Gamma}\boldsymbol{\Lambda}\boldsymbol{\Gamma}'$, with $\boldsymbol{\Lambda} = diag(\lambda_1, \lambda_2, \ldots, \lambda_p)$ and $\boldsymbol{\Gamma} = [\boldsymbol{\gamma}_{(1)}\ \boldsymbol{\gamma}_{(2)}\ \cdots\ \boldsymbol{\gamma}_{(p)}]$, where $\boldsymbol{\gamma}_{(i)}$ is the eigenvector corresponding to eigenvalue λ_i (see Appendix A for the definition of spectral decomposition). We shall assume for later reference that $\lambda_1 \geq \lambda_2 \geq \ldots \geq \lambda_p > 0$. The direction cosine vector of the i-th axis of ellipsoid (7.26) is $\boldsymbol{\gamma}_{(i)}$. The length of the i-th axis is $2c\lambda_i^{1/2}$ (for the ellipse given by (7.23), $\lambda_1 = 0.068$ and $\lambda_2 = 0.012$).

For $i = 1, 2, \ldots, p$, let us consider now projections of the centered observation $\mathbf{x}$ on the ellipsoid's i-th axis,

$$y_i = \boldsymbol{\gamma}'_{(i)}(\mathbf{x} - \boldsymbol{\mu}). \tag{7.27}$$

One can show that

$$\mathrm{Var}(y_i) = \lambda_i, \tag{7.28}$$

$i = 1, 2, \ldots, p$, and that the y_i's are independent.

Taken together, the given p projections define a one-to-one transformation from $\mathbf{x}$'s to $\mathbf{y}$'s, where $\mathbf{y} = [y_1\ y_2\ \cdots\ y_p]'$:

$$\mathbf{y} = \boldsymbol{\Gamma}'(\mathbf{x} - \boldsymbol{\mu}). \tag{7.29}$$

In the above analysis, the normality assumption was needed only to have ellipsoids (7.26) be the contours of the underlying distribution and to prove that the y_i's are independent. In all other considerations, only the variance-covariance structure of the distribution was taken into account. In fact, for any continuous p-dimensional distribution, it can be shown that transformation (7.29), to be called the *principal component transformation*, which has components (7.27), the i-th of them to be called the i-th *principal component* of $\mathbf{x}$, has the following properties: (i) no linear combination $\mathbf{a}'\mathbf{x}$, where $\mathbf{a}$ is a vector of length 1, has variance larger than λ_1 and, thus, the first principal component, y_1, is that linear combination of the $\mathbf{x}$ variables (with the other vector standardized to have length 1) which has the largest variance; (ii) the second principal component, y_2, whose variance is λ_2, is that linear combination of the $\mathbf{x}$ variables (with the other vector standardized) which has the second greatest variance subject to the constraint that it is uncorrelated with the first principal component; (iii) the j-th principal component, y_j, $j \leq p$, whose variance is λ_j, is that linear combination of the $\mathbf{x}$ variables (with the other vector standardized) which has the j-th greatest variance subject to the constraint that it is uncorrelated with the first $j-1$ principal components.

We have noticed already that (7.29) is a one-to-one transformation. However, due to interdependencies between the $\mathbf{x}$ variables, it is usually the case that a few of the first λ's account for a large portion of the sum $\sum_i \lambda_i$ (put otherwise, usually a few λ's prove much larger than the remaining majority of λ's). Now, given that the y_i'a are uncorrelated, it is reasonable to consider the sum $\sum_i \lambda_i$ a measure of the overall variability hidden in the data and, hence, to claim that, under the condition just mentioned, a few first principal components account for most of the overall variability. Thus, it is then justified to reduce the original p-variate problem to one described by a much smaller number of principal components, as providing a good approximation in few dimensions to the original problem in many more dimensions. It is this reduced problem which is eventually subjected to SPC analysis.

It is another problem how to interpret out of control signals for principal components, and how to react to them. For a relatively early but thorough exposition of the problem, see Jackson [8]. However concise, an excellent and more recent survey has been given by MacGregor [14]. In that paper, a PCA approach particularly suited to dealing with batch processes has been also described (see also Nomikos and MacGregor [16]). Extensions of PCA approaches to the so-called dynamic biplots have been dealt with by Sparks et al. [18].

7.4   A Robust Estimate of "In Control" Location

We have seen in Chapter 3 how one might estimate location for one dimensional data in such a way that contaminants do not very much affect this estimate. The primary device used there was the median as an estimator for the center of the dominant "in control" distribution.

For higher dimensional data one needs more sophisticated trimming procedures. The following "King of the Mountain" algorithm of Lawera and Thompson [13] appears to be promising:

"King of the Mountain" Trimmed Mean Algorithm

1. Set the counter M equal to the historical proportion of bad lots times the number of lots.

2. For N lots compute the vector sample means of each lot $\{\bar{\mathbf{X}}_i\}_{i=1}^N$.

3. Compute the pooled mean of the means $\bar{\bar{\mathbf{X}}}$.

4. Find the two sample means farthest apart in the cloud of lot means.

5. From these two sample means, discard the farthest from $\bar{\bar{\mathbf{X}}}$.

6. Let $M = M - 1$ and $N = N - 1$.

7. If the counter is still positive, go to 1, otherwise exit and print out $\bar{\bar{\mathbf{X}}}$ as $\bar{\bar{\mathbf{X}}}_T$.

To examine the algorithm, we examine a mixture distribution of lot means

$$\gamma N(\mathbf{0}, \mathbf{I}) + (1 - \gamma)\mathcal{N}(\boldsymbol{\mu}, \mathbf{I}). \tag{7.30}$$

Here we assume equal slippage in each dimension, i.e.,

$$(\boldsymbol{\mu}) = (\mu, \mu, \ldots, \mu). \tag{7.31}$$

Let us compare the trimmed mean procedure $\bar{\bar{X}}_T$ with the customary procedure of using the untrimmed mean $\bar{\bar{X}}$. In Tables 7.2 and 7.3, we show, for dimensions two, three, four and five, the average MSEs of the two estimators when $\gamma = .70$, for simulations of size 1,000.

Table 7.2.MSEs for 50 Lots.$\gamma = .7$.								
	d=2	d=2	d=3	d=3	d=4	d=4	d=5	d=5
μ	$\bar{\bar{X}}_T$	$\bar{\bar{X}}$	$\bar{\bar{X}}_T$	$\bar{\bar{X}}$	$\bar{\bar{X}}_T$	$\bar{\bar{X}}$	$\bar{\bar{X}}_T$	$\bar{\bar{X}}$
1	0.40	0.94	0.43	1.17	0.46	1.42	0.54	1.72
2	0.21	1.24	0.17	1.62	0.17	2.00	0.18	2.37
3	0.07	1.85	0.09	2.67	0.12	3.49	0.15	4.33
4	0.06	3.01	0.09	4.48	0.12	5.93	0.14	7.41
5	0.05	4.61	0.09	6.88	0.11	9.16	0.15	11.52
6	0.06	6.53	0.08	9.85	0.12	13.19	0.14	16.50
7	0.06	8.94	0.09	13.41	0.11	17.93	0.14	22.33
8	0.06	11.58	0.08	17.46	0.11	23.27	0.14	28.99
9	0.06	14.71	0.08	22.03	0.11	29.40	0.15	36.67
10	0.06	18.13	0.08	27.24	0.11	36.07	0.15	45.26

	d=2	d=2	d=3	d=3	d=4	d=4	d=5	d=5
μ	$\bar{X}_T$	$\bar{X}$	$\bar{X}_T$	$\bar{X}$	$\bar{X}_T$	$\bar{X}$	$\bar{X}_T$	$\bar{X}$
1	0.28	0.71	0.30	0.94	0.32	1.16	0.34	1.34
2	0.05	0.88	0.06	1.30	0.07	1.74	0.08	2.10
3	0.03	1.72	0.05	2.58	0.06	3.37	0.07	4.25
4	0.03	2.95	0.04	4.41	0.06	5.86	0.07	7.34
5	0.03	4.56	0.04	8.25	0.06	9.13	0.08	11.39
6	0.03	6.56	0.04	8.31	0.06	13.08	0.07	16.35
7	0.03	8.87	0.04	13.28	0.06	17.67	0.07	22.16
8	0.03	11.61	0.04	17.32	0.06	23.17	0.07	28.86
9	0.03	14.67	0.04	21.98	0.06	29.28	0.07	36.61
10	0.03	18.04	0.04	27.12	0.06	36.06	0.07	45.13

Table 7.3. MSEs for 100 Lots. $\gamma = .7$.

In Tables 7.4 and 7.5 we show the MSEs of the trimmed mean and the customary pooled sample mean for the case where $\gamma = .95$.

	d=2	d=2	d=3	d=3	d=4	d=4	d=5	d=5
μ	$\bar{X}_T$	$\bar{X}$	$\bar{X}_T$	$\bar{X}$	$\bar{X}_T$	$\bar{X}$	$\bar{X}_T$	$\bar{X}$
1	0.05	0.10	0.07	0.15	0.09	0.18	0.12	0.24
2	0.05	0.11	0.07	0.16	0.09	0.20	0.11	0.26
3	0.04	0.11	0.06	0.17	0.08	0.22	0.11	0.28
4	0.04	0.13	0.06	0.19	0.08	0.26	0.11	0.33
5	0.04	0.16	0.06	0.24	0.08	0.32	0.11	0.41
6	0.04	0.19	0.06	0.29	0.09	0.40	0.10	0.49
7	0.04	0.24	0.06	0.35	0.08	0.48	0.11	0.61
8	0.04	0.29	0.06	0.42	0.08	0.58	0.10	0.71
9	0.04	0.33	0.06	0.51	0.08	0.68	0.11	0.86
10	0.042	0.39	0.06	0.59	0.08	0.79	0.11	1.01

Table 7.4. MSEs for 50 Lots. $\gamma = .95$.

	d=2	d=2	d=3	d=3	d=4	d=4	d=5	d=5
μ	$\bar{X}_T$	$\bar{X}$	$\bar{X}_T$	$\bar{X}$	$\bar{X}_T$	$\bar{X}$	$\bar{X}_T$	$\bar{X}$
1	0.02	0.05	0.04	0.09	0.05	0.12	0.06	0.14
2	0.02	0.07	0.03	0.10	0.05	0.14	0.05	0.16
3	0.02	0.09	0.03	0.14	0.04	0.18	0.05	0.23
4	0.02	0.13	0.03	0.20	0.05	0.27	0.06	0.34
5	0.02	0.18	0.03	0.27	0.04	0.37	0.06	0.46
6	0.02	0.24	0.03	0.38	0.04	0.49	0.05	0.60
7	0.02	0.31	0.03	0.47	0.04	0.62	0.06	0.80
8	0.02	0.40	0.03	0.60	0.04	0.80	0.05	1.00
9	0.02	0.50	0.03	0.74	0.04	1.00	0.05	1.21
10	0.02	0.60	0.03	0.90	0.05	1.20	0.05	1.50

Table 7.5. MSEs for 100 Lots. $\gamma = .95$.

If the level of contamination is substantial (e.g., if $1 - \gamma = .3$), the use of a trimming procedure to find a base estimate of the center of the "in control" distribution contaminated by observations from other distributions may be strongly indicated. For more modest, but still significant levels of contamination (e.g., if $1 - \gamma = .05$), then simply using $\bar{\bar{X}}$ may be satisfactory. We note that the trimming procedure considered here is computer intensive and is not realistic to be performed on the usual hand held scientific calculator. However, it is easily computed on a personal computer or workstation. Since the standards for rejecting the null hypothesis that a lot is in control are generally done by off-line analysis on a daily or weekly basis, we do not feel the increase in computational complexity should pose much of a logistical problem.

7.5 A Rank Test for Location Slippage

Let us consider next a test which might be used in detecting changes in location when the assumptions of normality are questionable. If we have a group of N lots, then our procedure for testing whether a new lot is "typical" or not is as follows

Rank Test for Location Shift

1. Compute the distances of the sample means of each of the N lots from each other.

2. Compute the average, on a per lot basis, of these distances $\{D_i\}_{i=1}^N$.

3. Compute the distances of the mean of the new lot from each of the N lots.

4. Compute the average of these distances D_0.

5. Reject the hypothesis that the new lot is "typical" if $D_0 > Max_{i=1}^N\{D_i\}$.

It has been confirmed by extensive simulations that the significance level of this and other tests considered in this and the next section is approximately equal to $1/(N + 1)$. Let us now compare its performance with that of the parametric likelihood ratio test when we have as the generator of the "in control" data a p-variate normal distribution with mean $\mathbf{0}$ and covariance matrix $\mathbf{I}$, the identity. We will consider as alternatives "slipped" normal distributions, each with covariance matrix $\mathbf{I}$ but with a translated mean each of whose components is equal to the "slippage" μ. In Figure 7.10, using 20,000 simulations of 50 lots of size 5 per slippage value to obtain the base information, we compute the efficiency of the rank test to detect a shifted 51st lot relative to that of the likelihood ratio test in (7.5), i.e., the ratio of the power of the rank test to that of the $\chi^2(p)$ test (where p is the dimensionality of the data set).

In other words, here, we assume that both the base data and the lots to be tested have identity covariance matrix and that this matrix is known. We note that the efficiency of the rank test here, in a situation favorable to the likelihood ratio test, is close to one, with generally improving performance as the dimensionality increases. Here, we have used the critical values from tables of the χ^2 distribution. For such a situation, the χ^2 is the likelihood ratio test, so in a sense this is a very favorable case for the parametric test.

Next, in Figure 7.11, we note the relative efficiency when the data are drawn from a multivariate shifted **t** distribution with 3 degrees of freedom. We generate such a random variable in p dimensions, say, in the following manner. First, we generate a χ^2 variable v with 3 degrees of freedom. Then we generate p independent univariate normal variates $\mathbf{X}' = (X_1, X_2, \ldots, X_p)$ from a normal distribution with mean 0 and variance 1. If we wish to have a mean vector $\boldsymbol{\mu}$ and covariance matrix $\boldsymbol{\Sigma}$, we then find a linear transformation

$$\mathbf{Z} = \mathbf{A}\mathbf{X} = \boldsymbol{\Sigma}^{\frac{1}{2}}\mathbf{X} \tag{7.32}$$

so that $\mathbf{Z}$ has the desired covariance matrix. Then,

$$\mathbf{t} = \frac{\mathbf{Z}}{\sqrt{v/3}} + \boldsymbol{\mu} \tag{7.33}$$

will have a shifted **t** distribution with 3 degrees of freedom. To find an appropriate linear transformation, we can employ spectral decomposition techniques (see Appendix A) or, less elegantly, simply use

$$Z_1 \quad = \quad a_{11}X_1 \tag{7.34}$$
$$Z_2 \quad = \quad a_{21}X_1 + a_{22}X_2 \tag{7.35}$$
$$\ldots \qquad \ldots \tag{7.36}$$
$$Z_p \quad = \quad a_{p1}X_1 + a_{p2}X_2 + \ldots + a_{pp}X_p. \tag{7.37}$$

If $\boldsymbol{\Sigma}$ is given by

$$\boldsymbol{\Sigma} = \begin{pmatrix} \sigma_{11} & \sigma_{12} & \cdots & \sigma_{1p} \\ \sigma_{12} & \sigma_{22} & \cdots & \sigma_{2p} \\ \cdots & \cdots & \cdots & \cdots \\ \sigma_{1p} & \sigma_{2p} & \cdots & \sigma_{pp} \end{pmatrix}, \tag{7.38}$$

then, writing down the covariance for each Z_j, we note that

$$
\begin{aligned}
\sigma_{11} &= a_{11}^2 \\
\sigma_{12} &= a_{11}a_{21} \\
\cdots &\quad \cdots \\
\sigma_{1p} &= a_{11}a_{p1} \\
\sigma_{22} &= a_{21}^2 + a_{22}^2 \\
\cdots &\quad \cdots \\
\sigma_{pp} &= a_{p1}^2 + a_{p2}^2 + \ldots + a_{pp}^2.
\end{aligned}
\tag{7.39}
$$

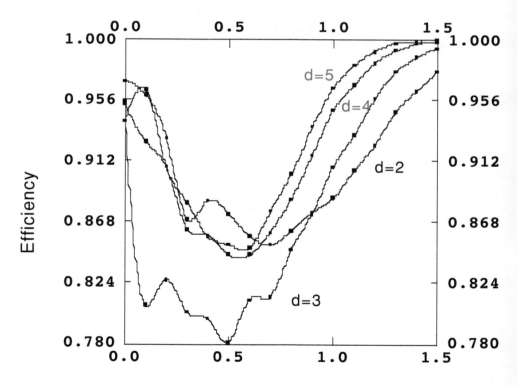

Figure 7.10. Simulated Efficiencies.

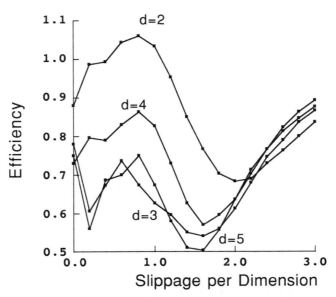

Figure 7.11. Simulated Efficiencies with t(3) Data.

Thus, we can simply go down the list in obvious fashion, solving for the a_{ij}. To generate the χ^2 variate with 3 degrees of freedom, we simply generate three independent $\mathcal{N}(0, 1)$ variates, square them, and take their sum. In Figure 7.11, we compute the efficiency of the rank test compared with that of a "χ^2" test *where the critical values have been picked to have the same Type I error as that of the rank test.* In other words, we have attempted to be unusually fair to the parametric test. The critical values for the noncentrality are much larger than those under the assumption that the data are normal.

At first glance, it appears that the rank test has actually lost utility when compared to the χ^2 test as we move from normal to more taily data. However, we note that the rank test we have described above always has a Type I error equal to $1/(N + 1)$, regardless of the form of the underlying distribution. The critical level for the noncentrality to obtain $1/(N + 1)$ as the critical value using the Hotelling T^2 test for $\mathbf{t}(3)$ data is much greater than that when the data are normal. We note that, as was the case with normal data, the powers of the rank test and that of the parametric test are comparable. But note that we have actually given the parametric test the advantage of determining the critical value assuming the data are $\mathbf{t}(3)$. Had we rather used that which would normally be utilized, namely the critical value assuming the data are normal, we would have obtained a test which very frequently declared the data to be out of control, when, in fact, it was in control.

As a practical matter, perhaps the major advantage of the rank test is that we have a critical region determined naturally and independently of the functional form of the underlying density.

7.6   A Rank Test for Change in Scale and/or Location

Having observed some utility in a rather simple "ranking" procedure, we now suggest a somewhat more complex algorithm. Basically, in statistical process control, we are looking for a difference in the distribution of a new lot, anything out of the ordinary. That might seem to indicate a nonparametric density estimation based procedure. But the general ability to look at averages in statistical process control indicates that for most situations, the Central Limit Theorem enables us to use procedures which point to distributions somewhat close to the normal distribution as the standards. In the case where data are truly normal, the functional form of the underlying density can be based exclusively on the mean vector and the covariance matrix.

Let us now suppose that we have a base sample of N lots, each of size n, with the dimensionality of the data being given by p. For each of these lots, we compute the sample mean $\bar{X}_i$ and sample covariance matrix S_i. Next, we compute the average of these N sample means, say, $\bar{\bar{X}}$, and the average of the sample covariance matrices $\bar{S}$. Then, we use the transformation

$$\mathbf{Z} = \bar{\mathbf{S}}^{-\frac{1}{2}}(\mathbf{X} - \bar{\bar{\mathbf{X}}}) \qquad (7.40)$$

which transforms $\bar{\bar{\mathbf{X}}}$ into a variate with approximate mean $\mathbf{0}$ and approximate covariance matrix $\mathbf{I}$. Next, we apply this transformation to each of the N lot means in the base sample. For each of the transformed lots, we compute the transformed mean and covariance matrix, $\bar{\mathbf{Z}}_i$ and $\mathbf{S}_{\mathbf{Z}_i}$, respectively. For each of these, we apply, respectively, the location norm

$$||\bar{\mathbf{Z}}_i||^2 = \sum_{j=1}^{p} \bar{Z}_{i,j}^2, \qquad (7.41)$$

and scale norm

$$||\mathbf{S}_i||^2 = \sum_{j=1}^{p} \sum_{l=1}^{p} S_{i,j,l}{}^2. \qquad (7.42)$$

If a new lot has location norm higher than any of those in the base sample, we flag it as untypical. If its scale norm is greater than those of any lot in the base sample, we flag it as untypical. The Type I error of either test is given by $1/(N+1)$; that of the combined test is given very closely by

$$1 - [1 - \frac{1}{N+1}]^2 = \frac{2N+1}{(N+1)^2}. \qquad (7.43)$$

In Figure 7.12, we apply the location test only for the data simulation in Figure 7.10. We note that the performance compares favorably to the parametric test, even for normal data.

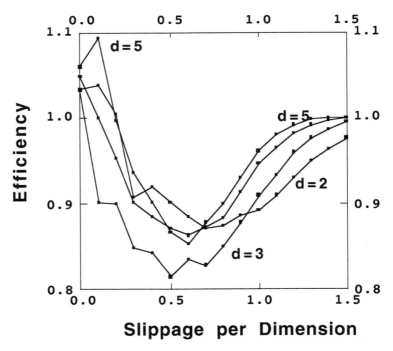

Figure 7.12. Simulated Efficiencies for Second Location Test.

Next, we consider applying the second rank test for location only to the $t(3)$ data of Figure 7.11.

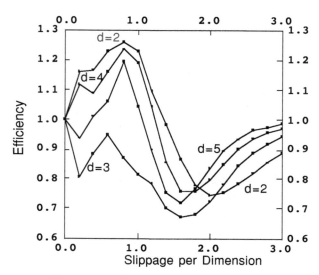

Figure 7.13. Simulated Efficiencies for t(3) Test.

Once again, the rank test performs well when its power is compared to that of the parametric test *even though we have computed the critical value for the parametric test assuming the data are known to be* **t**(3), an assumption very unlikely to be valid in the real world.

Up to this point, we have been assuming that both the base lots and the new lots were known to have identity covariance matrices. In such a case, the appropriate parametric test is χ^2 if the data are normal, and, if the data are not, we have employed simulation techniques to find appropriate critical values for the distribution in question. Now, however, we shift to the situation where we believe that the covariance matrices of the new lots to be sampled may not be diagonal. In such a situation, the appropriate test is the Hotelling T^2 test with degrees of freedom equal to $(p, n - p)$, i.e.,

$$T^2 = \frac{p(n-1)}{n-p} F_{p,n-p}(\alpha).$$

We have been assuming that the base lots (each of size 5) are drawn from $\mathcal{N}(\mathbf{0}, \mathbf{I})$. The sampled (bad) lot is drawn from $\mathcal{N}(\boldsymbol{\mu}, \boldsymbol{\Sigma})$ where

$$\boldsymbol{\mu} = \begin{pmatrix} \mu \\ \mu \\ \vdots \\ \mu \end{pmatrix} \qquad (7.44)$$

and

$$\Sigma = \begin{pmatrix} 1 & .8 & .8 & \dots & .8 \\ .8 & 1 & .8 & \dots & .8 \\ .8 & .8 & 1 & \dots & .8 \\ \vdots & \vdots & \vdots & \vdots & \vdots \\ 8 & .8 & .8 & \dots & \vdots \end{pmatrix}. \tag{7.45}$$

Thus, we are considering the case where the lot comes from a multivariate normal distribution with equal slippage in each dimension and a covariance matrix which has unity marginal variances and covariances .8. In Figure 7.14, we note the relative power of the "location" rank test when compared to that of the Hotelling T^2 procedure. The very favorable performance of the rank test is largely due to the effect that it picks up not only changes in location but also departures in the covariance matrix of the new lot from that of the base lots.

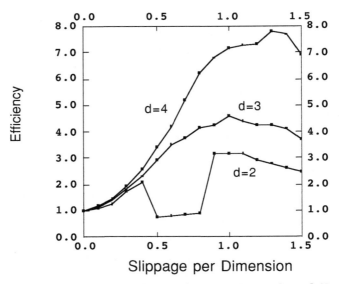

Figure 7.14. Simulated Efficiencies for Correlated Data

The basic setting of statistical process control lends itself very naturally to the utilization of normal distribution theory, since computation of lot averages is so customary. But, as we have seen, for modest lot sizes it is possible to run into difficulty if the underlying distributions have heavy tails. An attractive robust alternative to rank testing is that of "continuous resampling" based tests via such procedures as SIMDAT [19], [20].

References

[1] Alam, K. and Thompson, J.(1973). "On selecting the least probable multinomial event," *Annals of Mathematical Statistics, 43,* pp. 1981-1990.

[2] Alam, K. and Thompson, J.(1973). "A problem of ranking and estimation with Poisson processes," *Technometrics, 15,* pp. 801-808.

[3] Alt, F.B. (1985). "Multivariate quality control," in *Encyclopedia of Statistical Sciences,* (vol. 6), eds. S. Kotz and N. Johnson. New York: John Wiley & Sons.

[4] Alt, F.B. (1986). "SPC of dispersion for multivariate data," *ASQC Annual Quality Congress Transactions,* pp. 248-254.

[5] Alt, F.B., Goode, J.J., and Wadsworth, H.M. (1976). *Ann. Tech. Conf. Trans., ASQC,* pp. 754-759.

[6] Andrews, D.F., Bickel, P.J., Hampel, F.R., Huber, P.J., Rogers, W.H., and Tukey, J.W. (1972). *Robust Estimates of Location.* Princeton: Princeton University Press.

[7] Bridges, E., Ensor, K.B., and Thompson, J.R. (1992). "Marketplace competition in the personal computer industry," *Decision Sciences,* pp. 467-477.

[8] Jackson, J.E. (1991). *A User's Guide to Principal Components.* New York: John Wiley & Sons.

[9] Johnson, R.A. and Wichern, D.W. (1988). *Applied Multivariate Statistical Analysis.* Englewood Cliffs: Prentice Hall.

[10] Jolliffe, I.T. (1986). *Principal Component Analysis.* New York: Springer-Verlag.

[11] Kendall, M.G. and Stuart, A.(1958). *The Advanced Theory of Statistics, II.* New York: Hafner.

[12] Krzanowski, W.J. (1988). *Principles of Multivariate Analysis.* Oxford: Clarendon Press.

[13] Lawera, M. and Thompson, J.R. (1993). "Multivariate strategies in statistical process control," *Proceedings of the Thirty-Eighth Conference on the Design of Experiments in Army Research Development and Testing*, pp. 99-126

[14] MacGregor, J.F. (1997). "Using on-line process data to improve quality," *Int. Statist. Rev., 65*, pp.309-323.

[15] Mosteller, F. and Tukey, J.W. (1977). *Data Analysis and Regression.* Reading: Addison-Wesley.

[16] Nomikos, P. and MacGregor, J.F. (1995). "Multivariate SPC charts for monitoring batch processes," *Technometrics, 37*, pp.41-59.

[17] Ryan, T.P. (1989). *Statistical Methods for Quality Improvement.* New York: John Wiley & Sons.

[18] Sparks, R., Adolphson, A. and Phatak, A. (1997). "Multivariate process monitoring using the dynamic biplot." *Int. Statist. Rev., 65*, pp. 325-349.

[19] Taylor, M.S. and Thompson, J.R. (1986). "A data based algorithm for the generation of random vectors," *Computational Statistics and Data Analysis, 4*, pp. 93-101.

[20] Taylor, M.S. and Thompson, J.R. (1992). "A nonparametric based resampling algorithm," *Exploring the Limits of the Bootstrap*, Billard, L. and LePage, R., eds., pp. 397-403. New York: John Wiley & Sons.

[21] Wierda, S.J. (1994). "Multivariate statistical process control – recent results and directions for future research," *Statistica Neerlandica, 48*, pp. 147-168.

Problems

Problem 7.1. In the following table, the lot of 20 bivariate measurement data is given. Provided that the process is in control, it can be assumed that the lot comes from a normal distribution with mean $\mu_0 = (90.0, 85.1)'$.

a. Find the p-value of the Hotelling's T^2 test of the hypothesis that the process' mean is equal to μ_0.

b. Find the p-value of the t test (see (B.172) in the Appendix) of the hypothesis that the mean of the first variate, X_1, is 90.

c. Find the p-value of the t test of the hypothesis that the mean of the second variate, X_2, is 85.1.

d. Use the $\bar{X}$ charts to verify whether the two univariate sets of data, X_1 and X_2, can be considered (separately) to be in control.

e. Try the Page CUSUM charts to verify whether the two univariate sets of data, X_1 and X_2, can be considered (separately) to be in control.

f. Comment on the results.

Lot	x_1	x_2	Lot	x_1	x_2
1	89.987	85.150	11	90.018	85.081
2	90.016	85.073	12	89.910	85.057
3	89.962	85.179	13	89.853	85.113
4	89.976	85.122	14	89.896	85.212
5	90.137	85.229	15	89.999	85.105
6	89.993	85.281	16	89.900	84.992
7	89.937	85.147	17	90.120	85.320
8	89.831	84.930	18	90.044	85.092
9	90.239	85.419	19	90.067	85.144
10	90.112	85.098	20	89.809	85.054

Problem 7.2. In the following table, 20 lots of size 4 of bivariate measurement data are given. Provided that the process is in control, it can be assumed that the lots come from a bivariate normal distribution.

Lot	x_1	x_2	Lot	x_1	x_2	Lot	x_1	x_2
1	10.133	10.151	8	10.047	10.203	15	10.092	10.168
	10.012	10.106		10.099	10.074		9.815	9.929
	9.774	9.862		9.937	10.127		10.033	10.162
	10.012	10.017		9.877	9.945		9.918	10.080
2	9.915	10.065	9	9.920	9.977	16	10.032	10.107
	9.790	10.069		9.822	10.053		10.059	10.115
	10.293	10.343		9.872	10.073		10.055	10.179
	10.010	10.087		9.965	10.107		10.212	10.273
3	10.023	10.092	10	9.848	10.014	17	10.009	9.984
	9.854	9.932		10.100	10.118		9.978	10.149
	9.886	10.087		9.902	10.141		10.063	10.091
	10.028	10.070		9.905	10.116		10.116	10.235
4	9.965	10.128	11	10.251	10.237	18	9.991	10.025
	9.978	10.128		10.077	10.173		10.024	9.808
	10.118	10.101		9.896	9.984		10.253	9.762
	9.943	10.137		9.908	10.155		10.073	9.821
5	10.090	10.140	12	10.005	10.011	19	9.863	9.975
	9.953	10.108		10.225	10.269		9.978	10.069
	9.966	10.037		9.932	9.996		10.095	10.196
	10.081	10.101		9.972	10.048		10.110	10.199
6	9.908	10.093	13	10.038	10.059	20	9.824	10.044
	10.006	9.952		9.947	10.080		9.854	9.950
	10.108	10.116		9.957	10.057		10.207	10.282
	9.930	9.879		9.989	10.161		10.170	10.226
7	10.081	10.071	14	9.955	10.100			
	10.088	10.157		10.066	10.099			
	9.942	10.066		10.068	10.127			
	10.179	10.201		10.039	10.027			

a. Use the suitable version of the Hotelling's T^2 test to verify the hypothesis that the process' mean is $\boldsymbol{\mu}_0 = (10.0, 10.1)'$.

b. Use the $\bar{X}$ charts to verify whether the two univariate sets of data, X_1 and X_2, can be considered (separately) to be in control.

Problem 7.3. It is conjectured that a simplified heat treatment of a metal casting, suggested by a plant's foundry, may lead to a deterioration of the casting's quality. One of the cross-sections of the castings should be circular with diameter 285 mm. A trial sample of 20 castings has been made. A test consists in measuring two perpendicular diameters of the cross-section of each of the 20 castings. The data set obtained is given in the following table.

Lot	x_1	x_2	Lot	x_1	x_2
1	284.963	285.000	11	285.022	285.049
2	285.057	285.041	12	285.058	285.041
3	285.020	284.979	13	285.028	285.037
4	284.979	285.014	14	285.004	284.938
5	284.936	284.991	15	285.011	284.858
6	284.939	284.948	16	284.958	284.644
7	284.925	284.788	17	285.028	285.039
8	285.023	284.997	18	284.992	284.939
9	285.004	284.992	19	285.057	284.946
10	284.994	285.052	20	284.918	284.972

Perform the compound test to verify whether the process is in control. Comment on the results.

Problem 7.4. Consider Problem 7.2. Delete the out-of-control lots observed (if any), and recompute $\bar{\bar{x}}$ and $\mathbf{S}$. Verify whether the five new lots, given in the following table, are in control.

Lot	x_1	x_2
1	9.874	10.035
	9.987	9.953
	9.921	9.995
	9.969	10.023
2	10.239	10.435
	9.981	10.091
	10.088	10.224
	9.943	10.049
3	9.999	10.251
	9.847	10.080
	9.948	10.149
	10.222	10.285
4	10.060	10.149
	10.107	10.117
	9.968	9.994
	10.037	10.091
5	9.912	10.138
	10.056	10.211
	9.961	10.102
	9.861	10.066

Problem 7.5. The following are 20 measurements of four dimensional data. Provided that the process is in control, it can be assumed that the data set comes from a normal distribution with mean $\boldsymbol{\mu}_0 = (0, .1, .5, .3)'$. Perform the compound test to see whether the process is in control.

Lot	x_1	x_2	x_3	x_4
1	0.057	0.182	0.363	0.331
2	0.094	0.258	0.621	0.176
3	0.032	0.125	0.625	0.197
4	0.057	0.037	0.593	0.263
5	0.078	0.245	0.514	0.237
6	0.107	0.173	0.426	0.406
7	0.024	0.109	0.464	0.338
8	-0.007	0.131	0.637	0.253
9	0.135	0.193	0.616	0.306
10	0.052	0.135	0.425	0.471
11	0.102	0.141	0.612	0.024
12	0.183	-0.187	0.478	0.385
13	0.006	0.235	0.322	-0.008
14	0.115	0.220	0.501	0.334
15	0.054	0.125	0.534	0.308
16	-0.070	0.201	0.528	0.246
17	0.065	0.214	0.508	0.318
18	0.082	0.234	0.480	0.493
19	-0.085	0.000	0.480	0.270
20	-0.004	0.087	0.505	0.240

Appendix A

A Brief Introduction to Linear Algebra

A.1   Introduction

Frequently, the subject of interest cannot be adequately described by a single number but, rather, by a set of numbers. For example, the standard operational controls of a chemical process may involve not only the temperature, but the pressure as well. In each such case, suitable measurements yield a multivariate characteristic of the subject of interest.

Convenient analytical tools for dealing with multivariate data are vectors and matrices. In this appendix, only some basic concepts from vector and matrix algebra are briefly introduced or recalled, namely those which are of particular interest from the point of view of statistical multivariate analysis.

An n-tuple, or an array, $\mathbf{x}$ of n real numbers (*elements*) $x_1, x_2, \ldots, x_n$ arranged in a column, is called a *column vector* and it is written as

$$\mathbf{x} = \begin{bmatrix} x_1 \\ x_2 \\ \vdots \\ x_n \end{bmatrix}. \tag{A.1}$$

Analogously, an n-tuple of real numbers $x_1, x_2, \ldots, x_n$ arranged in a row is called a *row vector* and it is written as

$$\mathbf{x}' = [x_1, x_2, \ldots, x_n], \tag{A.2}$$

where the prime denotes the operation of *transposing* a column into a row. The reason that we define the row vector as the *transposition* of the column vector is that usually in the literature *vectors* are written as column vectors. Accordingly, whenever we write "vector" we mean a column vector. In the sequel, vectors are denoted by boldface lower case letters.

Elements x_i, $i = 1, \ldots, n$, of a vector $\mathbf{x}$ are often called *components* of $\mathbf{x}$. A *scalar* is a vector consisting of just one element, i.e., it is a single real number.

A more general concept is that of a *matrix*. An $n \times p$ matrix $\mathbf{A}$ is a rectangular array of real numbers (*elements*) arranged in n rows and p columns, and it is written as

$$\mathbf{A}_{(n \times p)} = \begin{bmatrix} a_{11} & a_{12} & \ldots & a_{1p} \\ a_{21} & a_{22} & \ldots & a_{2p} \\ \vdots & \vdots & & \vdots \\ a_{n1} & a_{n2} & \ldots & a_{np} \end{bmatrix} = (a_{ij}) \begin{matrix} i = 1, \ldots, n \\ j = 1, \ldots, p \end{matrix} ; \quad \text{(A.3)}$$

a_{ij}, referred to as the (i,j)th element of the matrix, denotes the element in the ith row and jth column of the matrix $\mathbf{A}$, $i = 1, \ldots, n$, $j = 1, \ldots, p$. Sometimes, we write $(\mathbf{A})_{ij}$ for a_{ij}. If it does not lead to ambiguity, we write $\mathbf{A}$ for $\mathbf{A}_{(n \times p)}$ and

$$(a_{ij}) \quad \text{for} \quad (a_{ij}) \begin{matrix} i = 1, \ldots, n \\ j = 1, \ldots, p \end{matrix} . \quad \text{(A.4)}$$

In what follows, we denote matrices by boldface upper case letters. A matrix, all of whose elements are zeros, is called a zero matrix; otherwise, a matrix is *nonzero*. If $\mathbf{A}$ is an $n \times p$ matrix, we say that it is of *order* (or *dimension*) $n \times p$. Clearly, vector $\mathbf{x}$ is a matrix with n rows and one column, i.e., it is a matrix of order $n \times 1$; for short, n-element vectors are said to be of *dimension* n. Correspondingly, row vector $\mathbf{x}'$ is a matrix of order $1 \times n$.

The *transpose* of an $n \times p$ matrix $\mathbf{A} = (a_{ij})$, denoted by $\mathbf{A}'$, is the $p \times n$ matrix obtained from $\mathbf{A}$ by interchanging the rows and columns; i.e., $\mathbf{A}'$ is the matrix with elements a_{ji}, $j = 1, \ldots, p$, $i = 1, \ldots, n$,

$$\mathbf{A}' = \begin{bmatrix} a_{11} & a_{21} & \ldots & a_{n1} \\ a_{12} & a_{22} & \ldots & a_{n2} \\ \vdots & \vdots & & \vdots \\ a_{1p} & a_{2p} & \ldots & a_{np} \end{bmatrix} = (a_{ji}). \quad \text{(A.5)}$$

Example: For $\mathbf{A}$ of order 3×4,

$$\mathbf{A} = \begin{bmatrix} 1 & 2 & 3 & 4 \\ 0 & 2 & -1 & -7 \\ 11 & -5 & 4 & 1 \end{bmatrix}, \tag{A.6}$$

we have

$$\mathbf{A}' = \begin{bmatrix} 1 & 0 & 11 \\ 2 & 2 & -5 \\ 3 & -1 & 4 \\ 4 & -7 & 1 \end{bmatrix}. \tag{A.7}$$

Clearly, for all matrices $\mathbf{A}$, we have

$$(\mathbf{A}')' = \mathbf{A}. \tag{A.8}$$

Note that our definition of the row vector is consistent with the general definition of the transpose of a matrix.

A matrix is *square* if it has the same number of rows and columns. For short, square matrices of order $n \times n$ are often said to be of order n. A square matrix $\mathbf{A}$ is *symmetric* if $\mathbf{A} = \mathbf{A}'$, i.e., if $a_{ij} = a_{ji}$ for all i and j. It is diagonal if $a_{ij} = 0$ for all $i \neq j$, i.e., if it can have nonzero elements on the main diagonal only, where the main diagonal of an $n \times n$ matrix $\mathbf{A}$ consists of elements $a_{11}, a_{22}, \ldots, a_{nn}$. A diagonal matrix $\mathbf{A}$ of order n is often written as $diag(a_1, a_2, \ldots, a_n)$. A diagonal matrix with all elements of the main diagonal equal to one is called the *identity matrix*, and is denoted by I or I_n when its order $n \times n$ is to be given explicitly,

$$\mathbf{I} = \begin{bmatrix} 1 & 0 & \cdots & 0 \\ 0 & 1 & \cdots & 0 \\ \vdots & \vdots & & \vdots \\ 0 & 0 & \cdots & 1 \end{bmatrix}. \tag{A.9}$$

All multivariate data dealt with in this book can be represented by vectors. However, performing suitable transformations of the data requires some matrix algebra.

A.2 Elementary Arithmetic

Two matrices $\mathbf{A} = (a_{ij})$ and $\mathbf{B} = (b_{ij})$ of the same order $n \times p$ are said to be *equal* if and only if $a_{ij} = b_{ij}$, $i = 1, 2, \ldots, n$, $j = 1, 2, \ldots, p$. Matrices of the same order can be added. The *sum* of two matrices $\mathbf{A} = (a_{ij})$ and $\mathbf{B} = (b_{ij})$ of the same order $n \times p$ is the $n \times p$ matrix $\mathbf{C} = (c_{ij})$, where

$$c_{ij} = a_{ij} + b_{ij}, \ i = 1, 2, \ldots, n, \ j = 1, 2, \ldots, p; \qquad (A.10)$$

we then write $\mathbf{C} = \mathbf{A} + \mathbf{B}$.

Example:

$$\begin{bmatrix} 2 & 3 \\ 1 & 2 \\ 4 & -1 \end{bmatrix} + \begin{bmatrix} -7 & 8 \\ 2 & 1 \\ 3 & 1 \end{bmatrix} = \begin{bmatrix} -5 & 11 \\ 3 & 3 \\ 7 & 0 \end{bmatrix}. \qquad (A.11)$$

Clearly, for all matrices $\mathbf{A}$, $\mathbf{B}$ and $\mathbf{C}$ of the same order, we have

(i) $\qquad \mathbf{A} + \mathbf{B} \ = \ \mathbf{B} + \mathbf{A}$

(ii) $\quad \mathbf{A} + (\mathbf{B} + \mathbf{C}) \ = \ (\mathbf{A} + \mathbf{B}) + \mathbf{C}$

(iii) $\qquad (\mathbf{A} + \mathbf{B})' \ = \ \mathbf{A}' + \mathbf{B}'$.

Matrices can be multiplied by a constant. The *scalar multiple* $c\mathbf{A}$ of the matrix $\mathbf{A}$ is obtained by multiplying each element of $\mathbf{A}$ by the scalar c. Thus, for $\mathbf{A} = (a_{ij})$ of order $n \times p$,

$$c\mathbf{A} = \begin{bmatrix} ca_{11} & ca_{12} & \cdots & ca_{1p} \\ ca_{21} & ca_{22} & \cdots & ca_{2p} \\ \vdots & \vdots & & \vdots \\ ca_{n1} & ca_{n2} & \cdots & ca_{np} \end{bmatrix}. \qquad (A.12)$$

Combining the definitions of matrix addition and scalar multiplication enables us to define *matrix subtraction*. The *difference* between two matrices $\mathbf{A} = (a_{ij})$ and $\mathbf{B} = (b_{ij})$ of the same order $n \times p$ is the $n \times p$ matrix $\mathbf{C} = (c_{ij})$, given by

$$\mathbf{C} = \mathbf{A} - \mathbf{B} = \mathbf{A} + (-1)\mathbf{B}; \qquad (A.13)$$

i.e., $c_{ij} = a_{ij} + (-1)b_{ij} = a_{ij} - b_{ij}, \ i = 1, 2, \ldots, n, \ j = 1, 2, \ldots, p$.

If the number of columns of a matrix $\mathbf{A}$ is equal to the number of rows of a matrix $\mathbf{B}$, one can define the *product* $\mathbf{AB}$ of $\mathbf{A}$ and $\mathbf{B}$. The product $\mathbf{AB}$ of an $n \times p$ matrix $\mathbf{A}$ and a $p \times m$ matrix $\mathbf{B}$ is the $n \times m$ matrix $\mathbf{C}$ whose (i, j)th element is given by

$$c_{ij} = \sum_{k=1}^{p} a_{ik}b_{kj}, \ i = 1, 2, \ldots, n, \ j = 1, 2, \ldots, m. \qquad (A.14)$$

Example:

$$\begin{bmatrix} 2 & 1 & 2 \\ 3 & 4 & 1 \end{bmatrix} \begin{bmatrix} 1 & 0 \\ 0 & -1 \\ 2 & 4 \end{bmatrix} = \begin{bmatrix} 6 & 7 \\ 5 & 0 \end{bmatrix}. \tag{A.15}$$

It is not hard to check that, for all matrices $\mathbf{A}$, $\mathbf{B}$ and $\mathbf{C}$ of orders conforming in a suitable sense,

(i) $\mathbf{A}(\mathbf{BC}) = (\mathbf{AB})\mathbf{C}$
(ii) $\mathbf{A}(\mathbf{B} + \mathbf{C}) = \mathbf{AB} + \mathbf{AC}$
(iii) $(\mathbf{B} + \mathbf{C})\mathbf{A} = \mathbf{BA} + \mathbf{CA}$
(iv) $(\mathbf{AB})' = \mathbf{B}'\mathbf{A}'.$

It should be noted that the order of multiplication is important. In order to see this consider an $n \times p$ matrix $\mathbf{A}$ and a $p \times m$ matrix $\mathbf{B}$. Now, if n is not equal to m, the product $\mathbf{AB}$ is well-defined while the product $\mathbf{BA}$ does not exist. Further, if $n = m$ but n and m are not equal to p, both products are well-defined but the matrices $\mathbf{AB}$ and $\mathbf{BA}$ are of different orders ($n \times n$ and $p \times p$, respectively). Finally, even if $\mathbf{A}$ and $\mathbf{B}$ are square matrices, $n = m = p$, the matrices $\mathbf{AB}$ and $\mathbf{BA}$ are of the same order but, usually, are different.

Example:

$$\begin{bmatrix} 2 & 1 \\ 3 & -4 \end{bmatrix} \begin{bmatrix} 1 & 2 \\ 0 & 1 \end{bmatrix} = \begin{bmatrix} 2 & 5 \\ 3 & -4 \end{bmatrix} \tag{A.16}$$

$$\begin{bmatrix} 1 & 2 \\ 0 & 1 \end{bmatrix} \begin{bmatrix} 2 & 1 \\ 3 & -4 \end{bmatrix} = \begin{bmatrix} 8 & -7 \\ 3 & -4 \end{bmatrix}. \tag{A.17}$$

Let us observe in passing that, for any square matrix $\mathbf{A}$ of order n and the identity matrix $\mathbf{I}_n$, $\mathbf{AI}_n = \mathbf{I}_n\mathbf{A} = \mathbf{A}$; hence the name of the matrix $\mathbf{I}_n$.

The concept of a matrix can, of course, be viewed as a generalization of that of a real number. In particular, a scalar is a 1×1 matrix. Similarly, the matrix algebra can be seen as a generalization of the algebra of real numbers. However, as the notion of the matrix product shows, there are important differences between the two algebras. One such difference has already been discussed. Another is that the product of two nonzero matrices need not be a nonzero matrix, whereas for all scalars c and d, whenever $cd = 0$, then either $c = 0$ or $d = 0$.

Example:

$$\begin{bmatrix} 2 & 3 & -1 \\ 4 & 6 & 2 \\ -6 & -9 & 7 \end{bmatrix} \begin{bmatrix} 6 \\ -4 \\ 0 \end{bmatrix} = \begin{bmatrix} 0 \\ 0 \\ 0 \\ 0 \end{bmatrix}. \tag{A.18}$$

Let us now introduce the concept of the trace of a square matrix. The *trace* of an $n \times n$ matrix $\mathbf{A}$, denoted by $tr(\mathbf{A})$, is defined as the sum of the diagonal elements of $\mathbf{A}$, $tr(\mathbf{A}) = \sum_{i=1}^{n} a_{ii}$. It can be shown that, for all matrices $\mathbf{A}$, $\mathbf{B}$, $\mathbf{C}$, $\mathbf{D}$ of conforming orders and a scalar γ,

(i) $tr(\mathbf{A} \pm \mathbf{B}) \;=\; tr(\mathbf{A}) \pm tr(\mathbf{B})$
(ii) $tr(\gamma \mathbf{A}) \;=\; \gamma tr(\mathbf{A})$
(iii) $tr(\mathbf{CD}) \;=\; tr(\mathbf{DC})$.

Since vectors are special cases of matrices, there is no need to consider separately the sum, scalar multiple and the difference of vectors. It is, however, useful to introduce two types of products for suitably transposed vectors.

Let $\mathbf{x}$ and $\mathbf{y}$ be two vectors of the same dimension. The *inner product* (or *dot product* or *scalar product*) of $\mathbf{x}$ and $\mathbf{y}$ is defined as $\mathbf{x'y}$. That is, for n-dimensional vectors

$$
\mathbf{x} = \begin{bmatrix} x_1 \\ x_2 \\ \vdots \\ x_n \end{bmatrix} \quad \text{and} \quad \mathbf{y} = \begin{bmatrix} y_1 \\ y_2 \\ \vdots \\ y_n \end{bmatrix}, \tag{A.19}
$$

the inner product $\mathbf{x'y}$ is given by

$$
x_1 y_1 + x_2 y_2 + \ldots + x_n y_n \,. \tag{A.20}
$$

Note that the inner product is always a scalar and that $\mathbf{x'y} = \mathbf{y'x}$.

The square root of the inner product of $\mathbf{x}$ with itself,

$$
\sqrt{\mathbf{x'x}} = \sqrt{x_1^2 + x_2^2 + \ldots + x_n^2} \,, \tag{A.21}
$$

is called the *length* of the vector $\mathbf{x}$.

The *outer product* of $\mathbf{x}$ and $\mathbf{y}$ is defined as $\mathbf{xy'}$. That is, for the given n-dimensional vectors $\mathbf{x}$ and $\mathbf{y}$, the outer product is an $n \times n$ matrix

$$
\begin{bmatrix} x_1 y_1 & x_1 y_2 & \cdots & x_1 y_n \\ x_2 y_1 & x_2 y_2 & \cdots & x_2 y_n \\ \vdots & \vdots & & \vdots \\ x_n y_1 & x_n y_2 & \cdots & x_n y_n \end{bmatrix}. \tag{A.22}
$$

Statistical analysis of multivariate data, in particular when the data are normally distributed, relies heavily on the concepts and calculations of the inverse and the square root of a matrix. The concepts themselves are straightforward but effective calculations of inverses and square roots of matrices require some additional considerations. To be honest, one should add that calculating the inverse and square root can be done on a computer using standard programs, and so the reader who is not interested in algebraic details can come directly to the sections on inverses and square roots, and refer to the skipped sections only to find a few necessary definitions.

A.3 Linear Independence of Vectors

Let $\mathbf{x}_1, \mathbf{x}_2, \ldots, \mathbf{x}_k$ be k vectors of the same dimension. The vector

$$\mathbf{y} = a_1 \mathbf{x}_1 + a_2 \mathbf{x}_2 + \ldots + a_k \mathbf{x}_k , \qquad (A.23)$$

where $a_1, a_2, \ldots, a_k$ are any fixed constants, is said to be a *linear combination* of the vectors $\mathbf{x}_1, \mathbf{x}_2, \ldots, \mathbf{x}_k$.

A set of vectors $\mathbf{x}_1, \mathbf{x}_2, \ldots, \mathbf{x}_k$ is said to be *linearly independent* if it is impossible to write any one of them as a linear combination of the remaining vectors. Otherwise, vectors $\mathbf{x}_1, \mathbf{x}_2, \ldots, \mathbf{x}_k$ are said to be *linearly dependent*.

It is worthwhile to mention that linear independence may be equivalently defined by first defining linear dependence of vectors. Namely, vectors $\mathbf{x}_1, \mathbf{x}_2, \ldots, \mathbf{x}_k$ can be said to be linearly dependent if there exist k numbers $a_1, a_2, \ldots, a_k$, not all zero, such that

$$a_1 \mathbf{x}_1 + a_2 \mathbf{x}_2 + \ldots + a_k \mathbf{x}_k = \mathbf{0} , \qquad (A.24)$$

where $\mathbf{0}$ denotes the zero vector. Now, if it is not the case, the set of vectors is linearly independent.

A particular case of linearly independent vectors is that of the *orthogonal* vectors. Vectors $\mathbf{x}$ and $\mathbf{y}$ are said to be orthogonal (or *perpendicular*) if $\mathbf{x}'\mathbf{y} = 0$. It can indeed be shown that a set of mutually orthogonal vectors is the set of linearly independent vectors (but not necessarily conversely!).

The concept of linear independence of vectors enables one to define the *rank* of a matrix. The *row rank* of a matrix is the maximum number of linearly independent rows of the matrix (the rows of the matrix are considered here as row vectors). Analogously, the *column rank* of a matrix is the maximum number of linearly independent columns of the matrix (the columns are considered here as the column vectors). It can be proved that the row rank and the column rank of a matrix are always equal. Thus, we can define either of them to be the *rank* of a matrix.

Let $\mathbf{A}$ be a matrix of order $n \times p$ and denote the rank of $\mathbf{A}$ by $r(\mathbf{A})$. It follows from the definition that $r(\mathbf{A})$ is a nonnegative number not larger than $\min\{n, p\}$ and that $r(\mathbf{A}) = r(\mathbf{A}')$. If $n = p$ and $r(\mathbf{A}) = n$, $\mathbf{A}$ is said to be of *full rank*. It can be shown that, for all matrices $\mathbf{A}$, $\mathbf{B}$ and $\mathbf{C}$ of conforming orders,

$$(i) \quad r(\mathbf{A} + \mathbf{B}) \;\leq\; r(\mathbf{A}) + r(\mathbf{B})$$
$$(ii) \quad r(\mathbf{AB}) \;\leq\; \min\{r(\mathbf{A}), r(\mathbf{B})\}$$
$$(iii) \quad r(\mathbf{A}'\mathbf{A}) \;=\; r(\mathbf{AA}') = r(\mathbf{A})$$
$$(iv) \quad r(\mathbf{BAC}) \;=\; r(\mathbf{A}) \text{ if } \mathbf{B} \text{ and } \mathbf{C} \text{ are of full rank.}$$

The concept of orthogonality can be extended to square matrices. A square matrix $\mathbf{A}$ is said to be *orthogonal* if $\mathbf{AA}' = \mathbf{I}$. In order to see how this last concept is related to the orthogonality of vectors, assume that $\mathbf{A}$ is of order $n \times n$ and denote the ith row of $\mathbf{A}$ by $\mathbf{a}_i'$, $i = 1, 2, \ldots, n$; hence, $\mathbf{a}_i$ denotes the ith column of $\mathbf{A}'$. Now, the condition $\mathbf{AA}' = \mathbf{I}$ can be written as

$$\begin{bmatrix} \mathbf{a}_1' \\ \mathbf{a}_2' \\ \vdots \\ \mathbf{a}_n' \end{bmatrix} \begin{bmatrix} \mathbf{a}_1 & \mathbf{a}_2 & \cdots & \mathbf{a}_n \end{bmatrix} = \begin{bmatrix} \mathbf{a}_1'\mathbf{a}_1 & \mathbf{a}_1'\mathbf{a}_2 & \cdots & \mathbf{a}_1'\mathbf{a}_n \\ \mathbf{a}_2'\mathbf{a}_1 & \mathbf{a}_2'\mathbf{a}_2 & \cdots & \mathbf{a}_2'\mathbf{a}_n \\ \vdots & \vdots & & \vdots \\ \mathbf{a}_n'\mathbf{a}_1 & \mathbf{a}_n'\mathbf{a}_2 & \cdots & \mathbf{a}_n'\mathbf{a}_n \end{bmatrix} = \mathbf{I}$$

$$(A.25)$$

and it follows that the rows of an orthogonal matrix $\mathbf{A}$ are mutually orthogonal and have unit lengths (indeed, $\mathbf{a}_i'\mathbf{a}_j = 0$ if $i \neq j$ and $\mathbf{a}_i'\mathbf{a}_i = 1$ for $i = 1, 2, \ldots, n$). We shall see in Section A.5 that, provided $\mathbf{A}$ is orthogonal, $\mathbf{AA}' = \mathbf{A}'\mathbf{A} = \mathbf{I}$. Thus, the columns of $\mathbf{A}$ are also orthogonal and have unit lengths; to verify this, it suffices to write $\mathbf{A}$ as $\begin{bmatrix} \mathbf{a}_{(1)} & \mathbf{a}_{(2)} & \cdots & \mathbf{a}_{(n)} \end{bmatrix}$, where $\mathbf{a}_{(i)}$ denotes the ith column of $\mathbf{A}$, and use this form of $\mathbf{A}$ to write the condition $\mathbf{A}'\mathbf{A} = \mathbf{I}$.

A.4 Determinants

Important Remark: Throughout the rest of this appendix we consider only square matrices.

The determinant, although seemingly neither intuitive nor operational, is one of the most useful concepts of matrix algebra. In order to avoid possible frustration of the reader, let us announce already here that one can calculate determinants in a simple way, without directly using their rather complicated definition. That simple way of evaluating determinants is given by (A.28)-(A.35) and the Sarrus' diagram below. Accordingly, the reader may decide to skip definition (A.26) and come directly to the formulas mentioned. The reader should then consider formulas (A.28)-(A.35) to be the defining properties of determinant (A.27).

The definition of the determinant has to be preceded by introducing the notion of the inversion of an ordered pair. Let $\alpha_1, \alpha_2, \ldots, \alpha_n$ be a given permutation (i.e., arrangement) of the numbers $1, 2, \ldots, n$, and consider all ordered pairs formed from this permutation:

$$(\alpha_1, \alpha_2), \ (\alpha_1, \alpha_3), \ \ldots, \ (\alpha_1, \alpha_n), \ (\alpha_2, \alpha_3), \ (\alpha_2, \alpha_4), \ \ldots, \ (\alpha_2, \alpha_n),$$

$$\ldots, \ (\alpha_{n-1}, \alpha_n).$$

The pair is said to form the *inversion* if $\alpha_i > \alpha_k$ and $i < k$. The *determinant* $|\mathbf{A}|$ of a square matrix $\mathbf{A}$ of order n is the sum

$$\sum (-1)^N a_{1j_1} a_{2j_2} \cdots a_{nj_n}, \tag{A.26}$$

where the summation is taken over all possible permutations

$$j_1, j_2, \ldots, j_n$$

of the numbers $1, 2, \ldots, n$, and N is the total number of inversions of the permutation $j_1, j_2, \ldots, j_n$; for instance, for $\mathbf{A}$ of order 4 and the particular permutation $4, 1, 3, 2$, $N = 4$.

It is often convenient to write determinants in a "more explicit" form:

$$|\mathbf{A}| = \begin{vmatrix} a_{11} & a_{12} & \cdots & a_{1n} \\ a_{21} & a_{22} & \cdots & a_{2n} \\ \vdots & \vdots & & \vdots \\ a_{n1} & a_{n2} & \cdots & a_{nn} \end{vmatrix}. \tag{A.27}$$

Obviously, for $\mathbf{A}$ of order 1×1, a scalar $\mathbf{A} = a_{11}$, we have

$$|\mathbf{A}| = a_{11}, \tag{A.28}$$

while for $\mathbf{A}$ of order 2×2, we have

$$|\mathbf{A}| = a_{11}a_{22} - a_{12}a_{21}. \tag{A.29}$$

It is also easy although a bit tedious to check that for $\mathbf{A}$ of order 3×3 we have

$$\begin{aligned} |\mathbf{A}| \ = \ & a_{11}a_{22}a_{33} + a_{12}a_{23}a_{31} + a_{13}a_{21}a_{32} \\ & - \ a_{13}a_{22}a_{31} - a_{12}a_{21}a_{33} - a_{11}a_{23}a_{32}. \end{aligned} \tag{A.30}$$

In fact, for $n = 3$, there are exactly six permutations: 1,2,3; 2,3,1; 3,1,2; 1,3,2; 2,1,3; 3,2,1; and the numbers of their inversions are, respectively, 0, 2, 2, 1, 1, 3.

An easy and almost automatic way of calculating determinants of matrices of order 3 is to use the following Sarrus' diagram:

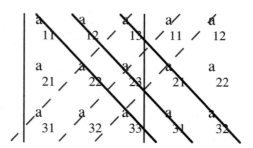

Namely, the determinant is obtained via adding the first two columns on the right hand side of the matrix, summing the products of elements along the solid (NW-SE) lines and subtracting the products along the dashed (NE-SW) lines.

With the order of $\mathbf{A}$ increasing, computing the determinant becomes an apparently hopeless task. However, one can then use a different method of computation. In order to introduce it, one has to define first the minor and the cofactor of an element of a matrix.

The *minor* of a_{ij}, the (i,j)th element of $\mathbf{A}$, is the value of the determinant obtained after deleting the ith row and the jth column of $\mathbf{A}$. The *cofactor* of a_{ij}, denoted by A_{ij}, is equal to the product of $(-1)^{i+j}$ and the minor of a_{ij}. For instance, for $\mathbf{A}$ of order 3,

$$A_{11} = \begin{vmatrix} a_{22} & a_{23} \\ a_{32} & a_{33} \end{vmatrix} = a_{22}a_{33} - a_{23}a_{32}, \tag{A.31}$$

$$A_{12} = -\begin{vmatrix} a_{21} & a_{23} \\ a_{31} & a_{33} \end{vmatrix} = a_{23}a_{31} - a_{21}a_{33}, \tag{A.32}$$

$$A_{13} = \begin{vmatrix} a_{21} & a_{22} \\ a_{31} & a_{32} \end{vmatrix} = a_{21}a_{32} - a_{22}a_{31}. \tag{A.33}$$

One can prove that, for $\mathbf{A}$ of order $n \times n$ not less than 2×2, we have

$$|\mathbf{A}| = \sum_{j=1}^{n} a_{ij} A_{ij} \tag{A.34}$$

for any i and

$$|\mathbf{A}| = \sum_{i=1}^{n} a_{ij} A_{ij} \qquad (A.35)$$

for any j.

For instance, for $\mathbf{A}$ of order 3, we can write

$$|\mathbf{A}| = a_{11} A_{11} + a_{12} A_{12} + a_{13} A_{13}. \qquad (A.36)$$

Therefore, when dealing with matrices of higher orders, we can use either of the given formulas to reduce successively the orders of cofactors which have to be effectively computed. Whatever the order of the matrix whose determinant is to be calculated, this procedure requires in effect effective computation of determinants of arbitrarily low order. And, as we mentioned already, we can let a computer do the whole job.

The following results hold for determinants of any matrices $\mathbf{A}$ and $\mathbf{B}$ of the same order:

(i) $\quad |\mathbf{A}| \quad = \quad |\mathbf{A}'|$
(ii) $\quad |c\mathbf{A}| \quad = \quad c^n |\mathbf{A}|$, where c is a scalar and $\mathbf{A}$ is of order n
(iii) $|\mathbf{AB}| \quad = \quad |\mathbf{A}| |\mathbf{B}|$.

Also, if $\mathbf{A}$ is a diagonal matrix of order n,

$$|\mathbf{A}| = a_{11} a_{22} \cdots a_{nn}. \qquad (A.37)$$

Note that properties (i) and (iii) imply that $|\mathbf{A}| = 1$ for an orthogonal matrix $\mathbf{A}$.

A matrix is called *singular* if its determinant is zero; otherwise, it is *non-singular*. Note that a scalar, which is a matrix of order 1, is singular if and only if it is zero. One should be warned, however, that in general a matrix need not be the zero matrix to be singular.

Example:

$$\begin{vmatrix} 1 & 3 & 2 \\ -2 & 4 & -1 \\ 2 & 6 & 4 \end{vmatrix} = 0. \qquad (A.38)$$

A matrix is non-singular if and only if it is of full rank.

A.5 Inverses

The *inverse* of $\mathbf{A}$ is the unique matrix $\mathbf{A}^{-1}$ such that

$$\mathbf{A}\mathbf{A}^{-1} = \mathbf{A}^{-1}\mathbf{A} = \mathbf{I}. \qquad (A.39)$$

The inverse exists if and only if $\mathbf{A}$ is non-singular or, equivalently, of full rank.

We shall show that an orthogonal matrix $\mathbf{A}$ has $\mathbf{A}'$ as its inverse. By definition, $\mathbf{A}\mathbf{A}' = \mathbf{I}$. Denote the product $\mathbf{A}'\mathbf{A}$ by $\mathbf{C}$. Now,

$$\mathbf{A}\mathbf{C}\mathbf{A}' = \mathbf{A}\mathbf{A}'\mathbf{A}\mathbf{A}'.$$

Hence, $\mathbf{A}\mathbf{C}\mathbf{A}' = \mathbf{A}\mathbf{A}'$ and

$$\mathbf{A}^{-1}\mathbf{A}\mathbf{C}\mathbf{A}'(\mathbf{A}')^{-1} = \mathbf{A}^{-1}\mathbf{A}\mathbf{A}'(\mathbf{A}')^{-1},$$

but this implies that $\mathbf{C} = \mathbf{I}$ and the required result follows. Thus, in the case of orthogonal matrices computing inverses is easy. In general, the following result holds for non-singular $\mathbf{A} = (a_{ij})$:

$$\mathbf{A}^{-1} = \frac{1}{|\mathbf{A}|}(A_{ij})', \tag{A.40}$$

where the matrix $(A_{ij})'$ is the transpose of (A_{ij}), the matrix with the (i,j)th element A_{ij} being the cofactor of a_{ij}.

Example: For

$$\mathbf{A} = \begin{bmatrix} 3 & 2 & 1 \\ 0 & 1 & 2 \\ 2 & 0 & 1 \end{bmatrix}, \tag{A.41}$$

we have $|\mathbf{A}| = 9$,

$$(A_{ij}) = \begin{bmatrix} 1 & 4 & -2 \\ -2 & 1 & 4 \\ 3 & -6 & 3 \end{bmatrix} \tag{A.42}$$

and, hence,

$$\mathbf{A}^{-1} = \begin{bmatrix} 1/9 & -2/9 & 1/3 \\ 4/9 & 1/9 & -2/3 \\ -2/9 & 4/9 & 1/9 \end{bmatrix}. \tag{A.43}$$

One can show that for all non-singular matrices $\mathbf{A}$ and $\mathbf{B}$ of the same order and for all nonzero scalars c, we have

$$
\begin{aligned}
(i) \quad & |\mathbf{A}|\,|\mathbf{A}^{-1}| &=& \quad 1 \\
(ii) \quad & (c\mathbf{A})^{-1} &=& \quad c^{-1}\mathbf{A}^{-1} \\
(iii) \quad & (\mathbf{A}^{-1})' &=& \quad (\mathbf{A}')^{-1} \\
(iv) \quad & (\mathbf{A}\mathbf{B})^{-1} &=& \quad \mathbf{B}^{-1}\mathbf{A}^{-1}.
\end{aligned}
$$

As was mentioned previously, computations of inverses can be transferred to a computer. One should be warned, however, that if a matrix is "nearly singular" (i.e., its determinant is close to zero), such a transfer becomes risky: round-off errors may then prove unacceptably large. Hence, it is recommended to check whether the product of the matrix and its obtained inverse is indeed equal to $\mathbf{I}$.

A.6   Definiteness of a Matrix

Important Remark: Throughout the rest of this appendix all matrices are assumed to be symmetric.

Let $\mathbf{x}$ be a p-dimensional vector and $\mathbf{A}$ be a symmetric matrix of order $p \times p$. The scalar

$$\mathbf{x}'\mathbf{A}\mathbf{x} = \sum_{i=1}^{p} \sum_{j=1}^{p} a_{ij} x_i x_j \tag{A.44}$$

is called a *quadratic form* in the vector $\mathbf{x}$.

A symmetric matrix $\mathbf{A}$ (and a quadratic form associated with it) is called
  (a) *positive definite* if $\mathbf{x}'\mathbf{A}\mathbf{x} > 0$ for all $\mathbf{x} \neq \mathbf{0}$;
  (b) *positive semi-definite* if $\mathbf{x}'\mathbf{A}\mathbf{x} \geq 0$ for all $\mathbf{x}$;
  (c) *negative definite* if $\mathbf{x}'\mathbf{A}\mathbf{x} < 0$ for all $\mathbf{x} \neq \mathbf{0}$;
  (d) *negative semi-definite* if $\mathbf{x}'\mathbf{A}\mathbf{x} \leq 0$ for all $\mathbf{x}$;
we write then $\mathbf{A} > \mathbf{0}$, $\mathbf{A} \geq \mathbf{0}$, $\mathbf{A} < \mathbf{0}$, $\mathbf{A} \leq \mathbf{0}$, respectively. Otherwise, $\mathbf{A}$ is called *indefinite*. Sometimes, positive semi-definite matrices are called *nonnegative definite* while negative semi-definite matrices are called *nonpositive definite*.

Note that positive (negative) definiteness of a matrix implies its positive (negative) semi-definiteness. It can be proved that positive- and negative- definite matrices are necessarily non-singular.

A.7   Eigenvalues and Eigenvectors

Eigenvalues and eigenvectors provide an extremely useful decomposition of symmetric matrices.

For a symmetric matrix $\mathbf{A}$ of order p and a scalar λ, the determinant $|\mathbf{A} - \lambda\mathbf{I}|$ (where, of course, $\mathbf{I} = \mathbf{I}_p$) is a pth order polynomial in λ. The p roots of the equation

$$|\mathbf{A} - \lambda\mathbf{I}| = 0 \tag{A.45}$$

are called the *eigenvalues* (or *characteristic roots* or *latent roots*) of $\mathbf{A}$. The equation itself is called the *characteristic equation* of $\mathbf{A}$. The characteristic equation may have multiple roots, and then some of the eigenvalues $\lambda_1, \lambda_2, \ldots, \lambda_p$ are equal. But, as can be shown for any symmetric $\mathbf{A}$, all the λ_i, $i = 1, 2, \ldots, p$, are real numbers.

For each eigenvalue λ_i, $i = 1, 2, \ldots, p$, there exists a vector $\mathbf{x}_i$ satisfying the vector equation

$$\mathbf{A}\mathbf{x}_i = \lambda_i \mathbf{x}_i \quad (\text{or} \quad (\mathbf{A} - \lambda_i \mathbf{I})\mathbf{x}_i = \mathbf{0}). \tag{A.46}$$

Any vector satisfying the above equation is called the *eigenvector* (or *characteristic vector* or *latent vector*) associated with the eigenvalue λ_i. For obvious reasons, any solution of this equation may be *standardized* (or *normalized*) so that, after standardization, it has length one, $\mathbf{x}_i'\mathbf{x}_i = 1$.

The following results hold for the eigenvalues and eigenvectors of any symmetric matrix $\mathbf{A}$ of order p:

(i) $|\mathbf{A}| = \lambda_1 \lambda_2 \cdots \lambda_p$;

(ii) $tr(\mathbf{A}) = \sum_{i=1}^{p} \lambda_i$;

(iii) if $\mathbf{A}$ is diagonal, $\mathbf{A} = diag(a_1, a_2, \ldots, a_p)$, then $\lambda_i = a_i$ for all i, $i = 1, 2, \ldots, p$;

(iv) if $\mathbf{A}$ is positive (negative) definite, then all eigenvalues of $\mathbf{A}$ are positive (negative);

(v) if $\mathbf{A}$ is positive (negative) semi-definite of rank r, then exactly r of the eigenvalues of $\mathbf{A}$ are positive (negative) while the remaining eigenvalues are zero;

(vi) the rank of $\mathbf{A}$ is equal to the number of non-zero eigenvalues of $\mathbf{A}$;

(vii) if λ_i is an eigenvalue of $\mathbf{A}$ and no other eigenvalue of $\mathbf{A}$ equals λ_i, then the eigenvector associated with λ_i is unique up to a scalar factor (i.e., it is unique if it is assumed to be standardized);

(viii) eigenvectors associated with distinct eigenvalues of $\mathbf{A}$ are necessarily orthogonal one to another;

(ix) eigenvectors of $\mathbf{A}$ can always be chosen in such a way that they are mutually orthogonal.

From the point of view of applications, the most important result to be introduced in this section is the following *spectral decomposition* (or *Jordan decomposition*) of a symmetric matrix $\mathbf{A}$ of order p:

$$\mathbf{A} = \mathbf{\Gamma}\mathbf{\Lambda}\mathbf{\Gamma}' = \sum_{i=1}^{p} \lambda_i \mathbf{x}_i \mathbf{x}_i', \tag{A.47}$$

where $\mathbf{\Lambda}$ is a diagonal matrix of eigenvalues of $\mathbf{A}$,

$$\mathbf{\Lambda} = diag(\lambda_1, \lambda_2, \ldots, \lambda_p),$$

and $\boldsymbol{\Gamma}$ is an orthogonal matrix whose columns are standardized eigenvectors of $\mathbf{A}$, $\boldsymbol{\Gamma} = [\mathbf{x}_1, \mathbf{x}_2, \ldots, \mathbf{x}_p]$. In view of the previously presented results, this decomposition can always be done; it suffices to find the eigenvalues of $\mathbf{A}$ and choose corresponding eigenvectors in such a way that they be standardized and mutually orthogonal.

Let us use the concept of the spectral decomposition of a matrix to prove a useful property of the so-called idempotent matrices. A symmetric matrix $\mathbf{A}$ is called *idempotent* if $\mathbf{A} = \mathbf{A}^2$, where $\mathbf{A}^2 = \mathbf{A}\mathbf{A}$. By the spectral decomposition of a symmetric matrix $\mathbf{A}$, we have that $\mathbf{A} = \boldsymbol{\Gamma}\boldsymbol{\Lambda}\boldsymbol{\Gamma}'$ and $\mathbf{A}^2 = \boldsymbol{\Gamma}\boldsymbol{\Lambda}\boldsymbol{\Gamma}'\boldsymbol{\Gamma}\boldsymbol{\Lambda}\boldsymbol{\Gamma}' = \boldsymbol{\Gamma}\boldsymbol{\Lambda}^2\boldsymbol{\Gamma}'$. If $\mathbf{A}$ is idempotent, $\boldsymbol{\Lambda} = \boldsymbol{\Lambda}^2$ and, hence, $\lambda_i = 0$ or 1 for all i. Thus, by (vi) and (ii) above,

$$r(\mathbf{A}) = tr(\mathbf{A}) \qquad (A.48)$$

for any symmetric and idempotent $\mathbf{A}$.

We shall conclude this section with an example of a detailed derivation of the spectral decomposition of a matrix of order 3. Another problem is how to effectively find eigenvalues and eigenvectors of a matrix of order greater than 3. Finding the roots of a third-order polynomial is already cumbersome but, in general and without referring to iterative methods of numerical analysis, the task becomes unworkable if a polynomial is of order greater than four. Moreover, the greater is the order of a matrix the more tedious, although in principle routine, is the computation of the matrix' eigenvectors. No wonder, therefore, that the job is in fact always left to a computer.

Example: Let

$$\mathbf{A} = \begin{bmatrix} 18 & 6 & 2 \\ 6 & 23 & 3 \\ 2 & 3 & 15 \end{bmatrix}. \qquad (A.49)$$

Then $|\mathbf{A} - \lambda\mathbf{I}| = -\lambda^3 + 56\lambda^2 - 980\lambda + 5488$, and the eigenvalues of the characteristic equation are $\lambda_1 = 28$ and $\lambda_2 = \lambda_3 = 14$. The eigenvector associated with $\lambda_1 = 28$ is given by a solution of the following set of three equations, linear in components x_{11}, x_{21}, x_{31} of $\mathbf{x}$ and written below in the vector form:

$$\begin{bmatrix} -10 & 6 & 2 \\ 6 & -5 & 3 \\ 2 & 3 & -13 \end{bmatrix} \begin{bmatrix} x_{11} \\ x_{21} \\ x_{31} \end{bmatrix} = \mathbf{0}. \qquad (A.50)$$

It can be easily verified that only two of the three rows of the above matrix of coefficients are linearly independent. In order to do this, it suffices to: i) consider the rows as vectors, multiply the first row by 2, the second by 3, leave the third row unchanged and add so obtained vectors to get **0**; ii) observe that any two of the three rows are linearly independent. Thus, one of the three equations is redundant. We can arbitrarily set $x_{31} = 1$ and, say, from the first two equations obtain $x_{11} = 2$ and $x_{21} = 3$. Finally, $\mathbf{x}'_1 = [2, 3, 1]$.

As the eigenvectors associated with $\lambda_2 = 14$ and $\lambda_3 = 14$, we can readily choose linearly independent solutions of the following sets of equations (of course, this is in fact one set of equations, since the eigenvectors to be found are associated with equal eigenvalues):

$$
\begin{bmatrix} 4 & 6 & 2 \\ 6 & 9 & 3 \\ 2 & 3 & 1 \end{bmatrix} \begin{bmatrix} x_{12} \\ x_{22} \\ x_{32} \end{bmatrix} = \mathbf{0} \quad \text{and} \quad \begin{bmatrix} 4 & 6 & 2 \\ 6 & 9 & 3 \\ 2 & 3 & 1 \end{bmatrix} \begin{bmatrix} x_{13} \\ x_{23} \\ x_{33} \end{bmatrix} = \mathbf{0}. \quad \text{(A.51)}
$$

It is worth noting that the fact that one can find linearly independent eigenvectors $\mathbf{x}_2$ and $\mathbf{x}_3$ follows, without looking at the equations, from property (ix) in this Section.[1] Indeed, this time two of the three equations are redundant, i.e., the maximum number of linearly independent rows in the matrix of coefficients, or the rank of the matrix, is one. Arbitrarily setting $x_{12} = x_{22} = 1$, we obtain $x_{32} = -5$ and, hence, $\mathbf{x}'_2 = [1, 1, -5]$. Analogously, setting $x_{13} = 0$ and $x_{23} = 1$ yields $x_{33} = -3$, i.e., $\mathbf{x}'_3 = [0, 1, -3]$. One can easily check that, in accordance with property (viii), $\mathbf{x}'_1\mathbf{x}_2 = 0$ and $\mathbf{x}'_1\mathbf{x}_3 = 0$. On the other hand, although linearly independent, the vectors $\mathbf{x}_2$ and $\mathbf{x}_3$ are not orthogonal.

In the last section of this appendix, we give a surprisingly simple procedure of replacing linearly independent eigenvectors of a matrix by vectors which are orthogonal and are still the eigenvectors of the matrix under scrutiny. In particular, the procedure mentioned leads to the replacement of vector $\mathbf{x}_3$ of the example by vector $\tilde{\mathbf{x}}'_3 = [-16, 11, -1]$. It is a trivial matter to verify that $\tilde{\mathbf{x}}_3$ is an eigenvector of $\mathbf{A}$ as well, and that $\mathbf{x}_1$, $\mathbf{x}_2$ and $\tilde{\mathbf{x}}_3$ are mutually orthogonal.

In order to state the spectral decomposition of $\mathbf{A}$, and thus to conclude the example, it remains to standardize the eigenvectors obtained:

$$
\mathbf{y}'_1 = \frac{1}{\mathbf{x}'_1\mathbf{x}_1}\mathbf{x}'_1 = \left[2/\sqrt{14},\ 3/\sqrt{14},\ 1/\sqrt{14}\right],
$$

$$
\mathbf{y}'_2 = \frac{1}{\mathbf{x}'_2\mathbf{x}_2}\mathbf{x}'_2 = \left[1/\sqrt{27},\ 1/\sqrt{27},\ -5/\sqrt{27}\right],
$$

---

[1] Actually, property (ix) implies more, namely that $\mathbf{x}_2$ and $\mathbf{x}_3$ may be chosen to be mutually orthogonal.

$$\mathbf{y}_3' = \frac{1}{\tilde{\mathbf{x}}_3' \tilde{\mathbf{x}}_3} \tilde{\mathbf{x}}_3' = \left[-16/3\sqrt{42}, \ 11/3\sqrt{42}, \ -1/3\sqrt{42} \right].$$

Hence,

$$\mathbf{A} = \mathbf{\Gamma} \mathbf{\Lambda} \mathbf{\Gamma}' \qquad (A.52)$$

where $\mathbf{\Lambda} = diag(28, \ 14, \ 14)$ and

$$\mathbf{\Gamma} = \begin{bmatrix} 2/\sqrt{14} & 1/\sqrt{27} & -16/3\sqrt{42} \\ 3/\sqrt{14} & 1/\sqrt{27} & 11/3\sqrt{42} \\ 1/\sqrt{14} & -5/\sqrt{27} & -1/3\sqrt{42} \end{bmatrix}. \qquad (A.53)$$

A.8 Matrix Square Root

The matrix square root of a positive semi-definite matrix $\mathbf{A}$ of order p, denoted by $\mathbf{A}^{1/2}$, is the matrix satisfying the equality

$$\mathbf{A} = \mathbf{A}^{1/2} \mathbf{A}^{1/2}. \qquad (A.54)$$

$\mathbf{A}^{1/2}$ can easily be found using the spectral decomposition of $\mathbf{A}$,

$$\mathbf{A} = \mathbf{\Gamma} \mathbf{\Lambda} \mathbf{\Gamma}', \qquad (A.55)$$

where $\mathbf{\Lambda}$ is a diagonal matrix of eigenvalues of $\mathbf{A}$ and $\mathbf{\Gamma}$ is an orthogonal matrix whose columns are standardized eigenvectors of $\mathbf{A}$. Since $\mathbf{A}$ is positive semi-definite, all its eigenvalues λ_i are nonnegative and, hence, $\mathbf{\Lambda} = \mathbf{\Lambda}^{1/2} \mathbf{\Lambda}^{1/2}$, where $\mathbf{\Lambda}^{1/2} = diag(\lambda_1^{1/2}, \ \lambda_2^{1/2}, \ \ldots, \ \lambda_p^{1/2})$. Thus,

$$\mathbf{A} = \mathbf{\Gamma} \mathbf{\Lambda}^{1/2} \mathbf{\Lambda}^{1/2} \mathbf{\Gamma}' = \mathbf{\Gamma} \mathbf{\Lambda}^{1/2} \mathbf{\Gamma}' \mathbf{\Gamma} \mathbf{\Lambda}^{1/2} \mathbf{\Gamma}' \qquad (A.56)$$

since, due to orthogonality of $\mathbf{\Gamma}$, $\mathbf{\Gamma}' \mathbf{\Gamma} = \mathbf{I}$. Finally, therefore, we get that

$$\mathbf{A}^{1/2} = \mathbf{\Gamma} \mathbf{\Lambda}^{1/2} \mathbf{\Gamma}' \qquad (A.57)$$

for any positive semi-definite $\mathbf{A}$.

One can readily show that $\mathbf{A}^{1/2}$ is a symmetric matrix with the inverse

$$(\mathbf{A}^{1/2})^{-1} = \mathbf{\Gamma} \mathbf{\Lambda}^{-1/2} \mathbf{\Gamma}', \qquad (A.58)$$

where $\mathbf{\Lambda}^{-1/2} = diag(\lambda_1^{-1/2}, \ \lambda_2^{-1/2}, \ldots, \ \lambda_p^{-1/2})$.

A.9   Gram-Schmidt Orthogonalization

Let the characteristic equation of a symmetric matrix $\mathbf{A}$ of order p have a root of multiplicity k, $1 < k \leq p$. Thus, $\mathbf{A}$ has k equal eigenvalues, say, $\lambda_1 = \lambda_2 = \cdots = \lambda_k$. Assume that we are given k linearly independent eigenvectors $\mathbf{x}_1$, $\mathbf{x}_2$, ..., $\mathbf{x}_k$ associated with λ_1, λ_2, ..., λ_k, respectively, and note that $\mathbf{x}_1$, $\mathbf{x}_2$, ..., $\mathbf{x}_k$ are orthogonal to the remaining eigenvectors, $\mathbf{x}_{k+1}$, $\mathbf{x}_{k+2}$, ..., $\mathbf{x}_p$, of $\mathbf{A}$ (by property (viii) of Section A.7).

Replace vectors $\mathbf{x}_1$, $\mathbf{x}_2$, ..., $\mathbf{x}_k$ by the following vectors:

$$
\begin{aligned}
\tilde{\mathbf{x}}_1 &= \mathbf{x}_1 \\
\tilde{\mathbf{x}}_2 &= \mathbf{x}_2 - \frac{\mathbf{x}_2' \tilde{\mathbf{x}}_1}{\tilde{\mathbf{x}}_1' \tilde{\mathbf{x}}_1} \tilde{\mathbf{x}}_1 \\
&\vdots \\
\tilde{\mathbf{x}}_k &= \mathbf{x}_k - \frac{\mathbf{x}_k' \tilde{\mathbf{x}}_1}{\tilde{\mathbf{x}}_1' \tilde{\mathbf{x}}_1} \tilde{\mathbf{x}}_1 - \cdots - \frac{\mathbf{x}_k' \tilde{\mathbf{x}}_{k-1}}{\tilde{\mathbf{x}}_{k-1}' \tilde{\mathbf{x}}_{k-1}} \tilde{\mathbf{x}}_{k-1} .
\end{aligned}
\tag{A.59}
$$

It can be proved that $\tilde{\mathbf{x}}_1$, $\tilde{\mathbf{x}}_2$, ..., $\tilde{\mathbf{x}}_k$, $\mathbf{x}_{k+1}$, ..., $\mathbf{x}_p$ form a set of mutually orthogonal eigenvectors of $\mathbf{A}$. The process of orthogonalization given above is known as the Gram-Schmidt procedure.

Example: We shall apply the Gram-Schmidt procedure to the eigenvectors $\mathbf{x}_2' = [1,\ 1,\ -5]$ and $\mathbf{x}_3' = [0,\ 1,\ -3]$ of the example from Section A.7.

$$
\tilde{\mathbf{x}}_2 = \mathbf{x}_2 = \begin{bmatrix} 1 \\ 1 \\ -5 \end{bmatrix} \quad \text{and} \quad \tilde{\mathbf{x}}_3 = \mathbf{x}_3 - \frac{16}{27} \tilde{\mathbf{x}}_2 = \begin{bmatrix} -16/27 \\ 11/27 \\ -1/27 \end{bmatrix} .
$$

To simplify matters a bit, one can replace the last vector by the vector $[-16,\ 11,\ -1]'$, since the latter preserves the required properties of the former.

Appendix B

A Brief Introduction to Stochastics

B.1 Introduction

Probability calculus provides means to deal with chance, or random, phenomena in a mathematically rigorous way. Such phenomena can be either those which we passively observe or those which we purposely create. In the latter case, it is of course natural to speak of random experiments. However, it has become customary in the probabilistic and statistical literature to speak of random experiments in the former case as well; in fact, we may consider ourselves to be passive observers of the "experiments" performed by nature. We shall follow this convention in the sequel.

We say that an experiment is *random* if it can be performed repeatedly under essentially the same conditions, its outcome cannot be predicted with certainty, and the collection of all possible outcomes is known prior to the experiment's performance. When we say that the conditions under which the experiment is performed are "essentially" the same, we mean that we may assume that these conditions *are* the same and retain sufficient accuracy of our model of the actual experiment. For example, when tossing repeatedly a die on the beach, we can disregard the fact that some tosses are accompanied by gentle blows of a breeze. Of course, in the case of tossing a fair die, it is reasonable to assert unpredictability of the outcome. The third attribute of randomness is also fulfilled. The collection of all possible outcomes consists of all faces of the die.

The set of all possible outcomes of a random experiment is called a *sample space* and it will be denoted by S. An element of the sample space (i.e., a single outcome) is called a *sample point* or an *elementary event*. For instance, one can consider the time to the first failure of the hard disk of a computer to be a random experiment with $S = [0, \infty)$, where each real number $s \epsilon S$ is a sample point corresponding to the time to the first failure equal to s. One is often interested rather in the occurrence of any elementary event from a particular subset of S than in the occurrence of a particular elementary event. It is definitely so in the case of the time to the first failure. One can hardly be interested that this time be, say, exactly 2,557,440 minutes (or 5 years). At the same time, one can be interested that this time be not smaller than 5 years. In the case of tossing a six-sided die with faces 1 to 6, one can be interested in getting 6, but one can also be interested in getting an even outcome, 2 or 4 or 6. A subset of a sample space is called an *event*.

It is worth noting that even simple experiments require sometimes introducing quite involved sample spaces. Suppose for example that we are interested in the experiment consisting in tossing a coin until the head occurs for the first time. Denoting the head and tail occurrences by H and T, respectively, we easily find out that the sample space consists of infinitely many elementary events,

$$S = \{H, TH, TTH, TTTH, TTTTH, \ldots\}.$$

In practice, we are interested not in events alone but, at the same time, in chances, or probabilities, of their occurrence. Although the very term probability has several interpretations, we rarely have problems when assigning probabilities to "simple" events. The common understanding of the notion of probability in engineering sciences is well described by the so-called relative frequency approach. In order to present this approach, let us refer to the simplest possible example of tossing a fair coin once. In this example, $S = \{H, T\}$ and we ask what is the probability of, say, a head occurring. Imagine an experiment in which we toss the coin N times and take the ratio n/N, where n is the number of heads occurring in N tosses, as an approximation of the probability of head's occurrence in one toss. Had we let N tend to infinity, a conceivable but actually not possible experiment, we would have observed that the values of the ratio n/N, as N increases, stabilize. At the turn of the last century, Karl Pearson (1857-1936) performed the experiment with $N = 24,000$ and obtained $n/N = .5005$. More recently, computers were used to simulate our experiment with N assuming astronomically large values and with the ratio n/N being always practically equal to $1/2$. The problem with this approach is that although it is possible to prove that a limit value of the ratio in question exists, we cannot prove experimentally what this limit value is, since we cannot perform infinitely many experiments in a finite time. Still, it is undoubtedly reasonable to assign value $1/2$ to the probability of the head occurring in one toss of a fair coin. By the same token, we assign probability $1/6$ to obtaining any particular face in one toss of a fair, that is, balanced, six-sided die.

Note that the relative frequency approach imposes the following natural bounds on probabilities of events. The probability of a certain event must be equal to one, while the probability of an impossible event must be equal to zero. Formally, denoting the probability of an event $A \subseteq S$ by $P(A)$, we can write

$$P(S) = 1 \quad \text{and} \quad P(\emptyset) = 0, \tag{B.1}$$

where $\emptyset$ denotes the empty set. Indeed, S is the certain event, since the occurrence of S is equivalent to the occurrence of any of all elementary events, and one of these must of course occur. For example, if we toss a six-sided die, one of its faces must be shown or, if we have a PC, its hard disk must have some time to the first failure, this time being an element of $S = [0, \infty)$. On the other hand, an "empty event," which corresponds to the empty set, is an impossible event, since the random experiment has been constructed in such a way that some outcome from S must occur. It also follows from the relative frequency approach that, for all events $A \subseteq S$, probability $P(A)$ is defined to be a real number such that

$$0 \leq P(A) \leq 1. \tag{B.2}$$

In practice, we are most often interested in calculating probabilities of some combinations of events. For example, PC owners can be interested in maximizing the probability that their hard disks will break down either in the first year (i.e., during the warranty period) or after, say, five years from the purchase (practically nobody keeps a PC for more than five years). If we were given the probabilities of the events {time to failure is not greater than 1 year} and {time to failure is not smaller than 5 years}, we could calculate the probability of their union as

$$P(\{\text{time to failure} \leq 1 \text{ year}\} \cup \{\text{time to failure} \geq 5 \text{ years}\})$$
$$= P(\text{time to failure} \leq 1 \text{ year}) + P(\text{time to failure} \geq 5 \text{ years}),$$

where $C \cup D$ denotes the union of sets C and D, i.e., the set of elements which belong either to C or to D or to both. Similarly, the probability of obtaining 2 or 4 or 6 in one toss of a fair die with faces 1 to 6 is $1/2$. Let us consider one more example. Suppose we are interested in the experiment consisting in two successive tosses of a fair coin, and we wish to find the probability of obtaining a head in the first toss. In this experiment, the sample space contains four elementary events, $S = \{HH, HT, TH, TT\}$. Of course, we assume that all the four elementary events are equally likely (i.e., their probabilities are $1/4$). We have

$$P(\text{head in the first toss}) = P(HH \cup HT) = P(HH) + P(HT) = 1/2,$$

in accordance with what should be expected intuitively.

In all the above calculations of probabilities, we have applied the following general rule. If, for any fixed n, the events $A_1, A_2, \ldots, A_n$ are mutually exclusive (i.e., they are such that no two have an element in common, $A_i \cap A_j = \emptyset$ whenever $i \neq j$), then

$$P(A_1 \cup A_2 \cup \ldots \cup A_n) = P(A_1) + P(A_2) + \ldots + P(A_n). \qquad \text{(B.3)}$$

Conditions (B.1), (B.2) and (B.3) are known as the axioms of probability.

Whenever necessary, condition (B.3) can be generalized to the following one: If the events $A_1, A_2, A_3, \ldots$ are mutually exclusive, then

$$P(A_1 \cup A_2 \cup A_3 \cup \ldots) = P(A_1) + P(A_2) + P(A_3) + \ldots.$$

Here, infinitely many mutually exclusive events are taken into account.

As we have seen, condition (B.3) enables one to calculate probabilities of mutually exclusive events in a rigorous and consistent way. If events A and B are not exclusive, i.e., they have some elementary events in common, condition (B.3) can still be used after decomposing the union of the events, $A \cup B$, into two nonoverlapping sets. Actually, one can then show that, for any two events A and B,

$$P(A \cup B) = P(A) + P(B) - P(A \cap B), \qquad \text{(B.4)}$$

where $A \cap B$ denotes the intersection (i.e., the common part) of A and B.

Let us conclude this section with some more problems connected with games of chance. In fact, much of probability calculus in the past centuries was concerned with such problems. Whatever the reason for that historical fact, it is the easiest path to enter the world of probabilistic ideas. Note that our coin or die tossing examples were the problems of this kind. The reader, however, should not get the false impression that probability theory is good for nothing more than dealing with artificial games far removed from reality.

Let us first compute the number of ways that we can arrange in a distinctive order k objects selected without replacement from n, $n \geq k$, distinct objects. We easily see that there are n ways of selecting the first object, $n - 1$ ways of selecting the second object, and so on until we select $k - 1$ objects and note that the kth object can be selected in $n - k + 1$ ways. The total number of ways is called the *permutation* of n objects taken k at a time, $P(n, k)$, and is seen to be given by

$$P(n, k) = n(n - 1)(n - 2) \cdots (n - k + 1) = \frac{n!}{(n - k)!}, \qquad \text{(B.5)}$$

where $m! = m(m - 1)(m - 2) \cdots 2 \times 1$, $0! = 1$. In particular, there are $n!$ ways that we can arrange n objects in a distinctive order. Next, let us compute in how many ways we can select k objects from n objects when we are not concerned with the distinctive order of selection. This number of ways is called the *combination* of n objects taken k at a time, and is denoted by $C(n, k)$. We can find it by noting that $P(n, k)$ could be first computed by finding $C(n, k)$ and then multiplying it by the number of ways k objects could be distinctly arranged (i.e., $k!$). So we have

$$P(n, k) = C(n, k)P(k, k) = C(n, k)k!$$

and thus

$$\binom{n}{k} = C(n, k) = \frac{n!}{(n - k)!k!}. \qquad \text{(B.6)}$$

For example, the game of stud poker consists in the drawing of 5 cards from a 52 card deck (4 suits, 13 denominations). The number of possible hands is given by

$$C(52, 5) = \frac{52!}{47!5!} = 2,598,960.$$

We are now in a position to compute some basic probabilities which are slightly harder to obtain than, say, those concerning tossing a die. Each of the 2,598,960 possible poker hands is equally likely. To compute the probability of a particular hand, we simply evaluate

$$P(\text{hand}) = \frac{\text{number of ways of getting the hand}}{\text{number of possible hands}}.$$

Suppose we wish to find the probability of getting an all-spade hand. There are $C(13,5)$ ways of selecting 5 spades (without regard to their order) out of 13 spades. Hence,

$$P(\text{an all spade hand}) = \frac{C(13,5)}{C(52,5)}$$

$$= \frac{(9)(10)(11)(12)(13)}{(5!)(2,598,960)} = .0000495.$$

Finding the probability of getting four cards of a kind (e.g., four aces, four kings) is a bit more complicated. There are $C(13,1)$ ways of picking a denomination, $C(4,4)$ ways of selecting all the four cards of the same denomination, and $C(48,1)$ of selecting the remaining card. Thus,

$$P(\text{four of a kind}) = \frac{C(13,1)C(4,4)C(48,1)}{C(52,5)}$$

$$= \frac{(13)(1)(48)}{2,598,960} = .00024.$$

Similarly, to find the probability of getting two pairs, we have

$$P(\text{two pairs}) = \frac{C(13,2)C(4,2)C(4,2)C(44,1)}{C(52,5)}$$

$$= \frac{(78)(6)(6)(44)}{2,598,960} = .0475.$$

The next two Sections, as well as Sections B.5 and B.10, provide the reader with necessary elements of the mathematical framework of probability calculus. In Sections B.4, B.6 and B.12, the most widely used probabilistic models are briefly discussed. In Sections B.7-B.9, some specific and particularly powerful tools for dealing with random phenomena are introduced. Section B.11 deals with multivariate random phenomena. In the last Section, a brief exposition of statistical methods is given.

B.2   Conditional Probability

The probability of an event B occurring when it is known that some event A has occurred is called a *conditional probability* of B given that A has occurred or, shortly, the probability of B given A. The probability of B given A will be denoted by $P(B|A)$. Suppose, for instance, that B

is the event of having a hand of 5 spades in stud poker. It is clear that we should expect the "unconditional" probability $P(B)$ to be different from the conditional probability of having an all-spades hand given that we already know that we have at least four spades in the hand. Actually, we would rather expect $P(B|A)$ to be greater than $P(B)$. The fact that the occurrence of one event may influence the conditional probability of another event occurring is perhaps even more transparent in the following example. Suppose that we are interested in the event that the price of rye is higher tomorrow in Chicago than it is today. Since there is a transfer of grain use between rye and wheat, if we know that the price of wheat will go up, then the probability that the price of rye will go up is greater than if we did not have the information about the increase in the price of wheat.

In order to become able to evaluate conditional probabilities effectively, let us look into this concept more thoroughly. First, since event A has occurred, only those outcomes of a random experiment are of interest that are in A. In other words, the sample space S (of which A is a subset) has been reduced to A, and we are now interested in the probability of B occurring, relative to the new sample space A. Formally, we should postulate that $P(A|A) = 1$: indeed, the probability of A given A has occurred should be one. Second, since A is now the sample space, the only elements of event B that concern us are those that are also in A. In other words, we are now interested in the intersection (or the common part) of A and B. Formally, we should postulate that $P(B|A) = P(A \cap B|A)$, where $A \cap B$ denotes the set consisting of only those elements which belong to A and to B. And finally, upon noticing that $P(A \cap B)$ and $P(A \cap B|A)$ are the probabilities of the same event, although computed relative to different sample spaces, we should postulate that $P(A \cap B)$ and $P(A \cap B|A)$ be proportional. More precisely, the last postulate states that for all events B such that $P(A \cap B) > 0$ the ratio $P(A \cap B|A)/P(A \cap B)$ is equal to some constant independent of B.

It is not hard to show that the three postulates mentioned imply the following formal definition of the conditional probability of B given A (assuming $P(A)$ is not zero):

$$P(B|A) = \frac{P(A \cap B)}{P(A)}. \tag{B.7}$$

The conditional probability of an all-spade hand (event B) given that there are at least four spades in the hand (event A) can now be readily calculated. Note that the event A consists of all hands with four spades and all hands with five spades and, therefore, that $A \cap B = B$. Hence,

$$P(B|A) \;=\; \frac{P(B)}{P(A)}$$

$$= \frac{\binom{13}{5} / \binom{52}{5}}{\left[\binom{13}{4}\binom{39}{1} + \binom{13}{5}\right] / \binom{52}{5}} = .0441,$$

since, by condition (B.3), $P(A) = P(A_1 \cup A_2) = P(A_1) + P(A_2)$, where A_1 and A_2 denote the events that there are exactly four and exactly five spades in the hand, respectively.

Note that it follows from (B.7) that

$$P(A \cap B) = P(B|A)P(A) = P(A|B)P(B). \tag{B.8}$$

These identities are known as the *multiplicative rule* of probability.

It may happen that events A and B have no effect upon each other. Consider, for example, the experiment consisting in two successive tosses of a fair coin. Clearly, the probability of any of the two possible outcomes of the second toss is unaffected by the outcome of the first toss and, conversely, the first toss is independent of the second. In such situations, the information that some event A has occurred does not change the probability of B occurring and, conversely, the information that B has occurred does not change the probability of A occurring. Formally, we have then

$$P(B|A) = P(B) \quad \text{and} \quad P(A|B) = P(A). \tag{B.9}$$

By (B.8), property (B.9) is equivalent to the following one:

$$P(A \cap B) = P(A)P(B). \tag{B.10}$$

We say that events A and B are *stochastically independent* or, shortly, *independent* if condition (B.10) is fulfilled. We leave it to the reader to find out why mutually exclusive events (of positive probability) cannot be independent.

B.3   Random Variables

Examples of random experiments show that elementary events may but do not have to be expressed numerically. However, in order to handle events defined on a sample space in a mathematically convenient way, it is necessary to attach numerical value to every elementary event. When the elementary events are themselves numbers, it is, in fact, already done. In general, it is convenient to introduce the concept of a random variable. The *random variable* is a function that associates a real number with each element in the sample space. The way in which a particular random variable is determined depends on an experimenter's need.

For example, if we consider time to the first failure of a PC's hard disk, elementary events are themselves numbers from the interval $S = [0, \infty)$. On the other hand, if we consider coin tossing once, we may attach, say, value zero to tail occurring and value one to head occurring, thus obtaining the random variable associated with the experiment under scrutiny. Of course, we can, if we wish to, define another random variable on the sample space $S = \{H, T\}$ and work with that another definition as well; the only problem is that we "remember" our definition and properly interpret random variable's values.

A random variable, to be abbreviated r.v., is *discrete* if the range of its values forms a discrete set; that is, if the r.v. can take only values that are either integers or can be put into one-to-one correspondence with some subset of integers. Obviously, r.v.'s associated with coin tossing experiments are discrete. A simple generalization of the coin tossing and die tossing experiments is the following one. Suppose we are given a random variable X which assumes the values $x_1, x_2, \ldots, x_k$ ($x_i \neq x_j$ whenever $i \neq j$) with the same probability $1/k$. (From now on, we shall denote random variables by uppercase letters X, Y, Z, etc., and their values by corresponding lowercase letters x, y, z, etc.) We say that X has the *(discrete) uniform distribution* given by the *probability function*

$$P(X = x_i) = \frac{1}{k}, \quad x = x_1, x_2, \ldots, x_k. \qquad \text{(B.11)}$$

For the sake of brevity, we shall sometimes denote probability functions by $f(x)$ instead of writing explicitly $P(X = x)$. It is also convenient to define the *cumulative distribution function* (c.d.f.) $F(x)$ of a r.v. X as the probability of the event that the r.v. X assumes a value not greater than x,

$$F(x) = P(X \leq x), \quad -\infty < x < \infty. \qquad \text{(B.12)}$$

Note that $F(x)$ is defined for all real numbers x. By the definition, it is a nondecreasing function, bounded from below by zero and from above by one.

In the case of a uniformly distributed r.v. X,

$$F(x) = \sum_{x_i \leq x} P(X = x_i) = \sum_{x_i \leq x} f(x_i), \qquad \text{(B.13)}$$

a sum over all x_i's which are not greater than x, $-\infty < x < \infty$.

So far, we have discussed only the simplest possible discrete probability distribution. In the next section, we shall discuss four more distributions, namely the hypergeometric, binomial, geometric and Poisson distributions. Already a cursory look at the probability functions of these distributions, given by (B.30), (B.31), (B.32) and (B.36), respectively, makes it evident that a way of summarizing the information provided by the probability functions would be welcome. The problem is that probability functions have to be determined at many different points. In particular, the Poisson probability function is defined for all nonnegative integers! So it is natural to attempt to characterize the most important aspects of a probability distribution using as few number characteristics as possible. The most common of such summaries are the mean and variance of an r.v. (or, equivalently, of its distribution).

The *mean*, or *expected value*, of the r.v. X is defined as

$$\mu = E(X) = \sum x P(X = x) = \sum x f(x), \qquad \text{(B.14)}$$

where the summation is over all possible values x of the r.v. X. Thus, the mean is simply the weighted average of the values of X, with the weights being the probabilities of the values' occurrences. For the discrete uniform distribution in its general form, we have

$$\mu = E(X) = \frac{1}{k} \sum_{i=1}^{k} x_i. \qquad \text{(B.15)}$$

The mean provides some information about the "location," or the "center," of the probability distribution. However, it does provide no information about the "spread," or "variability," of the distribution about its mean value. It is the variance of the distribution which measures expected squared departures from μ. The *variance* of the r.v. X is given by

$$E[(X - \mu)^2] = \text{Var}(X) = \sigma^2 = \sum (x - \mu)^2 f(x), \qquad \text{(B.16)}$$

where the summation is over all possible values x of the r.v. X. The square root of the variance is called the *standard deviation*. For the discrete uniform distribution in its general form, we have

$$\text{Var}(X) = \sigma^2 = \frac{1}{k}\sum_{i=1}^{k}(x_i - \mu)^2. \tag{B.17}$$

Straightforward calculations show that if X is "rescaled" by a fixed factor a and "shifted" by a constant b, i.e., if it is transformed to the r.v. $aX + b$, then

$$E(aX + b) = a\mu + b \text{ and } \text{Var}(aX + b) = a^2\text{Var}(X). \tag{B.18}$$

Note that the above results are intuitively appealing: e.g., shifting a random variable changes its "location" but does not affect its "variability." More generally, if we are interested in the mean (or the expectation) of some function $g(X)$ of X, we have

$$E(g(X)) = \sum_{x} g(x)f(x), \tag{B.19}$$

where the summation is over all possible values x of X, and, for any fixed reals a and b, we obtain

$$E(ag(X) + b) = aE[g(X)] + b. \tag{B.20}$$

Applying (B.20) repeatedly, we obtain

$$\text{Var}(X) = E(X^2) - [E(X)]^2 \tag{B.21}$$

for any r.v. X.

It is often the case in engineering sciences that random phenomena are functions of more than one random variable. Thus, we have to answer the question how to characterize the joint distribution of several r.v.'s. We shall confine ourselves to the case of two r.v.'s only, since generalizations to more r.v.'s are obvious.

For an r.v. X which assumes the values x_i, $i = 1, 2, \ldots, k$, and an r.v. Y which assumes the values y_j, $j = 1, 2, \ldots, m$, where either k or m or both can be equal to infinity, it is natural to define the joint p.f. of X and Y via

$$f(x_i, y_j) = P(X = x_i, Y = y_j), \quad i = 1, 2, \ldots, k, \quad j = 1, 2, \ldots, m \tag{B.22}$$

and the joint c.d.f. via

$$F(x, y) = \sum_{x_i \leq x} \sum_{y_j \leq y} f(x_i, y_j). \tag{B.23}$$

Note that given the joint p.f. $f(x, y)$, we can immediately determine the p.f.'s of X and Y alone:

$$f_X(x_i) = \sum_{j=1}^{m} f(x_i, y_j), \quad i = 1, 2, \ldots, k \tag{B.24}$$

and

$$f_Y(y_j) = \sum_{i=1}^{k} f(x_i, y_j), \quad j = 1, 2, \ldots, m. \tag{B.25}$$

$f_X(x)$ and $f_Y(y)$ are called the *marginal* p.f.'s of X and Y, respectively. Note that it readily follows from (B.10) that the joint p.f. of two independent r.v.'s X and Y is given by

$$f(x, y) = f_X(x) f_Y(y). \tag{B.26}$$

Furthermore, we can calculate the expectation of a function $g(x, y)$ of the r.v.'s X and Y:

$$E(g(X, Y)) = \sum_{i=1}^{k} \sum_{j=1}^{m} g(x_i, y_j) f(x_i, y_j). \tag{B.27}$$

In particular, the *covariance* of X and Y, defined as

$$\mathrm{Cov}(X, Y) = E\{[X - E(X)][Y - E(Y)]\}, \tag{B.28}$$

can be seen as some measure of "co-dependence" between random variables. If X and Y "vary together," i.e., if they tend to assume large or small values simultaneously, then the covariance takes positive values. If, rather, large values of one random variable are associated with small values of the other one, the covariance takes negative values. In the former situation we may speak of "positive," while in the latter of "negative," dependence. Moreover, if X and Y are stochastically independent, we have

$$\mathrm{Cov}(X, Y) = E[X - E(X)]E[Y - E(Y)] = 0.$$

However, the reader should be warned that, in general, the converse is not true; it is possible to give examples of dependent r.v.'s whose covariance is zero.

The covariance, seen as the measure of co-dependence between r.v.'s, has the drawback of not being normalized. Namely, covariances can take values from $-\infty$ to ∞. A normalized version of this measure is provided by the *correlation* between X and Y, which is defined as

$$\rho(X,Y) = \frac{\text{Cov}(X,Y)}{\sqrt{\text{Var}(X)\text{Var}(Y)}}. \qquad (\text{B.29})$$

It is easy to prove that correlation $\rho(X,Y)$ takes values only between -1 and 1. To see this, let a be an arbitrary constant, and consider

$$\begin{aligned} 0 \;\leq\; & E\{[a(X - E(X)) - (Y - E(Y))]^2\} \\ = \;& a^2\sigma_X^2 + \sigma_Y^2 - 2a\text{Cov}(X,Y), \qquad (\text{B.30}) \end{aligned}$$

where σ_X^2 and σ_Y^2 are the variances of X and Y, respectively. Substituting in (B.30) $a = \text{Cov}(X,Y)/\sigma_X^2$ yields

$$\rho^2 \leq 1$$

and the desired result follows. Of course, $\rho(X,Y)$ preserves other properties of the covariance of X and Y. Moreover, one can readily show that $\rho(X,Y) = 1$ if $Y = aX + b$ for any positive a and any b, and $\rho(X,Y) = -1$ if $Y = aX + b$ for any negative a and any b. Thus, the correlation between random variables attains its maximum and minimum values when there is a linear relationship between the variables.

B.4 Discrete Probability Distributions

B.4.1 Hypergeometric Distribution

Suppose we are given a batch of 100 manufactured items. It is known that 5 of them are defective but we do not know which ones. We draw at random and without replacement 10 items. The question is what is the probability that we have drawn no defectives, what is the probability that we have drawn exactly 1 defective, etc. More generally, we can assume that we are given a population of N items, of which M are of type 1 and $N - M$ are of type 2. We draw without replacement n items and ask what is the probability distribution of the number of items of type 1 among the given n items. Now we can define the random variable X which assumes value k if k items of type 1 have been drawn. In order to find the probability distribution of X, let us note first that $P(X = k) = 0$ if $k > M$ or $k > n$. Moreover, if the total number of items of type 2, $N - M$, is smaller than n, then at least $n - (N - M)$ items of type 1 have to be drawn. Proceeding analogously as in Section B.1, we find, therefore, that

$$P(X = k) = \frac{C(M, k)C(N - M, n - k)}{C(N, n)} \qquad \text{(B.31)}$$

if $\max\{n - (N - M), 0\} \leq k \leq \min\{n, M\}$. This is the probability function of the *hypergeometric* distribution with parameters M, N and n. Of course, did we not assume that we know the number of defective items, the above problem would be the most typical problem of acceptance sampling in quality control. In practice, we avoid this obstacle using some estimate of the unknown parameter M.

B.4.2 Binomial and Geometric Distributions

A similar probabilistic model is provided by the following experiment. Consider a single random experiment in which there are only two possible outcomes and assume that we are interested in n independent repetitions of this experiment. Let us call the possible outcomes of each single experiment a success and a failure, respectively. Thus, the whole experiment consists in performing n independent repeated trials, and each trial results in an outcome that may be classified as a success or a failure. Let us assume that the probability of success, p, remains constant from trial to trial (thus, the same can be claimed about the probability of failure, $1 - p$). Such experiments are known as *binomial experiments*. The difference between this experiment and the one that led us to the hypergeometric distribution is that formerly we sampled without replacement, while the binomial experiment corresponds to sampling with replacement. Formerly, the probability that the next item drawn will be defective depended on the number of defectives obtained in the previous draws. In the simpler, binomial experiment, this probability does not change from trial to trial. It is intuitively obvious, however, that both models become practically equivalent if a population from which we draw items is large enough. In fact, it is the binomial model which is much more often used in quality control.

Define the random variable X as the number of successes in n independent trials of a binomial experiment. Clearly, X can assume values $0, 1, 2, \ldots, n$. We can ask what is the probability function (or the probability distribution) of the r.v. X. That is, we want $P(X = x)$, the probability of x successes and $(n - x)$ failures in n trials, where x runs over the values $0, 1, 2, \ldots, n$. In order to answer the question, fix x and note that each particular sequence of trials with x successes has the probability of occurring

$$\underbrace{pp\cdots p}_{x}\underbrace{(1-p)(1-p)\cdots(1-p)}_{n-x \text{ times}} = p^x(1-p)^{n-x},$$

since the trials are independent. Hence, it remains to verify that there are $C(n,x)$ different sequences of trials with x successes. Indeed, this result will follow if we number the trials from 1 to n and ask in how many ways we can select x numbers of trials from all n numbers of trials. Finally therefore,

$$P(X = x) = \binom{n}{x} p^x(1-p)^{n-x}, \quad x = 0, 1, 2, \ldots, n. \qquad \text{(B.32)}$$

This is the probability function of the *binomial distribution* with parameters n and p. The r.v. X is called the binomial random variable. In a natural way, the binomial distribution is used as a basis for acceptance sampling and acceptance-rejection control charts of quality control.

For the binomial distribution with parameters n and p, we have

$$\mu = E(X) = 0f(0) + 1f(1) + 2f(2) + \cdots + nf(n) = np,$$

and

$$\begin{aligned}
\text{Var}(X) = \quad \sigma^2 \quad &= (0 - np)^2 f(0) + (1 - np)^2 f(1) + (2 - np)^2 f(2) \\
&\quad + \cdots + (n - np)^2 f(n) = np(1 - p).
\end{aligned}$$

The above results are not simple when calculated directly, but we shall show in Section B.8 how to obtain them in a surprisingly easy way.

For $n = 1$, the binomial distribution becomes the *Bernoulli distribution*. The only possible values of the r.v. X are then 0 and 1, and the probability of success, $P(X = 1)$, is equal to p while $P(X = 0) = 1 - p$. The r.v. X is called the *Bernoulli random variable* and it corresponds to just one experiment. Note that for any fixed n, a binomial r.v. is in fact the sum of n independent and identically distributed Bernoulli r.v.'s. In other words, a binomial experiment consists in repeating a *Bernoulli experiment* n times.

Sometimes, an experiment consists in repeating Bernoulli experiments until the first success occurs. The number of Bernoulli experiments performed, to be denoted also by X, is then a random variable. The range of X is equal to the set of all nonnegative integers, with the probability of assuming any particular value, x,

$$P(X = x) = p(1 - p)^{x-1}, \ x = 1, 2, \ldots, \tag{B.33}$$

where p denotes the probability of success. The given distribution is called the *geometric distribution* and X is called the *geometric random variable*. The mean and variance of the geometric distribution can be easily obtained using the method of Section B.8. They can also be readily obtained using the following argument. We have, of course, that

$$\sum_{x=1}^{\infty} (1 - p)^{x-1} = \frac{1}{p}.$$

Now, differentiating both sides with respect to p, we obtain

$$\sum_{x=1}^{\infty} (x - 1)(1 - p)^{x-2} = \frac{1}{p^2}.$$

Hence,

$$E(X) = \sum_{x=1}^{\infty} xp(1 - p)^{x-1} = \frac{1}{p}.$$

Using the same differentiation trick again yields, after a little algebra,

$$E(X^2) = \frac{2 - p}{p^2}.$$

Thus, since $Var(X) = E(X^2) - [E(X)]^2$,

$$Var(X) = \frac{1 - p}{p^2}.$$

Let us note that the random variable equal to the number of trials before the first success occurs, $Y = X - 1$, has the distribution

$$P(Y = y) = p(1 - p)^y, \ y = 0, 1, 2, \ldots.$$

Some authors call Y, not X, a geometric r.v.

B.4.3   Poisson Distribution

The last discrete probability distribution we are going to discuss in this section is the distribution named after Simeon Poisson (1781-1840) who was the first to describe rigorously experiments of the following kind. Consider some events which happen at random instants. Let X be the random variable representing the number of events occurring in a given time interval, say, of length T. Assume that the number of events occurring in one time interval is independent of the number of events that occur in any other disjoint time interval,

$$
\begin{aligned}
P(k \text{ occurrences} & \text{ in } [t_1, t_2] \text{ and } m \text{ occurrences in } [t_3, t_4]) \\
& = P(k \text{ in } [t_1, t_2]) P(m \text{ in } [t_3, t_4]) \qquad \text{(B.34)}
\end{aligned}
$$

if $[t_1, t_2] \cap [t_3, t_4] = \emptyset$. Assume also that the probability that a single event will occur in a short time interval is proportional to the length of the time interval,

$$
P(1 \text{ occurrence in } [t, t + \varepsilon]) = \lambda \varepsilon, \qquad \text{(B.35)}
$$

where λ is a proportionality constant. Finally, assume that the probability that more than one event will occur in a short time interval is negligible,

$$
P(\text{more than 1 occurrence in } [t, t + \varepsilon]) = o(\varepsilon), \qquad \text{(B.36)}
$$

where $\lim_{\varepsilon \to \infty} o(\varepsilon)/\varepsilon = 0$. The experiment which fulfills these assumptions is called a *Poisson experiment*, and the r.v. X equal to the number of events occurring in a Poisson experiment is called a *Poisson random variable*.

In telecommunication studies, the number of incoming telephone calls in a given time interval is considered to be a Poisson r.v. Another example of a Poisson r.v. is the number of alpha particles that enter a prescribed region in a given time interval. Also, the number of failures of some devices up to a fixed time is assumed to be a Poisson r.v. This last fact may seem counterintuitive since, contrary to the properties of the Poisson experiment, one would rather expect that if, say, no failure has occurred up to an instant t, then the chances for a failure in an adjacent interval $(t, t + s]$ should increase. However, some devices can indeed be assumed to be "memoryless" in the sense that they "do not remember" that they have already been in use for a time t. For instance, many electronic components could be considered practically everlasting were they not subject to damage by external phenomena of random character.

The probability function of the Poisson r.v. X, representing the number of events occurring in a given time interval of length T, has the form

$$P(X = x) = \frac{e^{-\nu}\nu^x}{x!}, \quad x = 0, 1, 2, \ldots, \tag{B.37}$$

where $\nu = \lambda T$ is the parameter of the distribution, $\nu > 0$. Note that, under the assumptions stated, the number of events' occurrences cannot be bounded from above by some fixed positive integer. It can be shown, however, that the mean and variance of the Poisson distribution with parameter ν both have the value ν. In particular, therefore, ν is the average number of events that occur in the given time interval. Parameter λ may be called the average occurrence rate, since it is equal to the average number of events when $T = 1$. In a wider context, the Poisson p.f. will be derived in Section B.12.

Although we have confined ourselves to Poisson experiments which are connected with observing events in a given time interval, it should be clear that we could focus on events that occur in a specified region of size T. Namely, we can also speak of a Poisson experiment if: (i) the number of events occurring in one region is independent of the number of events in a disjoint region; (ii) the probability that a single event will occur in a small region is proportional to the size of the region; (iii) the probability that more than one event will occur in a small region is negligible. Simply, all that applied above to a given time interval applies now to a specified region.

An important property of the binomial and Poisson distributions is that the former may be approximated by the latter when the number of trials n is large, the probability p is small and $\lambda T = np$.

Taking all these properties into account, the Poisson distribution can be used in quality control, for example, to control the number of defects on a surface of a certain element or to control the number of nonconforming units in a batch.

B.5   More on Random Variables

Let us now turn to random variables whose range is "continuous" in the sense that it forms an interval, finite or not, of real numbers. For example, the random variable which is defined as the time to the first failure of a PC's hard disk has a "continuous" range, $[0, \infty)$. The simplest random variable of this type is a continuous counterpart of the uniformly distributed r.v. described in Section B.3. Consider the experiment consisting in choosing at random a point from the interval $[0, 1]$. Now, since each outcome is assumed equally likely and there are infinitely many possible outcomes, each particular outcome must have probability zero. Due to the infinite number of possible outcomes, it does not contradict the fact that some outcome must occur in the experiment. For the r.v. X assuming value x whenever point x has occurred, we should have

$$P(0 \leq X \leq 1) = 1$$

since the probability of a certain event is one, and, for any numbers $a, b, \ 0 \leq a \leq b \leq 1$, we should have

$$P(a \leq X \leq b) = \frac{\text{length of } [a, b] \text{ interval}}{\text{length of } [0, 1] \text{ interval}}$$

due to the assumed randomness of the choice of a point from $[0, 1]$. Thus,

$$
\begin{aligned}
P(a \leq X \leq b) &= \frac{\int_a^b 1 dx}{\int_0^1 1 dx} \\
&= \frac{b - a}{1} = b - a, \ \ 0 \leq a \leq b \leq 1,
\end{aligned}
$$

in accordance with the intuitive understanding of the experiment. The given probability distribution is called the *uniform*, or *rectangular* (continuous), distribution on the interval $[0, 1]$, and will be denoted by $U[0, 1]$.

A random variable is said to be *continuous* if its range forms an interval, finite or not, and the probability of assuming exactly any of its values is zero. In such cases, it is reasonable to ask only about the probability that an r.v. assumes a value from an interval or some combination of intervals. The given example suggests also a more specific definition of a continuous r.v. Namely, a random variable X is continuous if there exists a function $f(x)$, defined over the set of all reals R, such that

(a) $f(x) \geq 0$ for all $x \epsilon R$
(b) $\int_{-\infty}^{\infty} f(x) dx = 1$
(c) $P(a \leq X \leq b) = \int_a^b f(x) dx, \ \ a \leq b, \ \ a, b \epsilon R.$

The function $f(x)$ is called a *probability density function* (p.d.f.) for the r.v. X.

It follows from the definition that, for a continuous r.v. X with p.d.f. $f(x)$ and cumulative distribution function $F(x)$, $F(x) = P(X \leq x)$, we have

$$F(x) = \int_{-\infty}^{x} f(t)dt \text{ and } f(x) = \frac{dF}{dx} \tag{B.38}$$

for each $x \epsilon R$.

For the uniform distribution on $[0, 1]$, the p.d.f. is equal to one on the interval $[0, 1]$, and is zero elsewhere. (We leave it to the reader to show that the p.d.f. for the uniform distribution on an arbitrary fixed interval $[A, B]$ is equal to $1/(B - A)$ on that interval, and is zero elsewhere.) For later reference, let us observe also that the c.d.f. $F(x)$ of $U[0, 1]$ is zero for $x \leq 0$, $F(x) = x$ for $0 < x \leq 1$ and stays equal to one for all $x > 1$.

When graphed, probability density functions provide an information where and how the "probability mass" of the probability distribution is located on the real line R.

In analogy with the discrete case, we define the mean of an r.v. X with p.d.f. $f(x)$ as

$$E(X) = \mu = \int_{-\infty}^{\infty} x f(x)dx \tag{B.39}$$

and the variance as

$$\text{Var}(X) = \sigma^2 = \int_{-\infty}^{\infty} (x - \mu)^2 f(x)dx. \tag{B.40}$$

Replacing summation by integration and noting that

$$E[g(X)] = \int_{-\infty}^{\infty} g(x)f(x)dx, \tag{B.41}$$

we easily obtain that properties (B.18), (B.20) and (B.21) hold in the continuous case as well.

In turn, given two continuous random variables X and Y, it is natural to define their joint c.d.f. and p.d.f. via

$$F(x, y) = \int_{-\infty}^{x} \int_{-\infty}^{y} f(x, y)dxdy \tag{B.42}$$

and

$$f(x, y) = \frac{\partial^2 F(x, y)}{\partial x \partial y}, \tag{B.43}$$

respectively. Consequently, for any rectangle $[a, b] \times [c, d]$ in the xy plane,

$$P(X\epsilon[a, b], Y\epsilon[c, d]) = \int_a^b \int_c^d f(x, y)dxdy. \qquad \text{(B.44)}$$

Of course, computing probabilities over more involved regions in the xy plane is also possible. For instance, if we are interested in the probability distribution of the sum $V = X + Y$, we can calculate

$$P(X + Y \le v) = \int \int_{x+y\le v} f(x, y)dxdy. \qquad \text{(B.45)}$$

The expectation of a function $g(x, y)$ of the r.v.'s X and Y is given by

$$E(g(X, Y)) = \int_{-\infty}^{\infty} \int_{-\infty}^{\infty} g(x, y)f(x, y)dxdy, \qquad \text{(B.46)}$$

and, in particular, the covariance of X and Y, defined by (B.27), has the same properties as covariances of discrete random variables. Also, correlation ρ between random variables has the same properties in discrete and continuous cases. Finally, the *marginal* p.d.f.'s of X and Y are given by

$$f_X(x) = \int_{-\infty}^{\infty} f(x, y)dy \text{ and } f_Y(y) = \int_{-\infty}^{\infty} f(x, y)dx, \qquad \text{(B.47)}$$

and, just as in the discrete case, it readily follows from (B.10) and (B.42) that the joint p.d.f. of two independent continuous r.v.'s X and Y is given by

$$f(x, y) = f_X(x)f_Y(y). \qquad \text{(B.48)}$$

The case of more than two random variables is treated in some detail in Section A.11. Let us only mention here that p random variables $X_1, X_2, \ldots, X_p$ are said to be *mutually stochastically independent* or, shortly, *stochastically independent*, if and only if

$$f(x_1, x_2, \ldots, x_p) = f(x_1)f(x_2) \cdots f(x_p), \qquad \text{(B.49)}$$

where $f(x_1, x_2, \ldots, x_p)$ is the joint p.d.f. of $X_1, X_2, \ldots, X_p$ and $f(x_i)$ is the marginal p.d.f. of X_i, $i = 1, 2, \ldots, p$.

In the next section, we shall present some most widely used continuous probability distributions. Let us, however, conclude this section pointing out the surprising importance of the uniform $U[0, 1]$ distribution. Suppose we want to construct a random variable Y having a given c.d.f. $F(y)$, increasing between its boundary values 0 and 1. In order to achieve this, it suffices to verify that the random variable $Y = F^{-1}(X)$ has the required distribution if X is distributed uniformly on $[0, 1]$, and F^{-1} is the inverse of the function F. Indeed,

$$P(Y \leq y) = P(F^{-1}(X) \leq y) = P(X \leq F(y)) = F(y)$$

as was to be shown (the last equality is a direct consequence of the form of the c.d.f. corresponding to the $U[0, 1]$ distribution). Analogous argument shows that the random variable $F(Y)$, where $F(y)$ is the c.d.f. of Y, has the uniform distribution on $[0, 1]$. Thus, given an r.v. V with a c.d.f. $F_v(v)$, we can construct an r.v. Z with an arbitrary c.d.f. $F_z(z)$ by the composite transformation $Z = F_z^{-1}(F_v(V))$. The first transformation gives the uniformly distributed random variable, while the second transformation yields the desired result. Of course, computer simulations of random phenomena rely heavily on this property.

B.6   Continuous Probability Distributions

B.6.1   Normal and Related Distributions

Let us consider first the most widely used of all continuous distributions, namely the *normal*, or *Gaussian*, distribution. The probability density function of the normal random variable X is

$$f(x) = \frac{1}{\sqrt{2\pi\sigma^2}} \exp\left(-\frac{1}{2\sigma^2}(x - \mu)^2\right), \quad -\infty < x < \infty, \qquad (B.50)$$

where $\exp(y)$ denotes e raised to the power y, and μ and σ^2 are the parameters of the distribution, $-\infty < \mu < \infty$, $\sigma^2 > 0$. It can be shown that the constants μ and σ^2 are simply the mean and variance of the r.v. X, respectively (see Section B.8). The normal distribution with mean μ and variance σ^2 will be denoted by $\mathcal{N}(\mu, \sigma^2)$. It is easy to prove that if X is normally distributed with mean μ and variance σ^2, then the r.v.

$$Z = \frac{X - \mu}{\sigma}, \qquad (B.51)$$

where $\sigma = \sqrt{\sigma^2}$, is normally distributed with mean 0 and variance 1. Z is called the *standard normal* r.v.

If $X_1, X_2, \ldots, X_n$ are independent r.v.'s having normal distributions with means $\mu_1, \mu_2, \ldots, \mu_n$ and variances $\sigma_1^2, \sigma_2^2, \ldots, \sigma_n^2$, respectively, then the sum has the normal distribution with mean $\mu_1 + \mu_2 + \cdots + \mu_n$ and variance $\sigma_1^2 + \sigma_2^2 + \cdots + \sigma_n^2$. More generally, if $a_1, a_2, \ldots, a_n$ are n constants, then the linear combination $a_1 X_1 + a_2 X_2 + \cdots + a_n X_n$ has the normal distribution with mean $a_1\mu_1 + a_2\mu_2 + \cdots + a_n\mu_n$ and variance $a_1^2\sigma_1^2 + a_2^2\sigma_2^2 + \cdots + a_n^2\sigma_n^2$. (All the results on linear combinations of independent r.v.'s stated in this Section can easily be proved using the so-called moment-generating functions; see Section B.8.)

It is easily seen that the normal curve (i.e., the curve corresponding to the normal p.d.f.) is "bell shaped": it is positive for all $x \epsilon R$, smooth, attains its unique maximum at $x = \mu$, and is symmetric about μ.

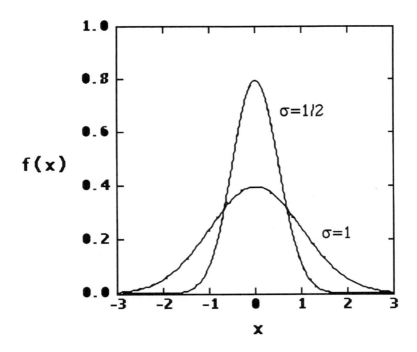

Figure B.1. Normal Densities With Zero Mean.

It was this shape of the normal curve that made many researchers of the 19th century believe that all continuous random phenomena must be of this type. And, although that was a vast exaggeration, we can indeed relatively often assume that a random phenomenon follows the normal law. The reason for this is twofold. First, we can quite often presume that the outcomes of an experiment are distributed symmetrically about some mean value and that the outcomes close to that mean are

more likely than those distant from it. In other words, we can consider
the outcomes to be at least approximately normally distributed. And
second, it turns out that if the observed outcomes are the averages of
some other independent random phenomena governed by an essentially
arbitrary probability distribution, then the probability distribution of
the outcomes is close to normal. This last result, which is, perhaps, the
most amazing result of probability theory, will be discussed in a greater
detail in Section B.9. Taking these facts into account, no wonder that
the normal distribution is also the most widely used distribution in statistical process control for quality improvement.

A distribution which is related to the normal distribution is the lognormal distribution. We say that the random variable X has the lognormal
distribution if its logarithmic transformation $\ln X$ is normally distributed.
The p.d.f. of X is given by

$$f(x) = \frac{1}{\sqrt{2\pi}\sigma x}\exp\left(-\frac{1}{2\sigma^2}(\ln x - \mu)^2\right) \quad \text{for} \ x > 0 \qquad (B.52)$$

and zero elsewhere. The constants μ and σ^2 are, respectively, the mean
and variance of the r.v. $\ln X$. The p.d.f. of the lognormal distribution
is not symmetric: it is zero for negative x and skewed to the right of its
maximum. However, the lognormal curve becomes "almost" symmetric
for small values of σ^2, say, for $\sigma^2 \leq .01$. Sometimes, the lognormal
distribution is used in statistical process control for quality improvement
and in representing distributions of the useful life (or the lifetime) of
various devices.

Another distribution related to the normal one is the *Cauchy* distribution. Let Z_1 and Z_2 be two independent standard normal random
variables. The random variable $X = Z_1/Z_2$ is said to have the Cauchy
distribution. Its p.d.f. is given by

$$f(x) = \frac{1}{\pi(1 + x^2)}, \quad -\infty < x < \infty. \qquad (B.53)$$

It is easy to see that a Cauchy r.v. does not have finite mean value.
Indeed, for large values of $|x|$, the product $x f(x)$ approaches the function
$1/(\pi x)$ and hence integral (B.39) does not exist. As this example shows,
not all random variables have finite means, let alone variances.

B.6.2   Gamma Distributions

In order to introduce an important family of the so-called gamma distributions, let us first define the *gamma function*:

$$\Gamma(\alpha) = \int_0^\infty x^{\alpha-1}e^{-x}dx \text{ for } \alpha > 0. \tag{B.54}$$

Integrating by parts, we obtain

$$\Gamma(\alpha) = (\alpha - 1)\Gamma(\alpha - 1) \text{ for } \alpha > 1. \tag{B.55}$$

When $\alpha = n$, with n a positive integer, repeated application of (B.55) yields

$$\Gamma(n) = (n - 1)!, \tag{B.56}$$

since $\Gamma(1) = \int_0^\infty e^{-x}dx = 1$. We say that the r.v. X has the *gamma distribution* with parameters α and β, if its p.d.f. is

$$f(x) = \frac{1}{\beta^\alpha \Gamma(\alpha)}x^{\alpha-1}e^{-x/\beta} \text{ for } x > 0 \tag{B.57}$$

and zero elsewhere, where both constants, α and β, are positive. The mean of X is $\alpha\beta$ and the variance of X is $\alpha\beta^2$ (see Section B.8).

The gamma distribution with parameter $\alpha = 1$ is called the *(negative) exponential distribution* with parameter β. That is, the exponential r.v. X has the p.d.f.

$$f(x) = \frac{1}{\beta}e^{-x/\beta} \text{ for } x > 0 \tag{B.58}$$

and zero elsewhere, where $\beta > 0$. It also follows from the above that X has the mean β and variance β^2.

If some events are occurring in time, independently one of another and according to a Poisson experiment with the average occurrence rate λ, then the inter-event times have the exponential distribution with mean $\beta = 1/\lambda$. For example, if the incoming telephone calls constitute the Poisson experiment, then the inter-arrival times between successive calls have the exponential distribution. By the same token, times between successive failures of a device and, in particular, time to the first failure (or the lifetime) can sometimes be modelled by the exponential distribution as well. (Poisson experiments will be reexamined within a more general framework in Section B.12.)

Let us mention that if $X_1, X_2, \ldots, X_n$ are independent r.v.'s having identical exponential distribution with parameter β, then the sum $X_1 +$

$X_2 + \cdots + X_n$ has the gamma distribution with parameters n and β. For example, time between n successive telephone calls can be modelled by this distribution. If $\beta = 1$, this distribution is called the *Erlang* distribution.

In turn, the gamma distribution with parameters $\alpha = \nu/2$ and $\beta = 2$, where ν is a positive integer, is called the *chi-square* (χ^2_ν for short) distribution with ν degrees of freedom. The chi-square r.v. X has the p.d.f.

$$f(x) = \frac{1}{2^{\nu/2}\Gamma(\nu/2)}x^{\nu/2-1}e^{-x/2} \text{ for } x > 0 \qquad (B.59)$$

and zero elsewhere. The r.v. X has the mean ν and variance 2ν.

If $X_1, X_2, \ldots, X_n$ are independent r.v.'s having chi-square distributions with $\nu_1, \nu_2, \ldots, \nu_n$ degrees of freedom, respectively, then the r.v. $X_1 + X_2 + \cdots + X_n$ has chi-square distribution with $\nu_1 + \nu_2 + \cdots + \nu_n$ degrees of freedom. If $X_1, X_2, \ldots, X_n$ are independent r.v.'s having identical normal distribution with mean μ and variance σ^2, then the r.v.

$$\sum_{i=1}^{n}\left(\frac{X_i - \mu}{\sigma}\right)^2$$

has the chi-square distribution with n degrees of freedom. In particular, the square of the standard normal r.v. Z has the chi-square distribution with one degree of freedom.

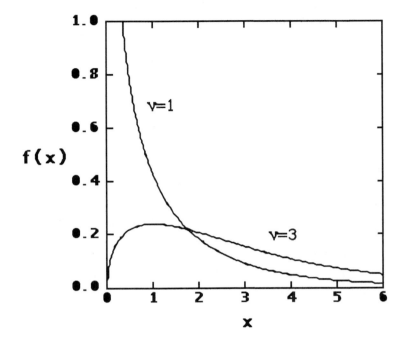

Figure B.2. Chi-Square Densities.

Let us mention in passing a distribution which is closely related to the chi-square distribution, although it does not belong to the gamma family. If $X_1, X_2, \ldots, X_n$ are independent r.v.'s having normal distributions with variance 1 and means $\mu_1, \mu_2, \ldots, \mu_n$, respectively, then the r.v.

$$\sum_{i=1}^{n} X_i^2$$

has the *noncentral* chi-square distribution with n degrees of freedom and noncentrality parameter $\lambda = \sum_{i=1}^{n} \mu_i^2$.

The importance of the chi-square distribution (as well as that of the t and F distributions which are discussed in the sequel) follows from the following fact. Given some random data $x_1, x_2, \ldots, x_n$, even if we know the type of the probability distribution from which the data are drawn, we very rarely know this distribution exactly. In almost all situations of practical interest, values of the parameters of the distribution are unknown. Hence, the inference about the data begins from estimating summary characteristics, such as mean and variance, of the underlying probability distribution. A natural sample counterpart of the mean is the average, or the *sample mean*, of the data

$$\bar{x} = \frac{1}{n}(x_1 + x_2 + \cdots + x_n), \qquad (B.60)$$

while that of the variance is the *sample variance*

$$s^2 = \frac{1}{n-1} \sum_{i=1}^{n} (x_i - \bar{x})^2; \qquad (B.61)$$

the reason for dividing the last sum by $n - 1$ and not, seemingly more naturally, by n will become apparent later. Note that prior to performing an experiment, i.e., prior to observing the data, the sample mean and sample variance are themselves random variables, since they are functions of random variables.

We shall prove in the next section that, whatever the underlying distribution of the random data, the sample mean seen as a random variable itself has the following properties:

$$E(\bar{X}) = E\left(\frac{X_1 + X_2 + \cdots + X_n}{n}\right) = \mu \qquad (B.62)$$

and

$$\mathrm{Var}(\bar{X}) = \mathrm{Var}\left(\frac{X_1 + X_2 + \cdots + X_n}{n}\right) = \frac{\sigma^2}{n}, \qquad (B.63)$$

where μ is the mean and σ^2 is the variance of the parent distribution of the data X_i, $i = 1, 2, \ldots, n$. For the sample variance, one can show that

$$E(S^2) = \sigma^2,$$

as one would wish it to be (hence the factor $1/(n-1)$, not $1/n$, in the definition of S^2). Unfortunately, no result of corresponding generality holds for the variance of S^2. If, however, the data come from the normal distribution with mean μ and variance σ^2, then the random variable

$$V = \frac{(n-1)S^2}{\sigma^2} \qquad (B.64)$$

can be shown to have chi-square distribution with $n - 1$ degrees of freedom. Hence, we have then

$$\mathrm{Var}(S^2) = \frac{2\sigma^4}{n-1}.$$

B.6.3 t and F Distributions

The other two probability distributions strictly connected with inferences about $\bar{X}$ and S^2 are the t and F distributions. The *(Student's) t* distribution with ν degrees of freedom is given by the p.d.f.

$$f(x) = \frac{\Gamma[(\nu + 1)/2]}{\Gamma(\nu/2)\sqrt{\pi\nu}} \left(1 + \frac{x^2}{\nu}\right)^{-(\nu+1)/2}, \quad -\infty < x < \infty. \qquad \text{(B.65)}$$

If the r.v. Z is standard normal and the r.v. V is chi-square distributed with ν degrees of freedom, and if Z and V are independent, then the r.v.

$$T = \frac{Z}{\sqrt{V/\nu}} \qquad \text{(B.66)}$$

has t distribution with ν degrees of freedom. The last result is very useful in the context of random sampling from a normal distribution with unknown mean and variance. Namely, since

$$Z = \frac{\bar{X} - \mu}{\sigma/\sqrt{n}} \qquad \text{(B.67)}$$

is standard normal and can be shown to be statistically independent of V given by (B.64), it follows from (B.66) that the r.v.

$$\frac{\bar{X} - \mu}{S/\sqrt{n}} \qquad \text{(B.68)}$$

has t distribution with $n - 1$ degrees of freedom. The fact that (B.68) has known distribution enables one to test certain hypotheses about the mean μ (see Section B.13).

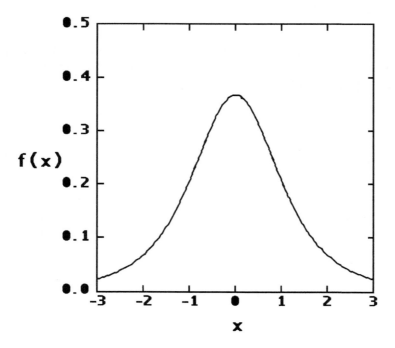

Figure B.3. t Density With 3 Degrees Of Freedom.

It is sometimes the case that we are given two independent sequences of data drawn from normal distributions with unknown means μ_1, μ_2 and variances σ_1^2, σ_2^2, respectively. The question is whether the underlying distributions are indeed different. The equality of means can be tested using our knowledge of the t distribution. In turn, the equality of variances can be tested using the F distribution with ν_1 and ν_2 degrees of freedom, which is defined as the distribution of the following ratio:

$$\frac{U/\nu_1}{V/\nu_2},\tag{B.69}$$

where U and V are independent chi-square distributed r.v.'s with ν_1 and ν_2 degrees of freedom, respectively. Now, if S_1^2 and S_2^2 are the sample variances of the first and second sequence of data, respectively, and if the true variances σ_1^2 and σ_2^2 are equal, then, by (B.64), the ratio S_1^2/S_2^2 is F distributed. More precisely, the ratio S_1^2/S_2^2 has F distribution with $n_1 - 1$ and $n_2 - 1$ degrees of freedom, where n_i, $i = 1, 2$, denotes the number of data in the ith sequence. The p.d.f. of the F distribution with ν_1 and ν_2 degrees of freedom is given by

$$f(x) = \frac{\Gamma((\nu_1+\nu_2)/2)(\nu_1/\nu_2)^{\nu_1/2}}{\Gamma(\nu_1/2)\Gamma(\nu_2/2)} \frac{x^{\nu_1/2-1}}{(1+\nu_1 x/\nu_2)^{(\nu_1+\nu_2)/2}} \qquad (B.70)$$

for $x > 0$ and zero elsewhere.

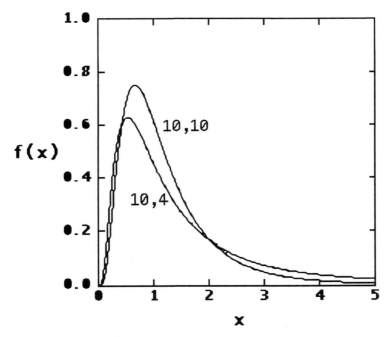

Figure B.4. F Densities With 10, 4 And 10, 10 Degrees of Freedom.

If U in the numerator of (B.69) has noncentral chi-square distribution with ν_1 degrees of freedom and noncentrality parameter λ, if V is the same as before, and U and V are independent, then the ratio (B.69) is said to have *noncentral F* distribution with ν_1 and ν_2 degrees of freedom and noncentrality parameter λ.

B.6.4   Weibull Distribution

It was mentioned in this section that, if certain assumptions are satisfied, the lifetimes (or times to a failure) can be modelled by the exponential distribution. Under different circumstances, for example, models based on the lognormal distribution can be used. However, it happens that, given some observed lifetimes, one cannot decide which of the models is adequate. Such a situation calls for a distribution which for some values of its parameters would look similarly to an exponential curve, and would resemble a lognormal curve for other values of the parameters. This sort of flexibility is provided by the *Weibull* distribution whose p.d.f. is given by

$$f(x) = c\beta^{-1}(x/\beta)^{c-1}\exp[-(x/\beta)^c], \quad x > 0 \qquad (B.71)$$

and zero elsewhere, $c > 0$ and $\beta > 0$. Clearly, the Weibull distribution reduces to the exponential when $c = 1$. For $c > 1$, it becomes similar to the lognormal distribution. One can show that the random variable $Y = (x/\beta)^c$ is exponential with mean one.

B.7   Laws of Large Numbers

Let us now consider the set of n data drawn from some probability distribution. Prior to the experiment which yields the data, they can be treated as a sequence of n independent and identically distributed (i.i.d.) random variables $X_1, X_2, \ldots, X_n$. Such sequence will be labeled as a *random sample* of size n. Suppose that the mean and variance of the underlying probability distribution are μ and σ^2, respectively. Otherwise, the probability distribution is unknown. We shall find the mean and variance of the sample mean (B.60) of the random sample.

It is easy to see that

$$
\begin{aligned}
\mu_{\bar{x}} &= \frac{E(X_1 + X_2 + \cdots + X_n)}{n} = \frac{E(X_1) + E(X_2) + \cdots + E(X_n)}{n} \\
&= \frac{\mu + \mu + \cdots + \mu}{n} = \mu.
\end{aligned}
$$

In this derivation, we have not used independence or the fact that all the r.v.'s have the same distribution, only the fact that they all have the same (finite) mean. We say that $\bar{X}$ is an *unbiased* estimator of μ.

Next we shall derive the variance of $\bar{X}$:

$$
\begin{aligned}
\sigma_{\bar{x}}^2 &= E[(\bar{X} - \mu)^2] \\
&= E\left[\left(\frac{(X_1 - \mu)}{n} + \frac{(X_2 - \mu)}{n} + \cdots + \frac{(X_n - \mu)}{n}\right)^2\right] \\
&= \sum_{i=1}^{n} \frac{E[(X_i - \mu)^2]}{n^2} + \text{terms like } E\left[\frac{(X_1 - \mu)(X_2 - \mu)}{n^2}\right].
\end{aligned}
$$

Now, by independence, the expectation of the cross-product terms is zero:

$$E[(X_1 - \mu)(X_2 - \mu)] = \int_{-\infty}^{\infty} (x_1 - \mu)(x_2 - \mu)f(x_1)f(x_2)dx_1dx_2$$
$$= E(X_1 - \mu)E(X_2 - \mu) = 0$$

(the argument for discrete distributions is analogous). Thus, we have

$$\sigma_{\bar{x}}^2 = \frac{\sigma^2}{n}.$$

We note that in the above derivation the fact that the X_i's are identically distributed has been superfluous. Only the facts that the r.v.'s are independent and have the same μ and σ^2 have been needed. The property that the variability of $\bar{X}$ about the true mean μ decreases as n increases is of key importance in experimental science. We shall develop this notion further below.

Let us begin by stating the celebrated *Chebyshev's inequality*. If Y is any r.v. with mean μ_y and variance σ_y^2, then for any $\varepsilon > 0$

$$P(|Y - \mu_y| > \varepsilon) \le \frac{\sigma_y^2}{\varepsilon^2}. \qquad (B.72)$$

As a practical approximation device, it is not a particularly useful inequality. However, as an asymptotic device, it is invaluable. Let us consider the case where $Y = \bar{X}$. Then, (B.72) gives us

$$P(|\bar{X} - \mu| > \varepsilon) \le \frac{\sigma^2}{n\varepsilon^2}, \qquad (B.73)$$

or equivalently

$$P(|\bar{X} - \mu| \le \varepsilon) > 1 - \frac{\sigma^2}{n\varepsilon^2}. \qquad (B.74)$$

Equation (B.74) is a form of the *weak law of large numbers*. The WLLN tells us that if we are willing to take a sufficiently large sample, then we can obtain an arbitrarily large probability that $\bar{X}$ will be arbitrarily close to μ.

In fact, even a more powerful result, the *strong law of large numbers*, is available. In order to make the difference between the WLLN and SLLN more transparent, let us denote the sample mean based on a sample of size n by $\bar{X}_n$, so that the dependence of $\bar{X}$ on n be emphasized. Now we can write the WLLN in the following way

$$\lim_{n \to \infty} P(|\bar{X}_n - \mu| \le \varepsilon) = 1 \qquad (B.75)$$

for each positive ε. On the other hand, the SLLN states that

$$P(\lim_{n\to\infty} |\bar{X}_n - \mu| = 0) = 1. \tag{B.76}$$

Loosely speaking, in the WLLN, the probability of $\bar{X}_n$ being closed to μ for only one n at a time is claimed, whereas in the SLLN, the closeness of $\bar{X}_n$ to μ for all large n simultaneously is asserted with probability one. The rather practical advantage of the SLLN is that if $g(x)$ is some function, then

$$P(\lim_{n\to\infty} |g(\bar{X}_n) - g(\mu)| = 0) = 1. \tag{B.77}$$

The WLLN and SLLN are particular cases of convergence in probability and almost sure convergence of a sequence of r.v.'s, respectively. Let $Y_1, Y_2, \ldots, Y_n, \ldots$ be an infinite sequence of r.v.'s. We say that this sequence of r.v.'s converges *in probability* or *stochastically* to a random variable Y if

$$\lim_{n\to\infty} P(|Y_n - Y| > \varepsilon) = 0$$

for each positive ε. We say that the sequence $Y_1, Y_2, \ldots, Y_n, \ldots$ converges *almost surely* or converges *with probability one* if

$$P(\lim_{n\to\infty} |Y_n - Y| = 0) = 1.$$

For brevity, the almost sure convergence is also called the a.s. convergence. In the case of the laws of large numbers, Y_n is equal to $\bar{X}_n$ and $Y = \mu$, that is, the limit is a real number or, equivalently, an r.v. which assumes only one value with probability one.

B.8   Moment-Generating Functions

Just as we defined the mean and variance of an r.v. X, we can define the kth *moment* of X,

$$E(X^k) \quad \text{for any} \quad k = 1, 2, \ldots, \tag{B.78}$$

and the kth *moment about the mean* μ of X,

$$E[(X - \mu)^k) \quad \text{for any} \quad k = 2, 3, \ldots. \tag{B.79}$$

For example, the kth moment of a continuous r.v. X with p.d.f. $f(x)$ can be computed as [1]

$$E(X^k) = \int_{-\infty}^{\infty} x^k f(x) dx. \tag{B.80}$$

---

[1]Note that an r.v. can have no or just a few finite moments. In particular, in the case of a Cauchy r.v., already the first moment does not exist.

Of course, moments of higher orders k provide additional information about the r.v. under scrutiny. For instance, it is easily seen that if an r.v. X is symmetric about its mean μ, then $E[(X - \mu)^{2r-1}] = 0$ for each integer r (provided these moments exist). It should be mentioned that the kth moment is a well-defined concept for any positive k, not only for integer k's. However, both in this section and in most of this book, our considerations are confined to kth moments with integer k.

Let us now define the *moment-generating function* $M_X(t)$ for an r.v. X via

$$M_X(t) = E(e^{tX}). \tag{B.81}$$

Assuming that differentiation with respect to t commutes with expectation operator E, we have

$$
\begin{aligned}
M_X'(t) &= E(Xe^{tX}) \\
M_X''(t) &= E(X^2 e^{tX}) \\
&\vdots \\
M_X^{(k)}(t) &= E(X^{(k)} e^{tX}).
\end{aligned}
$$

Setting t equal to zero, we see that

$$M_X^{(k)}(0) = E(X^k). \tag{B.82}$$

Thus, we see immediately the reason for the name moment-generating function (m.g.f.). Once we have obtained $M_X(t)$, we can compute moments of arbitrary order (assuming they exist) by successively differentiating the m.g.f. and setting the argument t equal to zero. As an example of this application, let us consider an r.v. distributed according to the binomial distribution with parameters n and p. Then,

$$
\begin{aligned}
M_X(t) &= \sum_0^n e^{tx} \binom{n}{x} p^x (1-p)^{n-x} \\
&= \sum_0^n \binom{n}{x} (pe^t)^x (1-p)^{n-x}.
\end{aligned}
$$

Now recalling the binomial identity

$$\sum_0^n \binom{n}{x} a^x b^{n-x} = (a+b)^n,$$

we have

$$M_X(t) = [pe^t + (1-p)]^n. \tag{B.83}$$

Next, differentiating with respect to t, we have

$$M'_X(t) = npe^t[pe^t + (1-p)]^{n-1}. \tag{B.84}$$

Then, setting t equal to zero, we have

$$E(X) = M'_X(0) = np. \tag{B.85}$$

Differentiating (B.84) again with respect to t and setting t equal to zero, we have

$$E(X^2) = M''_X(0) = np + n(n-1)p^2. \tag{B.86}$$

In order to calculate the variance, it suffices to recall that for any r.v. X we have

$$\mathrm{Var}(X) = E(X^2) - [E(X)]^2. \tag{B.87}$$

Thus, for the binomial X

$$\mathrm{Var}(X) = np(1-p). \tag{B.88}$$

Let us find also the m.g.f. of a normal variate with mean μ and variance σ^2.

$$
\begin{aligned}
M_X(t) &= \frac{1}{\sqrt{2\pi}\sigma} \int_{-\infty}^{\infty} e^{tx} \exp\left(-\frac{1}{2\sigma^2}(x-\mu)^2\right) dx \\
&= \frac{1}{\sqrt{2\pi}\sigma} \int_{-\infty}^{\infty} \exp\left(-\frac{1}{2\sigma^2}(x^2 - 2\mu x - 2\sigma^2 tx + \mu^2)\right) dx \\
&= \frac{1}{\sqrt{2\pi}\sigma} \int_{-\infty}^{\infty} \exp\left(-\frac{1}{2\sigma^2}(x^2 - 2x(\mu + t\sigma^2) + \mu^2)\right) dx \\
&= \frac{1}{\sqrt{2\pi}\sigma} \int_{-\infty}^{\infty} \exp\left(-\frac{1}{2\sigma^2}(x-\mu^*)^2\right) dx \exp\left(t\mu + \frac{t^2\sigma^2}{2}\right),
\end{aligned}
$$

where $\mu^* = \mu + t\sigma^2$. But recognizing that the integral is simply equal to $\sqrt{2\pi}\sigma$, we see that the m.g.f. of the normal distribution is given by

$$M_X(t) = \exp\left(t\mu + \frac{t^2\sigma^2}{2}\right). \tag{B.89}$$

It is now easy to verify that the mean and variance of the normal distribution are μ and σ^2, respectively.

The possible mechanical advantages of the m.g.f. are clear. One integration (summation) operation plus k differentiations yield the first k moments of a random variable. However, the moment-generating aspect of the m.g.f. pales in importance to some of its properties relating to the summation of independent random variables. Let us suppose, for example, that we have n independently distributed r.v.'s $X_1, X_2, \ldots, X_n$ with m.g.f.'s $M_1, M_2, \ldots, M_n$, respectively. Suppose that we wish to investigate the distribution of the r.v.

$$Y = c_1 X_1 + c_2 X_2 + \cdots + c_n X_n,$$

where $c_1, c_2, \ldots, c_n$ are fixed constants. Let us consider using the moment-generating functions to achieve this task. We have

$$M_Y(t) = E[\exp t(c_1 X_1 + c_2 X_2 + \cdots + c_n X_n)].$$

Using the independence of $X_1, X_2, \ldots, X_n$ we may write

$$
\begin{aligned}
M_Y(t) &= E[\exp t c_1 X_1] E[\exp t c_2 X_2] \cdots E[\exp t c_n X_n] \\
&= M_1(c_1 t) M_2(c_2 t) \cdots M_n(c_n t).
\end{aligned}
\tag{B.90}
$$

Given the density (or probability) function, we know what the m.g.f. will be. But it turns out that, under very general conditions, the same is true in the reverse direction; namely, if we know $M_X(t)$, we can compute a unique density (probability) function that corresponds to it. The practical implication is that if we find a random variable with an m.g.f. we recognize as corresponding to a particular density (probability) function, we know immediately that the random variable has the corresponding density (probability) function. Thus, in many cases, we are able to use (B.90) to give ourselves immediately the distribution of Y. Consider, for example, the sum

$$Y = X_1 + X_2 + \cdots + X_n$$

of n independent binomially distributed r.v.'s with the same probability of success p and the other parameter being equal to $n_1, n_2, \ldots, n_n$, respectively. Thus, the moment-generating function for Y is

$$
\begin{aligned}
M_Y(t) &= [pe^t + (1-p)]^{n_1} [pe^t + (1-p)]^{n_2} \cdots [pe^t + (1-p)]^{n_n} \\
&= [pe^t + (1-p)]^{n_1 + n_2 + \cdots + n_n}.
\end{aligned}
$$

We note that, not unexpectedly, this is the m.g.f. of a binomial r.v. with parameters $N = n_1 + n_2 + \cdots + n_n$ and p. Given (B.89), it is straightforward to give the corresponding result on the distribution of the sum (or a linear combination) of n independent normal r.v.'s with means $\mu_1, \mu_2, \ldots, \mu_n$ and variances $\sigma_1^2, \sigma_2^2, \ldots, \sigma_n^2$, respectively. Let us consider also the case of gamma distributed random variables. The m.g.f. of a gamma variate with parameters α and β can be computed in the following way

$$
\begin{aligned}
M(t) &= \frac{1}{\Gamma(\alpha)\beta^\alpha} \int_0^\infty e^{tx} x^{\alpha-1} e^{-x/\beta} dx \\
&= \frac{1}{\Gamma(\alpha)\beta^\alpha} \int_0^\infty x^{\alpha-1} e^{-x(1-\beta t)/\beta} dx;
\end{aligned}
$$

now this integral is finite only for $t < 1/\beta$ and substituting $y = x(1 - \beta t)/\beta$ yields

$$
\begin{aligned}
M(t) &= \frac{1}{\Gamma(\alpha)} \int_0^\infty y^{\alpha-1} e^{-y} dy \left(\frac{1}{1-\beta t}\right)^\alpha \\
&= \left(\frac{1}{1-\beta t}\right)^\alpha \quad for \ t < \frac{1}{\beta},
\end{aligned}
\tag{B.91}
$$

where the last equality follows from the form of the p.d.f. of the gamma distribution with parameters α and 1. In particular, the m.g.f. for a chi-square r.v. with ν degrees of freedom has the form

$$
M(t) = \left(\frac{1}{1-2t}\right)^{\nu/2} = (1-2t)^{-\nu/2}, \ t < 1/2.
\tag{B.92}
$$

It readily follows from (B.90) and (B.92) that the sum of n independent r.v.'s having chi-square distributions with $\nu_1, \nu_2, \ldots, \nu_n$ degrees of freedom, respectively, has chi-square distribution with $\nu_1 + \nu_2 + \cdots + \nu_n$ degrees of freedom.

B.9   Central Limit Theorem

We are now in a position to derive one version of the *central limit theorem*. Let us suppose we have a sample $X_1, X_2, \ldots, X_n$ of independently and identically distributed random variables with mean μ and variance σ^2. We wish to determine, for n large, the approximate distribution of the sample mean

$$
\bar{X} = \frac{X_1 + X_2 + \cdots + X_n}{n}.
$$

We shall examine the distribution of the sample mean when put into the standard form. Let

$$
Z = \frac{\bar{X} - \mu}{\sigma/\sqrt{n}} = \frac{X_1 - \mu}{\sigma\sqrt{n}} + \frac{X_2 - \mu}{\sigma\sqrt{n}} + \cdots + \frac{X_n - \mu}{\sigma\sqrt{n}}.
$$

Now, utilizing the independence of the X_i's and the fact that they are identically distributed with the same mean and variance, we can write

$$
\begin{aligned}
M_Z(t) &= E(e^{tZ}) = \prod_{i=1}^{n} E\left[\exp\left(t\frac{X_i - \mu}{\sigma\sqrt{n}}\right)\right] \\
&= \left\{E\left[\exp\left(t\frac{X_1 - \mu}{\sigma\sqrt{n}}\right)\right]\right\}^n \\
&= \left\{E\left[1 + t\frac{X_1 - \mu}{\sigma\sqrt{n}} + \frac{t^2}{2}\frac{(X_1 - \mu)^2}{\sigma^2 n} + o\left(\frac{1}{n}\right)\right]\right\}^n \\
&= \left(1 + \frac{t^2}{2n}\right)^n \rightarrow e^{t^2/2} \text{ as } n \rightarrow \infty.
\end{aligned}
\tag{B.93}
$$

But (B.93) is the m.g.f. of a normal distribution with mean zero and variance one. Thus, we have been able to show that the distribution of the sample mean of a random sample of n i.i.d. random variables with mean μ and variance σ^2 becomes "close" to the normal distribution with mean μ and variance σ^2/n as n becomes large.

Clearly, the CLT offers enormous conceptual and, in effect, computational simplifications. First and foremost, if the sample size is not too small, it enables us to approximate the distribution of the sample mean by a normal distribution regardless of the parent distribution of the random sample. Moreover, even if we knew that the parent distribution is, for instance, lognormal, computation of the exact distribution of $\bar{X}$ would be an enormous task. No wonder that the CLT is widely used in experimental sciences in general and in statistical process control in particular.

B.10 Conditional Density Functions

Let us return to questions of interdependence between random variables and consider briefly conditional distribution of one random variable given that another random variable has assumed a fixed value.

If two random variables X and Y are discrete and have a joint probability function (B.22), then, by (B.7) and (B.25), the *conditional probability function* of the r.v. X, given that $Y = y$, has the form

$$
f(x_i|y) = P(X = x_i|Y = y) = \frac{f(x_i, y)}{f_Y(y)},
\tag{B.94}
$$

where x_i runs over all possible values of X, and y is a fixed value from the range of Y. It is easy to verify that the conditional p.f. is indeed a probability function, i.e., that

$$f(x_i|y) > 0 \quad \text{for all} \quad x_i$$

and

$$\sum_{x_i} f(x_i|y) = 1.$$

Let us now suppose that r.v.'s X and Y are continuous and have joint p.d.f. (B.43). When deriving a formula for the conditional density function of X given $Y = y$, some caution is required since both random variables assume any particular value with probability zero. We shall use a type of a limit argument. Writing the statement of joint probability for small intervals in X and Y, we have by the multiplicative rule

$$P(x < X \le x + \varepsilon \cap y < X \le y + \delta)$$
$$P(y < Y \le y + \delta)P(x < X \le x + \varepsilon|y < Y \le y + \delta).$$

Now, exploiting the assumption of continuity of the density function, we can write

$$\int_x^{x+\varepsilon} \int_y^{y+\delta} f(x,y)dydx = \int_y^{y+\delta} f_Y(y)dy \int_x^{x+\varepsilon} f_{X|y}(x)dx$$
$$= \varepsilon\delta f(x,y) = \delta f_Y(y)\varepsilon f_{X|y}(x).$$

Here, we have used the terms f_Y and $f_{X|y}$ to denote the marginal density function of Y, and the *conditional density function* of X given $Y = y$, respectively. This gives us immediately (provided $f_Y(y) > 0$ for the given y)

$$f_{X|y}(x) = \frac{f(x,y)}{f_Y(y)}. \tag{B.95}$$

Note that this is a function of the argument x, whereas y is fixed; y is the value assumed by the random variable Y.

B.11   Random Vectors

B.11.1   Introduction

Sometimes we want to measure a number of quality characteristics of a single object simultaneously. It may be the case that the characteristics of interest are independent one of another, and can, therefore, be considered separately. More often, the characteristics are somehow related one to another, although this relationship cannot be precisely described. What we observe is not a set of independent random variables but a *random vector* whose elements are random variables somehow correlated one with another. Let $\mathbf{X} = [X_1, X_2, \ldots, X_p]'$ be such a random vector of any fixed dimension $p \geq 1$. By analogy with the univariate and two-variate cases, we define the cumulative distribution function of $\mathbf{X}$ as

$$F(\mathbf{X} \leq \mathbf{x}) = P(\mathbf{X} \leq \mathbf{x}) = P(X_1 \leq x_1, \ldots, X_p \leq x_p), \qquad \text{(B.96)}$$

where $\mathbf{x} = [x_1, \ldots, x_p]'$. All random vectors considered in this section will be assumed to be of continuous type. We can, therefore, define the (joint) probability density function $f(\mathbf{x})$ via

$$F(\mathbf{x}) = \int_{-\infty}^{x_1} \cdots \int_{-\infty}^{x_p} f(\mathbf{u}) du_1 \ldots du_p, \qquad \text{(B.97)}$$

where $\mathbf{u} = [u_1, \ldots, u_p]'$. Of course,

$$f(\mathbf{x}) = \frac{\partial^p F(\mathbf{x})}{\partial x_1 \ldots \partial x_p}. \qquad \text{(B.98)}$$

For any p-dimensional set $\mathcal{A}$,

$$P(\mathbf{X} \epsilon \mathcal{A}) = \int \cdots \int_{\mathcal{A}} f(\mathbf{x}) dx_1 \ldots dx_p, \qquad \text{(B.99)}$$

where the integration is over the set $\mathcal{A}$.

By analogy with the two-variate case, we can define marginal distribution of a subvector of $\mathbf{X}$. Consider the partitioned random vector $\mathbf{X} = [\mathbf{X}_1, \mathbf{X}_2]'$, where $\mathbf{X}_1$ has k elements and $\mathbf{X}_2$ has $p - k$ elements, $k < p$. The function

$$F(\mathbf{x}_1) = P(\mathbf{X}_1 \leq \mathbf{x}_1) = F(x_1, \ldots, x_k, \infty, \ldots, \infty), \qquad \text{(B.100)}$$

where $\mathbf{x}_1 = [x_1, \ldots, x_k]'$, is called the marginal c.d.f. of $\mathbf{X}_1$ and the function

$$f_1(\mathbf{x}_1) = \underbrace{\int_{-\infty}^{\infty} \cdots \int_{-\infty}^{\infty}}_{p-k \text{ times}} f(\mathbf{x}_1, \mathbf{x}_2) dx_{k+1} \ldots dx_p, \qquad \text{(B.101)}$$

where $f(\mathbf{x}) = f(\mathbf{x}_1, \mathbf{x}_2)$, is called the marginal p.d.f. of $\mathbf{X}_1$. Marginal distribution of any other subvector of $\mathbf{X}$ can be defined similarly. Also, analogously as in Section B.10, for a given value of $\mathbf{X}_2$, $\mathbf{X}_2 = \mathbf{x}_2$, the conditional p.d.f. of $\mathbf{X}_1$ can be defined as

$$f(\mathbf{x}_1|\mathbf{X}_2 = \mathbf{x}_2) = \frac{f(\mathbf{x}_1, \mathbf{x}_2)}{f_2(\mathbf{x}_2)}, \qquad (B.102)$$

where $f_2(\mathbf{x}_2)$ is the marginal p.d.f. of $\mathbf{X}_2$, which is assumed to be positive at $\mathbf{x}_2$. If the random subvectors $\mathbf{X}_1$ and $\mathbf{X}_2$ are stochastically independent, then

$$f(\mathbf{x}) = f_1(\mathbf{x}_1)f_2(\mathbf{x}_2). \qquad (B.103)$$

Recall that random variables $X_1, X_2, \ldots, X_p$, i.e., the elements of a random vector $\mathbf{X}$, are said to be (mutually) stochastically independent if and only if

$$f(\mathbf{x}) = f(x_1)f(x_2)\cdots f(x_p), \qquad (B.104)$$

where $f(\mathbf{x})$ is the joint p.d.f. of $\mathbf{X}$ and $f(x_i)$ is the marginal p.d.f. of X_i, $i = 1, 2, \ldots, p$.

B.11.2   Moment Generating Functions

For a random vector $\mathbf{X}$ of any fixed dimension p, with p.d.f. $f(\mathbf{x})$, the mean (or expectation) of a scalar-valued function $g(\mathbf{x})$ is defined as

$$E[g(\mathbf{X})] = \int_{-\infty}^{\infty} \cdots \int_{-\infty}^{\infty} g(\mathbf{x})f(\mathbf{x})dx_1 \ldots dx_p. \qquad (B.105)$$

More generally, the mean of a matrix

$$\mathbf{G}(\mathbf{X}) = (g_{ij}(\mathbf{X})),$$

that is, the mean of the matrix-valued function $\mathbf{G}(\mathbf{X})$ of the random vector $\mathbf{X}$, is defined as the matrix

$$E[\mathbf{G}(\mathbf{X})] = (E[g_{ij}(\mathbf{X})]). \qquad (B.106)$$

In particular, the vector

$$\boldsymbol{\mu} = E(\mathbf{X})$$

is the *mean vector* of $\mathbf{X}$ with components

$$\mu_i = \int_{-\infty}^{\infty} \cdots \int_{-\infty}^{\infty} x_i f(\mathbf{x})dx_1 \ldots dx_p, \quad i = 1, \ldots, p.$$

The properties of (B.106) are direct generalizations of those for the univariate and two-variate cases. For example, for any matrix of constants $\mathbf{A}_{(q \times p)}$ and any constant vector $\mathbf{b}_{(q \times 1)}$,

$$E(\mathbf{A}\mathbf{X} + \mathbf{b}) = \mathbf{A}E(\mathbf{X}) + \mathbf{b}. \qquad (B.107)$$

The mean vector plays the same role as the mean in the univariate case. It is the "location" parameter of $\mathbf{X}$. The "spread" of $\mathbf{X}$ is now characterized by the *covariance*, or *dispersion*, *matrix* which is defined as the matrix

$$\boldsymbol{\Sigma} = \mathbf{V}(\mathbf{X}) = E[(\mathbf{X} - \boldsymbol{\mu})(\mathbf{X} - \boldsymbol{\mu})'], \qquad (B.108)$$

where $\boldsymbol{\mu}$ is the mean vector of $\mathbf{X}$. Note that the mean vector and covariance matrix reduce to the mean and variance, respectively, when $p = 1$. The following properties of the covariance matrix $\boldsymbol{\Sigma}$ of order p are simple consequences of its definition:

$$\sigma_{ij} = \mathrm{Cov}(X_i, X_j) \; \; if \; \; i \neq j \qquad (B.109)$$

and

$$\sigma_{ii} = \mathrm{Var}(X_i), \; \; i = 1, \ldots, p, \qquad (B.110)$$

where σ_{ij} is the (i, j)th element of $\boldsymbol{\Sigma}$;

$$\boldsymbol{\Sigma} = E(\mathbf{X}\mathbf{X}') - \boldsymbol{\mu}\boldsymbol{\mu}'; \qquad (B.111)$$

$$\mathrm{Var}(\mathbf{a}'\mathbf{X}) = \mathbf{a}'\boldsymbol{\Sigma}\mathbf{a} \qquad (B.112)$$

for any constant vector $\mathbf{a}$ of dimension p; note that the left hand side of (B.112) cannot be negative and, hence, that $\boldsymbol{\Sigma}$ is positive semi-definite;

$$\mathbf{V}(\mathbf{A}\mathbf{X} + \mathbf{b}) = \mathbf{A}\boldsymbol{\Sigma}\mathbf{A}' \qquad (B.113)$$

for any constant matrix $\mathbf{A}_{(q \times p)}$ and any constant vector $\mathbf{b}_{(q \times 1)}$. If we are given two random vectors, $\mathbf{X}$ of dimension p and $\mathbf{Y}$ of dimension q, we can define the $p \times q$ matrix

$$\mathrm{Cov}(\mathbf{X}, \mathbf{Y}) = E[(\mathbf{X} - \boldsymbol{\mu})(\mathbf{Y} - \boldsymbol{\nu})'], \qquad (B.114)$$

where $\boldsymbol{\mu} = E(\mathbf{X})$ and $\boldsymbol{\nu} = E(\mathbf{Y})$. $\mathrm{Cov}(\mathbf{X}, \mathbf{Y})$ is called the covariance between $\mathbf{X}$ and $\mathbf{Y}$. The following properties of the covariance between two random vectors can easily be proved:

$$\mathrm{Cov}(\mathbf{X}, \mathbf{X}) = \mathbf{V}(\mathbf{X}); \qquad (B.115)$$

$$\text{Cov}(\mathbf{X}, \mathbf{Y}) = \text{Cov}(\mathbf{Y}, \mathbf{X})'; \qquad (\text{B.116})$$

if $\mathbf{X}_1$ and $\mathbf{X}_2$ are random vectors of the same dimension, then

$$\text{Cov}(\mathbf{X}_1 + \mathbf{X}_2, \mathbf{Y}) = \text{Cov}(\mathbf{X}_1, \mathbf{Y}) + \text{Cov}(\mathbf{X}_2, \mathbf{Y}); \qquad (\text{B.117})$$

if $\mathbf{X}$ and $\mathbf{Y}$ are of the same dimension, then

$$\mathbf{V}(\mathbf{X} + \mathbf{Y}) = \mathbf{V}(\mathbf{X}) + \text{Cov}(\mathbf{X}, \mathbf{Y}) + \text{Cov}(\mathbf{Y}, \mathbf{X}) + \mathbf{V}(\mathbf{Y}); \qquad (\text{B.118})$$

for any constant matrices $\mathbf{A}$ and $\mathbf{B}$ of orders $r \times p$ and $s \times q$, respectively,

$$\text{Cov}(\mathbf{AX}, \mathbf{BY}) = \mathbf{A}\text{Cov}(\mathbf{X}, \mathbf{Y})\mathbf{B}'; \qquad (\text{B.119})$$

finally, if $\mathbf{X}$ and $\mathbf{Y}$ are independent, then $\text{Cov}(\mathbf{X}, \mathbf{Y})$ is the zero matrix. All the above properties reduce to known properties of covariances between random variables when $p = q = 1$.

Let $\mathbf{X} = [X_1, \ldots, X_p]'$ be a random vector with mean vector $\boldsymbol{\mu}$. A moment of order k of the variables $X_{i_1}, X_{i_2}, \ldots, X_{i_m}$, $m \leq p$, is defined as

$$E[(X_{i_1} - \mu_{i_1})^{j_1} (X_{i_2} - \mu_{i_2})^{j_2} \cdots (X_{i_m} - \mu_{i_m})^{j_m}], \qquad (\text{B.120})$$

where $j_1, j_2, \ldots, j_m$ are positive integers such that $j_1 + j_2 + \cdots + j_m = k$. Note that many different moments of the r.v.'s $X_{i_1}, X_{i_2}, \ldots, X_{i_m}$ have the same order k. As in the univariate case, calculations of higher order moments are usually greatly facilitated by using the moment generating functions. The moment generating function $M_{\mathbf{X}}(\mathbf{t})$ for $\mathbf{X}$ is defined via

$$M_{\mathbf{X}}(\mathbf{t}) = E(e^{\mathbf{t}'\mathbf{X}}), \qquad (\text{B.121})$$

where, as usual, $\mathbf{t}'\mathbf{X}$ denotes the inner product of $\mathbf{t}$ and $\mathbf{X}$. The m.g.f. is, thus, a scalar function of the vector $\mathbf{t}$ of dimension p. Assuming that differentiation with respect to elements of $\mathbf{t}$ commutes with expectation operator E, we easily see that

$$E(X_1^{j_1} X_2^{j_2} \cdots X_p^{j_p}) = \left\{ \frac{\partial^{j_1 + j_2 + \cdots + j_p}}{\partial t_1^{j_1} \partial t_2^{j_2} \cdots \partial t_p^{j_p}} M_{\mathbf{X}}(\mathbf{t}) \right\}_{\mathbf{t}=0} \qquad (\text{B.122})$$

when this moment exists (in (B.120), some j_i's may be equal to zero).

B.11.3   Change of Variable Technique

The change of variable technique provides a powerful means for computing the p.d.f. of a transformed random vector given its original probability density. We shall use this technique in the next subsection to prove
an important property of multinormal random vectors. Let $\mathbf{X}$ be a random vector of dimension p having density $f(\mathbf{x})$, which is positive for $\mathbf{x}$
from a set $\mathcal{A}$ and is zero elsewhere (in particular, $\mathcal{A}$ may be equal to the
whole space of vectors $\mathbf{x}$, i.e., $f(\mathbf{x})$ may be positive for all values of the
vector $\mathbf{x}$). Let $\mathbf{Y} = \mathbf{u}(\mathbf{X})$ define a one-to-one transformation that maps
set $\mathcal{A}$ onto set $\mathcal{B}$, so that the vector equation $\mathbf{y} = \mathbf{u}(\mathbf{x})$ can be uniquely
solved for $\mathbf{x}$ in terms of $\mathbf{y}$, say, $\mathbf{x} = \mathbf{w}(\mathbf{y})$. Then the p.d.f. of $\mathbf{Y}$ is

$$f(\mathbf{w}(\mathbf{y}))J \text{ if } \mathbf{y}\epsilon\mathcal{B}, \qquad (B.123)$$

and is zero otherwise, where

$$J = \text{absolute value of } |\mathbf{J}| \qquad (B.124)$$

and $|\mathbf{J}|$ is the determinant of the $p \times p$ matrix of partial derivatives of
the inverse transformation $\mathbf{w}(\mathbf{y})$,

$$\mathbf{J} = \left(\frac{\partial x_i}{\partial y_j} \right) \begin{array}{l} i = 1, \ldots, p \\ j = 1, \ldots, p \end{array} . \qquad (B.125)$$

$|\mathbf{J}|$ is called the *Jacobian* of the transformation $\mathbf{w}(\mathbf{y})$. The above result
is a classical theorem of vector calculus and, although it is not hard to
be proved, its proof will be skipped. We shall see with what ease the
change of variable technique can be used.

B.11.4   Normal Distribution

The random vector $\mathbf{X}$ of dimension p is said to have *multivariate normal*
(or *p-dimensional multinormal* or *p-variate normal*) distribution if its
p.d.f. is given by

$$f(\mathbf{x}) = |2\pi\mathbf{\Sigma}|^{-1/2} \exp\{-\frac{1}{2}(\mathbf{x} - \boldsymbol{\mu})'\mathbf{\Sigma}^{-1}(\mathbf{x} - \boldsymbol{\mu})\}, \qquad (B.126)$$

where $\boldsymbol{\mu}$ is a constant vector and $\mathbf{\Sigma}$ is a constant positive definite matrix.
It can be shown that $\boldsymbol{\mu}$ and $\mathbf{\Sigma}$ are the mean vector and covariance
matrix of the random vector $\mathbf{X}$, respectively. For short, we write that $\mathbf{X}$

is $\mathcal{N}(\boldsymbol{\mu}, \boldsymbol{\Sigma})$ distributed. Comparing (B.50) and (B.126) we see that the latter is a natural multivariate extension of the former. Note that if the covariance matrix $\boldsymbol{\Sigma}$ is diagonal, $\boldsymbol{\Sigma} = diag(\sigma_{11}, \sigma_{22}, \ldots, \sigma_{pp})$, the p.d.f. (B.126) can be written as the product

$$f(\mathbf{x}) = \prod_{i=1}^{p} (2\pi\sigma_{ii})^{-1/2} \exp\{-\frac{1}{2}(x_i - \mu_i)\sigma_{ii}^{-1}(x_i - \mu_i)\}. \qquad (B.127)$$

Thus, the elements of $\mathbf{X}$ are then mutually independent normal random variables with means μ_i and variances σ_{ii}, $i = 1, 2, \ldots, p$, respectively. If the random vector $\mathbf{X}$ is multivariate normal, then the property that its elements are uncorrelated one with another (i.e., that $\mathrm{Cov}(X_i, X_j) = 0, i \neq j$) implies their mutual independence.

Let $\mathbf{X}$ be a p-variate normal random vector with mean vector $\boldsymbol{\mu}$ and covariance matrix $\boldsymbol{\Sigma}$, and let

$$\mathbf{Y} = \boldsymbol{\Sigma}^{-1/2}(\mathbf{X} - \boldsymbol{\mu}), \qquad (B.128)$$

where $\boldsymbol{\Sigma}^{-1/2}$ is the square root of the positive definite matrix $\boldsymbol{\Sigma}^{-1}$. Then $\mathbf{Y}$ is the p-variate normal random vector with zero mean vector and covariance matrix equal to the identity matrix, $\mathbf{I}$. In other words, $\mathbf{Y}$ is the random vector whose elements $Y_1, Y_2, \ldots, Y_p$ are mutually independent standard normal random variables. To see that it is indeed the case, observe first that

$$\mathbf{X} = \boldsymbol{\Sigma}^{1/2}\mathbf{Y} + \boldsymbol{\mu}. \qquad (B.129)$$

Transformation (B.129) defines the inverse transformation to the transformation (B.128) and its Jacobian (B.125) is

$$|\mathbf{J}| = J = |\boldsymbol{\Sigma}^{1/2}| = |\boldsymbol{\Sigma}|^{1/2}, \qquad (B.130)$$

where J is defined by (B.124), since $\boldsymbol{\Sigma}$ is positive definite and

$$|\boldsymbol{\Sigma}| = |\boldsymbol{\Sigma}^{1/2}\boldsymbol{\Sigma}^{1/2}| = |\boldsymbol{\Sigma}^{1/2}\boldsymbol{\Sigma}^{1/2}|$$

(see the properties of determinants). Hence, upon noting that

$$(\mathbf{x} - \boldsymbol{\mu})'\boldsymbol{\Sigma}^{-1}(\mathbf{x} - \boldsymbol{\mu}) = \mathbf{y}'\mathbf{y} \qquad (B.131)$$

and using (B.126) and (B.130), the p.d.f. (B.123) assumes the form

$$\prod_{i=1}^{p} (2\pi)^{-1/2} e^{-y_i^2/2}.$$

Thus, the proof is accomplished. Transformation (B.128) is very useful in practice. In principle, it enables one to replace a multivariate problem by a sequence of much simpler univariate problems. Usually, however, we neither know $\boldsymbol{\mu}$ nor we know $\boldsymbol{\Sigma}$ and, hence, some caution is needed here.

By (B.128),

$$(\mathbf{X} - \boldsymbol{\mu})'\boldsymbol{\Sigma}^{-1}(\mathbf{X} - \boldsymbol{\mu}) = \sum_{i=1}^{p} Y_i^2, \qquad (\text{B.132})$$

and it follows from the properties of the chi-square distribution that the random variable

$$(\mathbf{X} - \boldsymbol{\mu})'\boldsymbol{\Sigma}^{-1}(\mathbf{X} - \boldsymbol{\mu}) \qquad (\text{B.133})$$

has chi-square distribution with p degrees of freedom.

B.11.5   Quadratic Forms of Normal Vectors

Quadratic forms of normal random vectors play an important role in statistical inference. In Chapter 6, we use them for testing certain hypotheses about the so-called regression models.

Let us state first a version of the *Cochran's theorem*. Suppose $\mathbf{X} = [X_1, X_2, \ldots, X_p]'$ is a vector of p independent $\mathcal{N}(0,1)$ (i.e., standard normal) random variables. Assume that the sum of squares

$$\mathbf{X}'\mathbf{X} = \sum_{i=1}^{p} X_i^2$$

can be decomposed into k quadratic forms,

$$\mathbf{X}'\mathbf{X} = \sum_{j=1}^{k} Q_j,$$

where Q_j is a quadratic form in $\mathbf{X}$ with matrix $\mathbf{A}_j$ which has rank r_j, $Q_j = \mathbf{X}'\mathbf{A}_j\mathbf{X}$, $j = 1, 2, \ldots, k$. Then any of the following three conditions implies the other two:

(i) the ranks r_j, $j = 1, 2, \ldots, k$, add to p;

(ii) each of the quadratic forms Q_j has chi-square distribution with r_j degrees of freedom;

(iii) all the quadratic forms Q_j are mutually stochastically independent random variables.

We shall now prove a closely related result. Namely, if $\mathbf{X}$ is a vector of p independent $\mathcal{N}(0,1)$ random variables and $\mathbf{A}$ is a symmetric and idempotent matrix of order p, then the quadratic form $\mathbf{X}'\mathbf{A}\mathbf{X}$ has χ_r^2 distribution, where r is the rank of $\mathbf{A}$. By the spectral decomposition of $\mathbf{A}$ and the fact that all eigenvalues of $\mathbf{A}$, λ_i, are equal to 0 or 1 (see Section A.7 of the Appendix),

$$\mathbf{X}'\mathbf{A}\mathbf{X} = \sum_{i \in \mathcal{I}} \mathbf{X}'\boldsymbol{\gamma}_i \boldsymbol{\gamma}_i'\mathbf{X},$$

where $\boldsymbol{\gamma}_i$ is a standardized eigenvector of $\mathbf{A}$ corresponding to λ_i, $i \in \mathcal{I}$, and $\mathcal{I}$ is the set of indices of those eigenvalues λ_i which are equal to 1. We can write

$$\mathbf{X}'\mathbf{A}\mathbf{X} = \sum_{i \in \mathcal{I}} Y_i^2,$$

where $Y_i = \mathbf{X}'\boldsymbol{\gamma}_i = \boldsymbol{\gamma}_i'\mathbf{X}$, $i \in \mathcal{I}$. Each Y_i is the linear combination of standard normal r.v.'s and, hence, is a normally distributed r.v. with mean 0 and variance $\boldsymbol{\gamma}_i'\boldsymbol{\gamma}_i = 1$. Furthermore,

$$\begin{aligned}
\text{Cov}(Y_i Y_j) &= E(\mathbf{X}'\boldsymbol{\gamma}_i \mathbf{X}'\boldsymbol{\gamma}_j) \\
&= E(\boldsymbol{\gamma}_i'\mathbf{X}\mathbf{X}'\boldsymbol{\gamma}_j) \\
&= \boldsymbol{\gamma}_i' E(\mathbf{X}\mathbf{X}')\boldsymbol{\gamma}_j \\
&= 0 \ \text{if} \ i \neq j,
\end{aligned}$$

since $E(\mathbf{X}\mathbf{X}') = \mathbf{I}$ and $\boldsymbol{\gamma}_i'\boldsymbol{\gamma}_j = 0$ if $i \neq j$. Thus, the Y_i's are uncorrelated one with another and, since they are normally distributed, they are mutually independent. Finally, therefore, $\mathbf{X}'\mathbf{A}\mathbf{X}$ is chi-square distributed with the number of degrees of freedom equal to the number of elements of the set $\mathcal{I}$. By (ii) of Section A.7 and (A.48), it follows that this number is equal to the rank of $\mathbf{A}$.

B.11.6   Central Limit Theorem

Let us give the following multivariate extension of the central limit theorem. Let $\mathbf{X}_1, \mathbf{X}_2, \ldots, \mathbf{X}_n$ be a random sample of n independent random vectors having identical p-variate distribution with mean vector $\boldsymbol{\mu}$ and covariance matrix $\boldsymbol{\Sigma}$. Let

$$\bar{\mathbf{X}} = \frac{\mathbf{X}_1 + \mathbf{X}_2 + \cdots + \mathbf{X}_n}{n} \tag{B.134}$$

be the sample mean vector, and let

$$\mathbf{Z} = n^{1/2}\boldsymbol{\Sigma}^{-1/2}(\bar{\mathbf{X}} - \mu) = n^{-1/2}\boldsymbol{\Sigma}^{-1/2}\sum_{i=1}^{n}(\mathbf{X}_i - \mu). \tag{B.135}$$

Then, with n increasing, the distribution of $\mathbf{Z}$ approaches the p-variate normal distribution with zero mean vector and covariance matrix $\mathbf{I}$. The proof of this theorem is similar to that for the univariate case.

B.12   Poisson Process

The Poisson experiment, as described in Section B.4, was assumed to take place in a time interval of fixed length T. In fact, there is no need to restrict the experiment to any fixed time interval. To the contrary, already the examples discussed in Subsections B.4.3 and B.6.2 (following (B.58)) indicate that we are interested in continuing the experiment over time and observing the evolution of the process of interest. For example, if the incoming telephone calls constitute the Poisson experiment, we observe the *stochastic process* $X(t)$ as time t elapses, where $X(t)$ is the number of telephone calls up to current instant t. Also, in the context of quality control, items returned due to unsatisfactory performance usually form a Poisson process.

Let $X(t)$ denote the number of occurrences of an event from time 0 to time t, $t \geq 0$. Let $X(0) = 0$. The process $X(t)$ is said to be the *Poisson process* having rate λ, $\lambda \geq 0$, if the three postulates defining a Poisson experiment, (B.34)-(B.36), are fulfilled for each $t \geq 0$, and if

$$P(k \text{ in } [t_1, t_1 + s]) = P(k \text{ in } [t_2, t_2 + s]) \tag{B.136}$$

for all nonnegative t_1, t_2, and s. The fourth postulate implies that the the rate λ does not change with time.

Let $P(k, t) = P(k \text{ in } [0, t])$, i.e., $P(k, t)$ be the probability of k events up to time t. Then

$$
\begin{aligned}
P(k+1 \text{ in } [0, t+\varepsilon]) &= P(k+1 \text{ in } [0, t])P(0 \text{ in } [t, t+\varepsilon]) \\
&\quad + P(k \text{ in } [0, t])P(1 \text{ in } [t, t+\varepsilon]) + o(\varepsilon) \\
&= P(k+1, t)(1 - \lambda\varepsilon) + P(k, t)\lambda\varepsilon + o(\varepsilon).
\end{aligned}
$$

Thus,

$$\frac{P(k+1, t+\varepsilon) - P(k+1, t)}{\varepsilon} = \lambda[P(k, t) - P(k+1, t)] + \frac{o(\varepsilon)}{\varepsilon}.$$

Taking the limit as $\varepsilon \to 0$, we obtain

$$\frac{dP(k+1, t)}{dt} = \lambda[P(k, t) - P(k+1, t)]. \tag{B.137}$$

Now taking $k = -1$, we have

$$\frac{dP(0,t)}{dt} = -\lambda P(0,t), \tag{B.138}$$

since it is impossible for a negative number of events to occur. Hence

$$P(0,t) = \exp(-\lambda t). \tag{B.139}$$

Substituting (B.139) in (B.137) for $k = 0$, we have

$$\frac{dP(1,t)}{dt} = \lambda[\exp(-\lambda t) - P(1,t)], \tag{B.140}$$

and, hence,

$$P(1,t) = \exp(-\lambda t)(\lambda t). \tag{B.141}$$

Continuing in this way for $k = 1$ and $k = 2$, we can guess the general formula of the Poisson distribution:

$$P(k,t) = \frac{e^{-\lambda t}(\lambda t)^k}{k!}. \tag{B.142}$$

In order to verify that (B.142) satisfies (B.137), it suffices to substitute it into both sides of (B.137). Formula (B.139) gives us the probability of no event occurring up to time t. But this is equal to the probability that the first event occurs after time t and, thus,

$$P(\text{time of first occurrence} \leq t) = F(t) = 1 - \exp(-\lambda t). \tag{B.143}$$

$F(t)$ is the c.d.f. of the random variable defined as the time to the first event. Corresponding p.d.f. is equal to

$$f(t) = \frac{dF(t)}{dt} = \lambda e^{-\lambda t}, \tag{B.144}$$

i.e., it is the exponential density with parameter $\beta = 1/\lambda$, as was stated in Subsection B.6.2. A slightly more involved argument proves that the inter-event times are given by the same exponential distribution as well.

B.13   Statistical Inference

B.13.1   Motivation

The most obvious aim of quality inspection is to control the proportion of nonconforming items among all manufactured ones. It is too expensive, and often too time-consuming, to examine all of the items. Thus, we draw a sample of items from the population of all items and find out experimentally what is the proportion of nonconforming items in that particular sample. Now, the following interpretational problems arise. Assume that we know that, say, all of the 1000 items produced on a particular day were produced under the same conditions and using materials of the same quality. Still, if we drew not one but several different samples of items, we would most likely obtain different proportions of nonconforming items. Which of these experimentally obtained estimates of the true proportion of nonconforming items in the population are more informative, or more accurate, than the others? It is statistical inference which allows us to answer such questions. Statistical inference provides also an answer to a much more important question. Most of the changes in a production process are unintended and, hence, unknown in advance. As a matter of fact, the main reason for implementing the statistical process control is to discover these changes as quickly as possible. That is, we should never assume that no changes have occurred but, to the contrary, we should keep asking whether the samples we obtain for scrutiny do indeed come from the same population, characterized by the same, constant properties. If we are given several samples of items produced during one day, our main task is to verify whether the daily production can be considered as forming one, homogeneous, population or, rather, it is formed of at least two populations of different properties, in particular, of different defective rates. We are able to solve the problem using statistical means.

Acceptance sampling is the most primitive stage of quality control. But the problems outlined above are typical for the whole field of statistical process for quality. The most immediate aim of SPC is to systematically verify that both the mean value of a parameter of interest and the variablity of the parameter about this mean do not change in time. In other words, the problem is to verify that samples of the items measured come from one population, described by fixed mean value and variance. Note that it does not suffice to take samples of the items and measure their sample means and sample variances. From sample to sample, these measurements will be giving different results even if the production process is stable, that is, if the samples come from the same probability distribution. Thus, given different sample means and variances corresponding to different samples, we have to recognize if these values are typical for a stable production process or they point to a change in the process.

In what follows, we shall focus our attention on some basics of statistical inference about means and variances. We shall begin with getting some better insight into the properties of the sample mean and variance. Most of the considerations will be confined to the case of samples being drawn from normal populations. It should be emphasized, however, that our considerations give ground to statistical analysis of other parameters and other probability distributions in general as well.

B.13.2   Maximum Likelihood

Let us consider a random sample of n independent random variables $X_1, X_2, \ldots, X_n$ with common p.f. or p.d.f. $f(x)$. In other words, we can say that the random sample comes from a population characterized by a probability distribution $f(x)$. Constructing, for example, the sample mean $\bar{X}$ of the sample does not require any prior knowledge about the parent distribution $f(x)$. Moreover, provided only that $f(x)$ has finite mean and variance, we know that the expected value of $\bar{X}$ is equal to the true mean μ of the probability distribution $f(x)$ and that its variance is n times smaller than that of $f(x)$.

Another justification of using $\bar{X}$ as an estimator of unknown μ is provided by the following argument. If we cannot make any a priori assumptions about the parent distribution $f(x)$, the only available information about it is that contained in the random sample $X_1, X_2, \ldots, X_n$. Given that the following values of the random sample have been observed, $X_1 = x_1, X_2 = x_2, \ldots, X_n = x_n$, it is natural to identify the unknown probability distribution with the discrete uniform distribution (B.11) with $k = n$. But the expected value of this distribution is given by (B.15) and, thus, is equal to the sample mean (B.62). It is also worthwhile to note that the variance of the uniform distribution (B.11) with $k = n$ is equal to (B.17) with $k = n$ and, therefore, that it is equal to the sample variance (B.61) multiplied by factor $(n-1)/n$. In this way, the given argument justifies the use of a slightly modified sample variance as well (we shall encounter the modified sample variance in this subsection once more).

However, if we want to gain a deeper insight into the problem of estimating μ and σ^2, we have to make some more assumptions about the parent distribution $f(x)$. Usually, it is reasonable to assume that the parent distribution is of some known type, although the parameters of this distribution are unknown. For example, in the problem of controlling the mean of a production process, we usually assume that the parent distribution is normal with unknown mean μ and unknown variance σ^2.

In this last example, one is, of course, tempted to use $\bar{X}$ as the estimator of unknown μ and the sample variance S^2 as the estimator of an unknown variance σ^2. We shall now discuss one of possible general approaches to the problem of estimating unknown parameters of probability distributions which offers additional justification for using $\bar{X}$ and S^2 (or its modification, $((n-1)/n)S^2$ in the particular case mentioned. Each of the n random variables constituting the random sample drawn from a normal distribution has the same p.d.f. given by (B.50). In order to make its dependence on the parameters μ and σ^2 explicit, let us denote this density by

$$f(x; \theta), \qquad (B.145)$$

where θ denotes the set of parameters; in our case, $\theta = (\mu, \sigma^2)$. By (B.49), the joint p.d.f. of the whole random sample of n independent random variables is given by the product

$$L(x_1, x_2, \ldots, x_n; \theta) = f(x_1; \theta)f(x_2; \theta) \cdots f(x_n; \theta), \qquad (B.146)$$

where $-\infty < x_1 < \infty$, $-\infty < x_2 < \infty$, ..., $-\infty < x_n < \infty$. If we have observed n values of the random sample, $X_1 = x_1$, $X_2 = x_2$, ..., $X_n = x_n$, the p.d.f. assumes a fixed value, given also by formula (B.146). This value, however fixed, is unknown, since we do not know the true values of the parameters θ of the distribution. In a sense, (B.146) remains a function of the parameters θ. We can now ask the following question. Given that we have observed data x_1, x_2, ..., x_n, what values of parameters μ and σ^2 are "most likely"? To put it otherwise, we can think of $L(x_1, x_2, \ldots, x_n; \theta)$ as a function of unknown θ which measures how "likely" a particular θ is to have given the observed data x_1, x_2, ..., x_n. It is then natural to consider a θ that maximizes $L(x_1, x_2, \ldots, x_n; \theta)$ to be the most likely set of parameters of the joint density of the random sample.

Function $L(x_1, x_2, \ldots, x_n; \theta)$, considered as a function of θ, is called the *likelihood function*. The suggested method of finding estimators of parameters θ is called the *maximum likelihood method* and the estimators obtained are called the *maximum likelihood estimators*. In the case of normal joint density, (B.146) takes on the form

$$L(x_1, x_2, \ldots, x_n; \theta) = (2\pi\sigma^2)^{-n/2} \exp\left[\frac{\sum_{i=1}^{n}(x_i - \mu)^2}{2\sigma^2} \right], \qquad (B.147)$$

where $-\infty < \mu < \infty$ and $0 < \sigma^2 < \infty$. The maximum likelihood estimators of μ and σ^2 are those which maximize the likelihood function

(B.147). In practice, maximization of $L(x_1, x_2, \ldots, x_n; \theta)$ is replaced by a usually simpler maximization of the logarithm of $L(x_1, x_2, \ldots, x_n; \theta)$. This can indeed be done, since both functions achieve their maxima for the same θ. The logarithm of a likelihood function is called the *log-likelihood function*. In our case, we have

$$\ln L(x_1, x_2, \ldots, x_n; \theta) = -\frac{n(2\pi\sigma^2)}{2} - \frac{\sum_{i=1}^{n}(x_i - \mu)^2}{2\sigma^2}. \qquad \text{(B.148)}$$

Maximization of (B.148) with respect to μ and σ^2 yields the following maximum likelihood estimators of these parameters

$$\hat{\mu} = \bar{X} \quad \text{and} \quad \hat{\sigma}^2 = \frac{n-1}{n}S^2, \qquad \text{(B.149)}$$

respectively. Thus, in the case considered, the sample mean has occurred to be the maximum likelihood estimator of μ. Interestingly enough, S^2 has to be slightly modified if one wishes it to become the maximum likelihood estimator of σ^2. It follows that the expected value of $\hat{\sigma}^2$ is not equal to the true variance σ^2. From the practical point of view, however, the modification required is inessential.

The maximum likelihood approach is not confined to the normal case only. In fact, it is the most widely used approach to estimating unknown parameters of both discrete and continuous distributions. Note that in the case of a discrete parent distribution, the likelihood function $L(x_1, x_2, \ldots, x_n; \theta)$ given by (B.146) has a particularly clear interpretation. For the observed data $x_1, x_2, \ldots, x_n$, it is simply the probability of obtaining this set of data.

B.13.3   Confidence Intervals

The idea of constructing the so-called confidence intervals forms a basis for evaluating control limits for control charts. This is so, because the confidence intervals tell us, given the data observed, within which interval "we would expect" the true value of an unknown parameter to lie. From the point of view of statistical process control, the most important confidence intervals are those for the mean and variance of a parent distribution.

Let us consider a random sample of n independent random variables $X_1, \ldots, X_n$ with common normal p.d.f given by (B.50). Let, as usual, $\bar{X}$ denote the sample mean and assume that the variance σ^2 is known.

Then (see (B.67)),

$$Z = \frac{\bar{X} - \mu}{\sigma/\sqrt{n}} \qquad (B.150)$$

is a standard normal random variable. Let $z_{\alpha/2}$ be such a number that

$$P(Z > z_{\alpha/2}) = \alpha/2, \qquad (B.151)$$

where α is fixed, $0 < \alpha < 1$. For any given α, corresponding $z_{\alpha/2}$ can be found in the statistical tables for the standard normal distribution or can be provided by any computer statistical package. Now, since the standard normal p.d.f. is symmetric about zero,

$$P(Z < -z_{\alpha/2}) = \alpha/2 \qquad (B.152)$$

and, hence,

$$P(-z_{\alpha/2} \le Z \le z_{\alpha/2}) = 1 - \alpha. \qquad (B.153)$$

Thus,

$$P(-z_{\alpha/2} \le \frac{\bar{X} - \mu}{\sigma/\sqrt{n}} \le z_{\alpha/2}) = 1 - \alpha. \qquad (B.154)$$

But the last equation can be written in the following form

$$P(\bar{X} - z_{\alpha/2}\frac{\sigma}{\sqrt{n}} \le \mu \le \bar{X} + z_{\alpha/2}\frac{\sigma}{\sqrt{n}}) = 1 - \alpha. \qquad (B.155)$$

In this way, we have obtained the confidence interval for μ: If $\bar{x}$ is an observed value of the sample mean of a random sample $X_1, \ldots, X_n$ from the normal distribution with known variance σ^2, the $(1-\alpha)100\%$ *confidence interval* for μ is

$$\bar{x} - z_{\alpha/2}\frac{\sigma}{\sqrt{n}} \le \mu \le \bar{x} + z_{\alpha/2}\frac{\sigma}{\sqrt{n}}. \qquad (B.156)$$

That is, we have a $(1-\alpha)100\%$ chance that the given interval includes the true value of unknown μ. The number $1 - \alpha$ is called the *confidence coefficient*. Usually, we choose the confidence coefficient equal to 0.95.

In practice, the variance of a normal population is rarely known. We have, however, two easy ways out of the trouble. First, we know that the random variable given by (B.68) has t distribution with $n - 1$ degrees of freedom. The t distribution is symmetric about zero and, thus, proceeding analogously as before, we obtain the following $(1-\alpha)100\%$ confidence interval for μ if σ^2 is not known

$$\bar{x} - t_{\alpha/2}\frac{s}{\sqrt{n}} \le \mu \le \bar{x} + t_{\alpha/2}\frac{s}{\sqrt{n}}, \qquad (B.157)$$

where s is the observed value of the square root of the sample variance and $t_{\alpha/2}$ is such that $P(T > t_{\alpha/2}) = \alpha/2$ for a random variable T which has t distribution with $n - 1$ degrees of freedom. For any given α, corresponding $t_{\alpha/2}$ can be found in the statistical tables for the relevant t distribution or can be provided by any computer statistical package. The second solution is an approximate one. If the sample size n is not too small, we can substitute s for σ in (B.156) and use that former confidence interval. Intuitively, for large n, we can consider s to be a sufficiently accurate estimate of σ. More rigorously, t distribution tends to the standard normal distribution as n increases.

It is equally straightforward to construct a confidence interval for unknown variance of a normal population. We know that the random variable V defined by (B.64) has chi-square distribution with $n - 1$ degrees of freedom. Let $\chi^2_{1-\alpha/2}$ and $\chi^2_{\alpha/2}$ be such numbers that

$$P(V > \chi^2_{1-\alpha/2}) = 1 - \alpha/2 \ \text{ and } \ P(V > \chi^2_{\alpha/2}) = \alpha/2. \qquad \text{(B.158)}$$

For any given α, corresponding $\chi^2_{1-\alpha/2}$ and $\chi^2_{\alpha/2}$ can be found in the statistical tables for the relevant χ^2 distribution or can be provided by any computer statistical package. Now, we have

$$P\left(\chi^2_{1-\alpha/2} \leq \frac{(n-1)S^2}{\sigma^2} \leq \chi^2_{\alpha/2}\right) = 1 - \alpha \qquad \text{(B.159)}$$

and, hence,

$$P\left(\frac{(n-1)S^2}{\chi^2_{\alpha/2}} \leq \sigma^2 \leq \frac{(n-1)S^2}{\chi^2_{1-\alpha/2}}\right) = 1 - \alpha. \qquad \text{(B.160)}$$

Thus, if s^2 is an observed value of the sample variance of a random sample $X_1, \ldots, X_n$ from a normal distribution, the $(1 - \alpha)100\%$ confidence interval for the variance σ^2 is

$$\frac{(n-1)s^2}{\chi^2_{\alpha/2}} \leq \sigma^2 \leq \frac{(n-1)s^2}{\chi^2_{1-\alpha/2}}. \qquad \text{(B.161)}$$

This time, unlike in the two former cases, we had to choose two different limit points $\chi^2_{1-\alpha/2}$ and $\chi^2_{\alpha/2}$. The choice of only one point would be useless, since the χ^2 distribution is not symmetric about zero.

Let us conclude this discussion with the following observation. In all the cases, we constructed the "equal tailed" confidence intervals. That is, we required that

$$P(Z > z_{\alpha/2}) = P(Z < -z_{\alpha/2}) = \alpha/2,$$

$$P(T > t_{\alpha/2}) = P(T < -t_{\alpha/2}) = \alpha/2,$$

$$P(V > \chi^2_{\alpha/2}) = P(V < \chi^2_{1-\alpha/2}) = \alpha/2,$$

respectively. Most often, it is indeed most natural to construct the equal
tailed confidence intervals. Sometimes, however, it may be recommended
to use some other limit values for a confidence interval. The choice of
the limit values is up to an experimenter and the only strict requirement
is that a predetermined confidence level be achieved. Say, when constructing a confidence interval for the variance of a normal population,
we have to require that

$$P(\underline{v} \leq V \leq \overline{v}) = 1 - \alpha$$

for a given α, and any limit values $\underline{v}$ and $\overline{v}$ of the experimenter's choice.

B.13.4   Testing Hypotheses

A dual problem to the construction of confidence intervals is that of
testing hypotheses. Instead of constructing an interval which includes
the true value of a parameter of interest with predetermined confidence,
we can ask whether we have sufficient evidence to reject our hypothesis
about the true value of the parameter. In the context of SPC, we usually
ask whether, given the evidence coming from the observations taken, we
should not reject the hypothesis that the true value of the parameter of
interest remains unchanged. The parameters which are examined most
often are the mean and variance of some measurements of a production
process.

Consider a normal population with unknown mean μ and known variance σ^2. We want to test the *null* hypothesis that μ is equal to some
specified value μ_0,

$$H_0: \quad \mu = \mu_0, \tag{B.162}$$

against the general *alternative* hypothesis that it is not the case,

$$H_1: \quad \mu \neq \mu_0. \tag{B.163}$$

The conclusion has to be based on a suitable random experiment. Thus,
in order to solve the problem, two tasks have to be accomplished. First,
we have to introduce a *test statistic* whose range could be partitioned into
two nonoverlapping sets in such a way that the values from one set would
indicate that the null hypothesis should be rejected whereas the values
from the other set would indicate that the null hypothesis should be

accepted. Of course, rejection of H_0 implies acceptance of the alternative hypothesis, H_1. And second, a random sample of observations should be obtained, so that the test statistic could be assigned a value. Due to the random character of the whole experiment, we cannot hope for reaching a 100% certain conclusion. It is therefore crucial to propose a test statistic which is "capable of" discerning the two possibilities with maximum possible accuracy.

Let us put our problem into a general framework. Suppose a random sample $X_1, X_2, \ldots, X_n$ is drawn from a population with common p.d.f. or p.f. $f(x; \theta)$. We wish to test the null hypothesis $H_0 : \theta \epsilon \Theta_0$ against the alternative hypothesis $H_1 : \theta \epsilon \Theta_1$, where Θ_0 and Θ_1 are some fixed and disjoint subsets of the *parameter space* Θ, i.e., of the whole range of possible values of the parameters θ. In case of (B.162) and (B.163), just one parameter is tested, $\theta = \mu$, $\Theta = (-\infty, \infty)$, $\Theta_0 = \{\mu_0\}$, i.e., Θ_0 is the set consisting of only one point μ_0, and $\Theta_1 = (-\infty, \mu_0) \cup (\mu_0, \infty)$. If a hypothesis is determined by a set consisting of only one point in the parameter space, the hypothesis is called *simple*. Otherwise, it is called the *composite* hypothesis.

We shall consider the following, intuitively plausible, test statistic

$$\lambda(x_1, x_2, \ldots, x_n) = \frac{\sup\{L(x_1, x_2, \ldots, x_n; \theta) : \theta \epsilon \Theta_1\}}{\sup\{L(x_1, x_2, \ldots, x_n; \theta) : \theta \epsilon \Theta_0\}}, \qquad \text{(B.164)}$$

where $L(x_1, x_2, \ldots, x_n; \theta)$ is the likelihood function,

$$L(x_1, x_2, \ldots, x_n; \theta) = f(x_1; \theta) f(x_2; \theta) \cdots f(x_n; \theta). \qquad \text{(B.165)}$$

In the numerator of (B.164), the supremum of $L(x_1, x_2, \ldots, x_n; \theta)$ is taken over all $\theta \epsilon \Theta_1$, while in the denominator, the supremum is taken over all $\theta \epsilon \Theta_0$. Note that according to our interpretation of the likelihood function, the test statistic $\lambda(x_1, x_2, \ldots, x_n)$ should assume large values if H_1 is true, that is, if the true value of the parameters θ belongs to the set Θ_1. Analogously, $\lambda(x_1, x_2, \ldots, x_n)$ should assume small values if H_0 is true. Tests based on the statistic (B.164) are called the *likelihood ratio tests*. If the null hypothesis is simple, as is the case in example (B.162)-(B.163), formula (B.164) simplifies to the following one

$$\lambda(x_1, x_2, \ldots, x_n) = \frac{\sup\{L(x_1, x_2, \ldots, x_n; \theta) : \theta \epsilon \Theta_1\}}{L(x_1, x_2, \ldots, x_n; \theta_0)}, \qquad \text{(B.166)}$$

where θ_0 is given by the null hypothesis, $H_0 : \theta = \theta_0$.

It remains to determine the *critical value* of $\lambda(x_1, x_2, \ldots, x_n)$, which partitions the set of all possible values of the statistic into two complementary subsets of "small" and "large" values of the statistic. If the observed value of the test statistic (B.164) falls into the subset of "large" values, the null hypothesis is rejected. If the observed value falls into the subset of "small" values, the null hypothesis is accepted. The subset of values of the test statistic leading to rejection of H_0 is called the *critical region* of the test, whereas the other subset is called the *acceptance region* of the test. Let us first determine the critical value c of $\lambda(x_1, x_2, \ldots, x_n)$ for the case of a simple null hypothesis, $H_0 : \theta = \theta_0$. Prior to the experiment, a random sample $X_1, X_2, \ldots, X_n$ is, of course, a sequence of n random variables. Now, even if the null hypothesis is true, it may still happen that the test statistic will assume arbitrarily large value. We determine c in such a way that the probability that H_0 will be rejected, when it is true, be equal to a predetermined *significance level* α,

$$P_{\theta_0}[\lambda(X_1, X_2, \ldots, X_n) > c] = \alpha, \qquad (B.167)$$

where P_{θ_0} denotes the probability calculated under the assumption that the true value of the parameter is θ_0. A simple generalization of this requirement to the case of composite null hypothesis has the form

$$\sup\{P_\theta[\lambda(X_1, X_2, \ldots, X_n) > c] : \theta \epsilon \Theta_0\} = \alpha. \qquad (B.168)$$

The supremum is taken here over all $\theta \epsilon \Theta_0$. In other words, whatever value of θ from the set determining the null hypothesis is true, we require that the probability of rejecting H_0 not exceed the significance level α. Rejection of the null hypothesis when it is true is called the *type I error*. (Acceptance of the null hypothesis when it is false is called the *type II error*.) Observe that the rejection region has the form

$$\{x_1, x_2, \ldots, x_n : \lambda(x_1, x_2, \ldots, x_n) > c\}, \qquad (B.169)$$

that is, it is the set of such points $(x_1, x_2, \ldots, x_n)$ in the n-dimensional sample space that the function $\lambda(x_1, x_2, \ldots, x_n)$ assumes values greater than c. In practice, it is often convenient to replace statistic (B.164) by an equivalent statistic, which is in one-to-one correspondence with (B.164). We can thus describe the likelihood ratio test in terms of either of these statistics. In fact, all the test statistics introduced in the sequel are some equivalent statistics to those defined by (B.164).

Of course, we are interested in having tests which guarantee small probabilities of committing type II errors. This requirement is most often investigated using the concept of the power of a test. The *power function* of a test is the function that associates the probability of rejecting H_0 by the test with each θ from the parameter space $\boldsymbol{\Theta}$. Thus, for $\theta \in \boldsymbol{\Theta}_1$, the value of the power function at θ is equal to one minus the probability of committing the type II error when the true value of the parameter is θ. It can sometimes be proved that the likelihood ratio tests are most powerful in a certain sense and, even if such a proof is not available, it can usually be shown that these tests have desirable properties anyway.

Any choice of the significance level of a test is always somewhat arbitrary. Interpretation of results of testing a hypothesis is greatly facilitated by providing the so-called p-value. Given a test and the computed value of the test statistic, the p-*value* of the test is the smallest level of significance at which the null hypothesis is to be rejected. Thus, if the p-value proves small for a particular case under consideration (say, it is equal to .01), one is certainly inclined to reject H_0. We call the data used to compute a test statistic *significant*, if they lead to the rejection of H_0. The p-value says "how significant" the data are.

Let us return to problem (B.162)-(B.163). It can be shown that the likelihood ratio test, provided the parent probability distribution is normal and its variance is known, is given by the test statistic

$$\frac{|\bar{X} - \mu_0|}{\sigma/\sqrt{n}}. \qquad (B.170)$$

The critical value c is equal to $z_{\alpha/2}$ given by (B.151). Thus, the null hypothesis is rejected if the given statistic assumes a value which is greater than $z_{\alpha/2}$. This result is intuitively appealing, in particular in view of (B.155).

Sometimes, we want to test (B.162) against the hypothesis

$$H_1 : \mu > \mu_0. \qquad (B.171)$$

(For obvious reasons, hypothesis (B.163) is called *two-sided*, while hypothesis (B.171) is called *one-sided*.) In view of the preceding considerations, it is not surprising that the likelihood ratio test is given by the test statistic

$$\frac{\bar{X} - \mu_0}{\sigma/\sqrt{n}} \qquad (B.172)$$

and the critical value c is given by z_α, where z_α is such that

$$P(Z > z_\alpha) = \alpha, \qquad (B.173)$$

and Z is a standard normal random variable. We leave it to the reader
to guess the likelihood ratio test for testing (B.162) against the one-sided
hypothesis $H_1 : \mu < \mu_0$.

If the population is normal and we want to test (B.162) against (B.163),
but the population variance σ^2 is unknown, it can be shown that the likelihood ratio test is given by the test statistic

$$\frac{|\bar{X} - \mu_0|}{S/\sqrt{n}}, \tag{B.174}$$

where S is the square root of the sample variance. The critical value
c is equal to $t_{\alpha/2}$ from (B.157). It is worthwhile to note that, in the
problem just considered, H_0 was a composite hypothesis. Indeed, with
σ^2 unknown, H_0 corresponds to the set of parameters $\Theta_0 = \{\mu_0, 0 <
\sigma^2 < \infty\}$; i.e., although the mean μ is equal to μ_0 under H_0, the variance
σ^2 is arbitrary. Tests based on (B.174) are known as t-tests.

As in the case of the confidence interval for μ when σ^2 is unknown,
if n is not too small, we can replace $t_{\alpha/2}$ by $z_{\alpha/2}$ given by (B.151). We
leave it to the reader to guess the form of the likelihood ratio tests for
the one-sided alternative hypotheses $H_1 : \mu > \mu_0$ and $H_1 : \mu < \mu_0$.

Finally, let us consider the problem of testing the hypothesis that the
variance of a normal population is equal to a fixed number σ_0^2 against
the alternative hypothesis that the population variance is not equal to
σ^2. The population mean is unknown. That is, we wish to test

$$H_0 : \sigma^2 = \sigma_0^2 \quad \text{versus} \quad H_1 : \sigma^2 \neq \sigma_0^2. \tag{B.175}$$

It can be shown that the likelihood ratio test accepts the null hypothesis
if and only if

$$c_1 \leq \frac{(n-1)S^2}{\sigma_0^2} \leq c_2, \tag{B.176}$$

where c_1 and c_2 are such that

$$c_1 - c_2 = n \ln(c_1/c_2) \tag{B.177}$$

and

$$F(c_2) - F(c_1) = 1 - \alpha, \tag{B.178}$$

where F is the c.d.f. of the χ^2 distribution with $n-1$ degrees of freedom.
This time, perhaps somewhat surprisingly, the likelihood ratio test does
not correspond exactly to the equal tailed confidence interval for the
dual problem, given by (B.161). However, for sufficiently large n, $\chi^2_{1-\alpha/2}$

and $\chi^2_{\alpha/2}$, given by (B.158), approximately satisfy conditions (B.177) and (B.178). Thus, at least approximately, the given test and the confidence interval (B.161) correspond one to another.

Although we shall not discuss these cases, let us mention that it is easy to construct likelihood ratio tests when we are given two independent normal random samples, and we wish to test either equality of means (with variances known or unknown but equal) or equality of variances (with means unknown and arbitrary). In the latter case, as hinted in Subsection B.6.3, the test is based on the F statistic.

B.13.5   Goodness-of-Fit Test

Sometimes, we want to verify the hypothesis that a random sample comes from a specified probability distribution which does not have to belong to any family of distributions defined by some parameters. Thus, the problem posed does not reduce to that of testing a hypothesis about the parameters of the hypothesized distribution. A natural way to solve this problem is to fit somehow the data to the hypothesized distribution.

In order to introduce one such approach, developed by Karl Pearson, let us consider the following very simple task. A six-sided die with faces 1 to 6 is tossed 120 times. The null hypothesis is that the die is balanced, i.e.,

$$H_0: \quad f(x) = 1/6 \text{ for } x = 1, 2, \ldots, 6, \tag{B.179}$$

where $f(x)$ denotes the probability function of the outcomes. The alternative hypothesis is that $f(x)$ is not a p.f. of the uniform distribution.

If the die is balanced, we would expect each face to occur approximately 20 times. Equivalently, we would expect all the proportions of occurrences of each of the six faces to be equal to $1/6$. Having this information about the hypothesized distribution, we can base our inference on the comparison of the observed proportions with the corresponding expected proportions. Suppose the following proportions, denoted by p_i, have been observed

face	1	2	3	4	5	6
p_i	$\frac{1}{15}$	$\frac{1}{3}$	$\frac{1}{12}$	$\frac{2}{15}$	$\frac{2}{15}$	$\frac{1}{4}$

In the sequel, we shall refer to each possible outcome of a discrete-valued experiment as a bin. That is, we have six bins in our example. We shall also say that the outcomes fall into bins.

The criterion for verifying the fit of data proposed by Pearson is

$$\chi^2 = n \sum_{i=1}^{k} \frac{(p_i - \pi_i)^2}{\pi_i},\qquad\text{(B.180)}$$

where π_i is the expected proportion of data in the ith bin, $i = 1, 2, \ldots, k$, k is the number of bins and n is the number of data. In the case above,

$$\chi^2 = 120 \left(\frac{(\frac{1}{15}-\frac{1}{6})^2}{\frac{1}{6}} + \frac{(\frac{1}{3}-\frac{1}{6})^2}{\frac{1}{6}} + \frac{(\frac{1}{12}-\frac{1}{6})^2}{\frac{1}{6}} + 2\frac{(\frac{2}{15}-\frac{1}{6})^2}{\frac{1}{6}} + \frac{(\frac{1}{4}-\frac{1}{6})^2}{\frac{1}{6}} \right)$$
$$= 32.13.$$

Before we present the way of using criterion (B.180) to test the goodness of fit, let us note that the above example shows clearly how to obtain this criterion's value in case of an arbitrary hypothesized discrete distribution. If a hypothesized probability distribution is continuous, we have first to bin the data and then see if the proportions thus obtained fit the corresponding expected proportions. Bins are formed as adjacent intervals on the real line, and expected proportions are given by

$$\pi_i = \int_{i\text{th bin}} f(x)dx,\qquad\text{(B.181)}$$

where $f(x)$ is the hypothesized p.d.f. Binning of the data should be done in such a way that all the expected proportions be approximately the same.

Criterion (B.180) can be used effectively to test the goodness of fit due to the following fact. If a random sample is not too small, and if the null hypothesis is true, then χ^2 given by (B.180) is an observed value of a random variable whose probability distribution is well approximated by the χ^2 distribution with $k - 1$ degrees of freedom. In practice, we require that at least 5 data fall in each bin. Of course, if the observed proportions are close to the expected ones, the value of (B.180) is small, indicating good fit. Thus, at the α level of significance, the critical region should have the form

$$\chi^2 > \chi^2_\alpha,\qquad\text{(B.182)}$$

where the value of χ^2_α is provided by the statistical tables of χ^2 distribution with $k - 1$ degrees of freedom.

The value of χ^2 in our example was equal to 32.13. For $\alpha = 0.05$, $\chi^2_{0.05} = 11.07$ and, hence, the null hypothesis is rejected at 0.05 level.

B.13.6   Empirical Distribution Functions

We may also wish to estimate the parent probability distribution of a random sample directly, instead of verifying the hypothesis that the random sample comes from a specified distribution. As previously, we do not want to assume that the parent distribution belongs to a family of distributions defined by some parameters. The primary way of accomplishing this general task is to construct the so-called empirical distribution function (e.d.f.).

Let us consider a random sample of n independent random variables $X_1, X_2, \ldots, X_n$ with common c.d.f. $F(x)$. The *empirical distribution function* of the random sample is defined as

$$F_n(x) = \frac{1}{n} \sum_{i=1}^{n} I(X_i \leq x), \quad -\infty < x < \infty, \qquad (B.183)$$

where $I(X_i \leq x)$ is the *indicator function* of the event $\{X_i \leq x\}$,

$$I(X_i \leq x) = \begin{cases} 1 & \text{if } X_i \leq x \\ 0 & \text{if } X_i > x. \end{cases}$$

Equivalently,

$$F_n(x) = \frac{\text{number of sample points satisfying } X_i \leq x}{n}. \qquad (B.184)$$

The e.d.f. can be written in still another form, using the concept of order statistics. By the *order statistics* of the sample $X_1, X_2, \ldots, X_n$ we mean the sample rearranged in order from least to greatest. The ordered sample is written as $X_{(1)}, X_{(2)}, \ldots, X_{(n)}$, where $X_{(1)} \leq X_{(2)} \leq \ldots \leq X_{(n)}$, and $X_{(k)}$ is called the kth order statistic. If we want to emphasize that the kth order statistic comes from the sample of size n, we write $X_{k:n}$ instead of $X_{(k)}$. Now

$$F_n(x) = \begin{cases} 0 & \text{if } X_{1:n} > x \\ k/n & \text{if } X_{k:n} \leq x < X_{k+1:n}, \ k = 1, 2, \ldots, n-1 \\ 1 & \text{if } X_{n:n} \leq x. \end{cases} \qquad (B.185)$$

Thus, as is easily seen from the above definitions, the e.d.f. is a piecewise constant function with jumps at the observed data points $X_1, X_2, \ldots, X_n$. The jumps are of value $1/n$, unless more than one observation X_i assumes the same value (note that for a continuous parent distribution it happens with probability zero). One can object that, since the e.d.f. is necessarily a nonsmooth function, it is not an "elegant" estimator of any continuous parent c.d.f. $F(x)$. Yet, the closer examination of the properties of $F_n(x)$ shows that it is a good estimator of both discrete and continuous c.d.f.'s.

In order to appreciate the concept of the e.d.f., let us note first that, by (B.183) and for any fixed value of the argument x, $F_n(x)$ is the sample mean of n independent Bernoulli r.v.'s, each of which assumes value one with probability $P(X_i \leq x)$. But $P(X_i \leq x) = F(x)$, and it follows from the properties of the Bernoulli r.v.'s that

$$E(X_i) = F(x) \text{ and } \text{Var}(X_i) = F(x)(1 - F(x))$$

for each $i = 1, 2, \ldots, n$. Hence, by (B.62) and (B.63),

$$E(F_n(x)) = F(x) \text{ and } \text{Var}(F_n(x)) = \frac{F(x)(1 - F(x))}{n}. \qquad \text{(B.186)}$$

Moreover, the SLLN implies that $F_n(x)$ converges almost surely to $F(x)$ as n tends to infinity, while the CLT implies that, for large n, $F_n(x)$ is approximately normally distributed with mean $F(x)$ and variance $F(x)(1 - F(x))/n$.

Sometimes, we are interested in estimating not a whole c.d.f. but only the so-called quantiles of a parent distribution. For the sake of simplicity, let us assume for a moment that the random sample is governed by a continuous probability distribution with a c.d.f. $F(x)$ which is increasing between its boundary values 0 and 1. A value x_p such that $F(x_p) = p$ for any fixed $p\epsilon(0, 1)$ is called the *quantile* of order p of $F(x)$. Quantiles of orders .25 and .75 are called the *lower quartile* and *upper quartile*, respectively. Quantile of order .5 is called the *median*. Often, the quantile of order p is called the $(100p)$th *percentile*. It follows from (B.185) that a natural sample counterpart of the quantile of order p can be defined as

$$\tilde{x}_p = \begin{cases} X_{np:n} & \text{if } np \text{ is an integer} \\ X_{[np]+1:n} & \text{otherwise,} \end{cases} \qquad \text{(B.187)}$$

where, for any fixed a, $[a]$ denotes the largest integer not greater than a. Indeed,

$$F_n(\tilde{x}_p) = p$$

if np is an integer. If np is a fraction, $F_n(x)$ is never equal to p exactly. In fact, since $F_n(x)$ is a piecewise constant function, defining sample quantiles is always somewhat arbitrary. For example, as we shall see in the next subsection, a bit more natural definition of the sample median is available.

The properties discussed make the e.d.f. a widely used tool in statistical inference. It is another question, not to be dealt with here, that much work has been done to obtain smooth estimators of continuous c.d.f.'s (and p.d.f.'s).

B.13.7   Nonstandard Estimators of Mean and Variance

We know well from the previous considerations that the sample mean is a natural and good estimator of the population mean. It happens, however, that the sample we have at our disposal does not come exactly from the random phenomenon of our interest but, rather, that the probability distribution of the sample is contaminated by some other distribution. In such circumstances, we would wish to have an estimator which is likely to disregard the elements of the sample that come from the contaminating distribution. The sample mean usually does not perform this task well, while the so-called sample median quite often does. In particular, the concept of the sample median is very useful in the context of statistical process control. This point is briefly discussed below and elaborated at length in Chapter 3.

Throughout this subsection, we shall assume that the random sample is governed by a continuous probability distribution (with c.d.f. $F(x)$ and p.d.f. $f(x)$). For the sake of simplicity, we shall assume also that the c.d.f. $F(x)$ is increasing between its boundary values 0 and 1. The median of the distribution is defined as the number $\xi \equiv x_{.5}$ for which

$$F(\xi) = 1/2.$$

That is, half the area under the p.d.f. $f(x)$ is to the left of ξ and half is to the right of ξ (note that an additional condition is required to make ξ uniquely defined if $F(x)$ is not assumed to be increasing). It is easily seen that the mean and median are equal if the probability distribution is *symmetric*, i.e., if $F(x) = 1 - F(-x)$ or, equivalently, $f(x) = f(-x)$ for each x. Indeed, both are then equal to zero. It is also clear that the equality of the mean and median holds for probability distributions which arise from shifting a symmetric distribution by an arbitrary constant. We say that a probability distribution of the sample $X_1, X_2, \ldots, X_n$ is *symmetric about a constant* c if the common distribution of the r.v.'s $X_1 - c, X_2 - c, \ldots, X_n - c$ is symmetric. Of course, the mean and median are then equal to c. The most important family of such distributions is that of normal distributions. In general, however, the mean and median may differ (take, e.g., a unimodal skewed density).

A natural sample counterpart of the population median, slightly different from that implied by (B.187), is the following *sample median*

$$\tilde{\mu} = \begin{cases} X_{((n+1)/2)} & \text{if } n \text{ is odd} \\ 1/2(X_{(n/2)} + X_{(n/2+1)}) & \text{if } n \text{ is even,} \end{cases} \tag{B.188}$$

where n is the sample size and $X_{(k)}$ denotes the kth order statistic. Before we give the properties of the sample median, we have to derive some general results for arbitrary order statistics. In particular, the c.d.f. of the kth order statistic of a sample of size n, to be denoted $F_{(k)}(x)$, has the form

$$\begin{aligned} F_{(k)}(x) &= P(X_{(k)} \le x) \\ &= P(\text{at least } k \text{ of the } X_i\text{'s are not greater than } x) \\ &= \sum_{i=k}^{n} \binom{n}{i} F^i(x)(1 - F(x))^{n-i}, \end{aligned} \tag{B.189}$$

since each term in the summand is the binomial probability that exactly i of $X_1, X_2, \ldots, X_n$ are not greater than x. Hence, differentiating with respect to x, we obtain that the p.d.f. of the kth order statistic is given by

$$f_{(k)}(x) = \frac{n!}{(k-1)!(n-k)!} f(x)(F(x))^{k-1}(1 - F(x))^{n-k}. \tag{B.190}$$

Substituting the p.d.f. of the kth order statistic to (B.39) yields the mean of the statistic:

$$\mu_{k:n} = \frac{n!}{(k-1)!(n-k)!} \int_{-\infty}^{\infty} x f(x)(F(x))^{k-1}(1 - F(x))^{n-k} dx, \tag{B.191}$$

where $\mu_{k:n}$ denotes the mean of the kth order statistic $X_{k:n}$ of a sample of size n. After some tedious, and hence skipped, algebra one obtains that the following recurrence relation holds for the means of order statistics of samples from the same parent distribution:

$$(n - k)\mu_{k:n} + k\mu_{k+1:n} = n\mu_{k:n-1}, \tag{B.192}$$

where $k = 1, 2, \ldots, n - 1$.

Relations (B.191) and (B.192) enable us to prove that the mean of
the sample median is equal to the mean of a parent distribution of the
sample, provided that the parent distribution of the sample is symmetric
about some constant. In the proof, without loss of generality, we shall
confine ourselves to the case of $c = 0$, that is, of symmetric parent
distributions; by (B.18) and the definition of distribution's symmetry
about c, the proof will be accomplished for any c. Now, if the sample
size is odd, $\tilde{\mu} = X_{(n+1)/2:n}$ and, by (B.191), $\mu_{(n+1)/2:n} = 0$, since the
integrand in (B.191) is an odd function,

$$g(x) = -g(-x),$$

where $g(x) = xf(x)(F(x))^{(n-1)/2}(1 - F(x))^{(n-1)/2}$; indeed, x is the odd
function while, under the symmetry assumption, $f(x)$ and $[F(x)(1 -
F(x))]^{(n-1)/2}$ are even functions. But the mean of a symmetric distri-
bution is necessarily zero and, thus, the required result holds for odd n.
For n even and $k = n/2$, the result also readily follows, since the left
hand side of (B.192) is equal to $nE(\tilde{\mu})$, while the right hand side is equal
to zero by the same argument as previously, but applied to $\mu_{n/2:n-1}$.
Finally, therefore, the proof is concluded.

Given (B.190), it is obvious that calculation of the variance of $\tilde{\mu}$ for
particular parent distributions is a cumbersome task. We shall content
ourselves with citing a more general but only asymptotically valid result,
which says that, for large n, $\tilde{\mu}$ is approximately normally distributed
with mean μ and variance $1/(4nf^2(\xi))$, provided the parent p.d.f. is
continuous in a neighhborhood of ξ.

It readily follows from the last result that the sample median converges
in probability to the median, that is,

$$\lim_{n \to \infty} P(|\tilde{\mu}_n - \xi| > \varepsilon) = 0 \qquad (B.193)$$

for each positive ε, where $\tilde{\mu}_n$ denotes the sample median of a sample of
size n; more explicitly, (B.193) states that for each positive ε and δ
there exists an N such that

$$P(|\tilde{\mu}_n - \xi| > \varepsilon) < \delta$$

for all $n > N$. Without coming to technical details of a fully rigorous
proof, it suffices to observe that, in the limit, for $n = \infty$, the sample median is normally distributed with mean ξ and variance 0. In other words,
asymptotically, the probability distribution of $\tilde{\mu}_n$ becomes concentrated
at ξ, which implies (B.193).

The above proof of convergence in probability cannot be generalized in such a way as to answer the question of the almost sure convergence of $\tilde{\mu}_n$. Since convergence properties form an important ingredient of justifying the usefulness of any estimator, we shall now give another, simple and complete, proof of (B.193), which does not refer to any distributional properties of the sample median. A suitable modification of this latter proof answers the question of the a.s. convergence of $\tilde{\mu}_n$ as well.

For any fixed $\varepsilon > 0$, we have

$$F(\xi - \varepsilon) < F(\xi) = 1/2 < F(\xi + \varepsilon). \tag{B.194}$$

Let

$$\eta = \min\{1/2 - F(\xi - \varepsilon), F(\xi + \varepsilon) - 1/2\}. \tag{B.195}$$

Now, since the e.d.f. $F_n(x)$ converges almost surely to $F(x)$ for each x, we have

$$P(|F_n(\xi - \varepsilon) - F(\xi - \varepsilon)| < \eta/2) \geq 1 - \delta/2 \tag{B.196}$$

and

$$P(|F_n(\xi + \varepsilon) - F(\xi + \varepsilon)| < \eta/2) \geq 1 - \delta/2 \tag{B.197}$$

for any fixed δ and all sufficiently large n. Combining (B.195) - (B.197) yields

$$P(F_n(\xi - \varepsilon) < 1/2 < F_n(\xi + \varepsilon)) \geq 1 - \delta. \tag{B.198}$$

For n odd, it follows from (B.185), (B.198) and the definition of $\tilde{\mu}$ that

$$P(\xi - \varepsilon < \tilde{\mu} < \xi + \varepsilon) \geq 1 - \delta \tag{B.199}$$

for all sufficiently large odd n. A similar but more careful examination of (B.198) shows that property (B.199) holds for n even as well, which proves the desired result. Although we do not give it here, let us mention again that a slightly more refined argument shows that the sample median converges to the mean not only in probability but also almost surely.

In statistical process control, the following conceptual model is of interest. We are given a number of independent samples of random variables, each sample having the same size, say, n. Denote the number of samples by N. Within each sample, the random variables are independent and identically distributed. Ideally, the distribution of the random variables is the same for all N samples. Our goal is to estimate the mean of this parent, or "norm," distribution. Provided all N samples are indeed governed by the norm distribution, the given task is simple. The obvious way to solve it is to calculate the sample mean of all nN random variables. Equivalently, we can calculate N sample means of all the samples first, and then take the average of the sample means obtained. The average of the sample means can be viewed as the sample mean of the sample means for particular samples. Ideally, the estimate thus obtained is unbiased and, with N increasing, it converges to the true mean of the norm distribution if only the variance of the distribution is finite.

However, whenever we are interested in implementing statistical process control for quality improvement, it is because "sometimes something different from the norm is happening." Namely, the norm distribution of some samples is contaminated by another distribution. In order to take this fact into account, it is reasonable to assume that each sample is governed by the norm distribution only with some probability less than one, say, with probability p, and is governed by some other distribution with probability $1 - p$. The other distribution is a mixture of the norm distribution and a contaminating distribution. Assume that the mean of the norm distribution is μ_1, while that of the other distribution is μ_2, $\mu_1 \neq \mu_2$. Assume also that the variances of both distributions are finite. Now, the average of the sample means is an unbiased estimator of $p\mu_1 + (1 - p)\mu_2$ and, as N increases, it converges to $p\mu_1 + (1 - p)\mu_2$, although we would wish it to converge to μ_1. To the worse, it does not help if we can make the size of the samples, n, arbitrarily large. With n increasing, each sample mean converges to the true mean of the sample, which is either equal to μ_1 or to μ_2, and the average of the sample means still converges to $p\mu_1 + (1 - p)\mu_2$. The reason for this last fact is that, for N large, the sample means obtained are approximately equal to μ_1 about pN times and to μ_2 about $(1 - p)N$ times. At the same time, precisely for the same reason, the sample median of the sample means must converge to μ_1 as n and N both increase, provided only that $p > .5$. This important result follows from the fact that pN is greater than $(1 - p)N$, the sample median is chosen from N data and, in the limit (as n and N both approach infinity), pN data are equal to μ_1 while $(1 - p)N$ are equal to μ_2. Intuitively, for sufficiently large n and N, more than half the data are concentrated around μ_1 and are separated from the data concentrated around μ_2 with probability arbitrarily close to one. In view of the definition of the sample median, this implies the required result.

A rigorous argument is slightly more involved. For each fixed n and for each sample, sample mean $\bar{X}$ is, of course, a random variable. We

shall show first that $\bar{X}$ can be made arbitrarily close to μ_1 with probability greater than .5 by choosing a sufficiently large sample size. Given
that a sample is governed by the norm distribution, it follows from the
Chebyshev's inequality (see (B.74)) with $\varepsilon = 1/n^{1/4}$ that the probability
that

$$\bar{X} \; \epsilon [\mu_1 - 1/n^{1/4}, \mu_1 + 1/n^{1/4}] \qquad (B.200)$$

is greater than $1 - \sigma_1^2/\sqrt{n}$, where σ_1^2 denotes the variance of the norm
distribution. Now, since the probability that a sample is governed by
the norm distribution is p, the unconditional probability of event (B.200)
satisfies the following inequality

$$P(\bar{X} \; \epsilon [\mu_1 - 1/n^{1/4}, \mu_1 + 1/n^{1/4}]) > p(1 - \sigma_1^2/\sqrt{n}) \qquad (B.201)$$

and, for n sufficiently large, the right hand side of (B.201) is indeed
greater than .5, since p was assumed to be greater than .5. Hence,
we obtain by the definition of the median of a probability distribution
that, for all such n, the median of the (unconditional) distribution of the
sample mean belongs to the interval $[\mu_1 - 1/n^{1/4}, \mu_1 + 1/n^{1/4}]$. But this
implies in turn that the sample median of N sample means converges,
as $N \to \infty$, to a value in this interval as well. Finally, therefore, unlike
the sample mean, the sample median is indeed insensitive to the model's
departures from the norm in the sense described.

The only measure of "spread" or "variability" of a probability distribution discussed so far has been the variance of the distribution. The
sample counterpart of the variance is, of course, the sample variance.
Another natural candidate for becoming a measure of sample's spread is
the *range* of the sample, defined as the difference between the last and
the first order statistic,

$$R = X_{n:n} - X_{1:n}. \qquad (B.202)$$

No doubt, the range provides the simplest possible way of measuring the
spread of the parent probability distribution. The interesting question is
what is the relationship between the variance and the range or, since the
range is a sample characteristic and, hence, is itself a random variable,
between the variance and the mean of the range.

Just as we derived the p.d.f. of the kth order statistic, one can solve
the more ambitious task of giving the joint p.d.f. of two order statistics
of the same sample, $X_{k:n}$ and $X_{m:n}$, say. In particular, one can derive
the joint p.d.f. of the first and the last order statistic, $X_{1:n}$ and $X_{n:n}$,

respectively. Using the change-of-variable technique, one can then derive the p.d.f. of the range R. Let us omit all the tedious calculations and state only the form of the p.d.f. of the sample's range:

$$f_r(y) = n(n-1) \int_{-\infty}^{\infty} f(x)(F(x+y) - F(x))^{n-2} f(x+y) dx, \quad \text{(B.203)}$$

where, as usual, $F(x)$ and $f(x)$ denote the parent c.d.f. and p.d.f., respectively. Integrating (B.203) yields the c.d.f. of the range

$$\begin{aligned} F_r(y) &= \int_{-\infty}^{y} f_r(t) dt \\ &= n \int_{-\infty}^{\infty} f(x)(F(x+y) - F(x))^{n-1} dx. \end{aligned} \quad \text{(B.204)}$$

We are now in a position to prove that the mean of the range of a sample of normal r.v.'s with variance σ^2 is linearly related to the square root of σ^2, i.e., to the standard deviation of the sample,

$$E(R) = \bar{R} = \sigma/b, \quad \text{(B.205)}$$

where b depends on the sample size. It follows that, for normal samples, the square root of the sample variance can be replaced by a suitable multiple of the range. To prove (B.205), let us consider a sample of n normally distributed r.v.'s with mean μ and variance σ^2, $X_1, X_2, \ldots, X_n$. We have

$$F_r(y) = P(R \leq y) = P(R/\sigma \leq y/\sigma) = P(R' \leq y/\sigma), \quad \text{(B.206)}$$

where R is given by (B.202) and $R' = R/\sigma$. Observe that the transformed sample,

$$\frac{X_1 - \mu}{\sigma}, \frac{X_2 - \mu}{\sigma}, \ldots, \frac{X_n - \mu}{\sigma},$$

comes from the standard normal distribution and that R' is its range. By (B.206), if we want the c.d.f. of the range of the original sample, we can compute $P(R' \leq r/\sigma)$ for the transformed sample. Thus, by (B.204),

$$\begin{aligned} F_r(y) &= P(R' \leq y/\sigma) \\ &= n \int_{-\infty}^{\infty} \varphi(x)(\Phi(x + y/\sigma) - \Phi(x))^{n-1} dx, \end{aligned} \quad \text{(B.207)}$$

where $\varphi(x)$ and $\Phi(x)$ denote the p.d.f. and c.d.f. of the standard normal distribution, respectively. Upon differentiating (B.207) and substituting u for y/σ, we have

$$
\begin{aligned}
E(R) &= n(n-1)\sigma^{-1} \int_0^\infty \int_{-\infty}^\infty y\varphi(x)(\Phi(x+y/\sigma) - \Phi(x))^{n-2} \\
&\quad \times \varphi(x+y/\sigma)dxdy \\
&= n(n-1)\sigma \int_0^\infty \int_{-\infty}^\infty u\varphi(x)(\Phi(x+u) - \Phi(x))^{n-2} \\
&\quad \times \varphi(x+u)dxdu.
\end{aligned} \tag{B.208}
$$

Therefore, relation (B.205) is proved. Moreover, we have obtained that b is equal to the inverse of

$$
n(n-1) \int_0^\infty \int_{-\infty}^\infty u\varphi(x)(\Phi(x+u) - \Phi(x))^{n-2}\varphi(x+u)dxdu. \tag{B.209}
$$

The values of b for different sample sizes n have to be calculated numerically. They are given in Table 3.1 for sample sizes $2, 3, \ldots, 10$, 15 and 20. Of course, the sample range is easier for calculation than the sample variance. The price to be paid is that the variance of the former is greater than that of the latter. The two variances can be considered comparable only for small sample sizes, say, up to $n = 10$.

B.14 Bayesian Statistics

B.14.1 Bayes Theorem

Suppose that the sample space S can be written as the union of disjoint sets: $S = A_1 \cup A_2 \cup \cdots \cup A_n$. Let the event H be a subset of S which has non-empty intersections with some of the A_i's. Then

$$
P(A_i|H) = \frac{P(H|A_i)P(A_i)}{P(H|A_1)P(A_1) + P(H|A_2)P(A_2) + \cdots + P(H|A_n)P(A_n)}. \tag{B.210}
$$

To explain the conditional probability given by equation (B.210), consider a diagram of the sample space, S. Consider that the A_i's represent n disjoint states of nature. The event H intersects some of the A_i's.

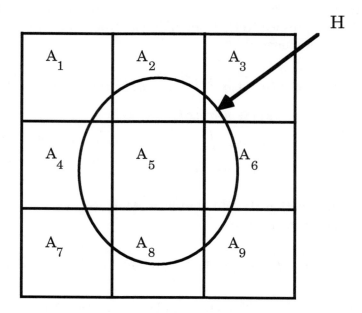

Figure B.5. Bayes Venn Diagram.

Then,

$$P(H|A_1) = \frac{P(H \cap A_1)}{P(A_1)} = \frac{P(A_1|H)P(H)}{P(A_1)}.$$

Solving for $P(A_1|H)$, we get

$$P(A_1|H) = \frac{P(H|A_1)P(A_1)}{P(H)},$$

and in general,

$$P(A_i|H) = \frac{P(H|A_i)P(A_i)}{P(H)}. \tag{B.211}$$

Now,

$$
\begin{aligned}
P(H) &= P((H \cap A_1) \cup (H \cap A_2) \cup \cdot \cup (H \cap A_n)) \\
&= \sum P(H \cap A_i), \text{ since the intersections } (H \cap A_i) \text{ are disjoint} \\
&= \sum P(H|A_i)P(A_i), \text{ where } j = 1, 2, \ldots, n.
\end{aligned}
$$

Thus, with (B.211) and $P(H)$ given as above, we get (B.210).

The formula (B.210) finds the probability that the true state of nature is A_i given that H is observed. Notice that the probabilities $P(A_i)$ must be known to find $P(A_i|H)$. These probabilities are called *prior* probabilities because they represent information prior to experimental data. The $P(A_i|H)$ are then *posterior* probabilities. For each $i = 1, 2, \ldots, n$, $P(A_i|H)$ is the probability that A_i was the state of nature in light of the occurrence of the event H.

B.14.2 A Diagnostic Example

Consider patients being tested for a particular disease. It is known from historical data that 5% of the patients tested have the disease, further, that 10% of the patients that have the disease test negative for the disease, and that 20% of the patients who do not have the disease test positive for the disease. Denote by D^+ the event that the patient has the disease, by D^- the event that the patient does not, and denote by T^+ the event the patient tests positive for the disease, and by T^- the event the patient tests negative.

If a patient tests positive for the disease, what is the probability that the patient actually has the disease? We seek the conditional probability, $P(D^+|T^+)$. Here, T^+ is the observed event, and D^+ may be the true state of nature that exists prior to the test. (We trust that the test does not cause the disease.) Using Bayes's theorem,

$$P(D^+|T^+) \quad = \quad \frac{P(T^+|D^+)P(D^+)}{P(T^+)} \qquad (B.212)$$

$$= \quad \frac{P(T^+|D^+)P(D^+)}{P(T^+|D^+)P(D^+) + P(T^+|D^-)P(D^-)}$$

$$= \quad \frac{.9 \times .05}{.9 \times .05 + .2 \times .95} = 0.1915.$$

Thus, there is nearly a 20% chance given a positive test result that the patient has the disease. This probability is the posterior probability, and if the patient is tested again, we can use it as the new prior probability. If the patient tests positive once more, we use equation (B.212) with an updated version of $P(D^+)$, namely, .1915. The posterior probability now is:

$$P(D^+|T^+) \quad = \quad \frac{P(T^+|D^+)P(D^+)}{P(T^+)}$$

$$= \frac{P(T^+|D^+)P(D^+)}{P(T^+|D^+)P(D^+) + P(T^+|D^-)P(D^-)}$$

$$= \frac{.9 \times .1915}{.9 \times .1915 + .2 \times .8085} = 0.5159.$$

Twice the patient tests positive for the disease and the posterior probability that the patient has the disease is now much higher. As we gather more and more information with further tests, our posterior probabilities will better and better describe the true state of nature.

In order to find the posterior probabilities as we have done, we needed to know the prior probabilities. A major concern in a Bayes application is the choice of priors, a choice which must be made sometimes with very little prior information. One suggestion made by Bayes is to assume that the n states of nature are equally likely (Bayes' Axiom). If we make this assumption in the example above, that is, that $P(D^+) = P(D^-) = .5$, then

$$P(D^+|T^+) = \frac{P(T^+|D^+)P(D^+)}{P(T^+|D^+)P(D^+) + P(T^+|D^-)P(D^-)}$$

$P(D^+)$ and $P(D^-)$ cancel, giving

$$P(D^+|T^+) = \frac{P(T^+|D^+)}{P(T^+|D^+) + P(T^+|D^-)}$$

$$= \frac{.9}{.9 + .2}$$

$$= 0.8182.$$

This is much higher than the accurate probability, .1912. Depending upon the type of decisions an analyst has to make, a discrepancy of this magnitude may be very serious indeed. As more information is obtained, however, the effect of the initial choice of priors will become less severe.

B.14.3 Prior and Posterior Density Functions

We have defined a parameter of a distribution to be a quantity that describes it. Examples include the mean μ and variance σ^2 of the normal distribution, and the number of trials n and the probability of success p of the binomial distribution. A great deal of statistical inference considers parameters to be fixed quantities, and sample data are used to estimate a parameter that is fixed but unknown. A Bayesian approach, however, supposes that a parameter of a distribution varies according to some distribution of its own. Observations taken on a random variable may then permit (better) estimates of the underlying true *distribution* of the parameter.

Suppose that the p.d.f. of X depends on the parameter θ. It is reasonable to write this as the conditional p.d.f. of X given θ: $f(x|\theta)$. Suppose further that θ has p.d.f. given by $g(\theta)$. The p.d.f. $g(\theta)$ is a density function for the *prior* distribution of θ, since it describes the distribution of θ before any observations on X are known. Bayes Theorem gives us a means to find the conditional density function of θ given X, $g(\theta|x)$. This is known as the *posterior* p.d.f. of θ, since it is derived after observation $X = x$ is taken. According to Bayes Theorem, we have for $f(x) > 0$

$$g(\theta|x) \;=\; \frac{f(x|\theta)g(\theta)}{f_1(x)} \;,\; \text{where} \tag{B.213}$$

$$f_1(x) = \int_{-\infty}^{\infty} f(x|\theta)g(\theta)d\theta$$

is the marginal (unconditional) p.d.f. of X. Note that $f(x|\theta)g(\theta)$ is the joint distribution of X and θ.

If a random sample of size n is observed, the prior distribution on θ can be updated to incorporate the new information gained from the sample. That is, the posterior density function of θ is derived from the sample data as a conditional density of θ given that the sample $(x_1, x_2, \ldots, x_n)$ is observed:

$$g(\theta|(x_1, x_2, \ldots, x_n)) = \frac{f((x_1, x_2, \ldots, x_n)|\theta)g(\theta)}{h(x_1, x_2, \ldots, x_n)}, \tag{B.214}$$

where the marginal density $h(x_1, x_2, \ldots, x_n)$ is found by integrating the numerator of equation (B.214) over θ. If the density of the random variable X depends upon more than one parameter, then we let $\Theta = (\theta_1, \theta_2, \ldots, \theta_k)$ be the vector of parameters governing the distribution of X, and equation (B.214) becomes (with $\mathbf{x} = (x_1, x_2, \ldots, x_n)$),

$$g(\Theta|\mathbf{x}) = \frac{f(\mathbf{x}|\Theta)g(\Theta)}{h(\mathbf{x})}.$$

We find the marginal $h(\mathbf{x})$ by

$$h(x_1, x_2, \ldots, x_n)$$
$$= \int_{\theta_1} \int_{\theta_2} \cdots \int_{\theta_k} f(\mathbf{x}|\Theta) g\Theta) d\theta_1 d\theta_2 \ldots d\theta_k.$$

The apparent difficulty in finding a posterior density from sample data is that the prior density function must be known. This is a problem similar to that addressed in example B.14.2. In that example, we found the posterior probability that a patient who tests positive for a disease actually has the disease. We used prior probabilities that were based on historical data, and so were *informed* priors. We considered a second case for which no historical data were available, so that *noninformed* priors had to be selected, and we got a posterior probability very different from that in the original (informed) case.

We are confronted by the same problem when looking for posterior densities. We must specify a prior distribution $g(\theta)$ that best describes the distribution of θ based on available knowledge. The available knowledge in fact may be only expert opinions and "best" guesses.

B.14.4   Example: Priors for Failure Rate of a Poisson Process

Suppose we have a device which fails at a rate of θ according to a Poisson process. If X is the number of failures in an interval of length t, then X is a Poisson random variable with probability function

$$P[X = x] = f(x) = \frac{e^{-\theta t}(\theta t)^x}{x!}.$$

In the Bayesian setting, we write this as the conditional probability function $f(x|\theta)$ and suppose that θ has prior density $g(\theta)$. What we seek in $g(\theta)$ is a density function that uses all of the prior knowledge (or prior belief) of the distribution of the failure rate, θ. If we had a pretty good idea that θ would not exceed some maximum value, say $\theta_{max} = b$, then we might assume that θ is distributed *uniformly* over the interval, $(0, b)$. Then

$$g(\theta) = \frac{1}{b} \text{ for } 0 < \theta < b$$
$$= 0 \text{ otherwise.}$$

Using (B.213),

$$g(\theta|x) = \frac{(e^{-\theta t}(\theta t)^x/x!)(1/\theta_{max})}{\int_0^b ((e^{-\theta t}(\theta t)^x)/x!)(1/b)d\theta}$$

$$= \frac{(e^{-\theta t}\theta^x)(1/b)}{(1/b)\int_0^b e^{-\theta t}\theta^x d\theta}$$

$$= \frac{(e^{-\theta t}\theta^x)}{\int_0^b e^{-\theta t}\theta^x d\theta}.$$

If we make the change of variable $y = \theta t$ in the denominator, we get

$$= \frac{(e^{-\theta t}\theta^x)}{\int_0^{bt} e^{-y}\frac{y^x}{t^x}\frac{dy}{t}}$$

$$= \frac{e^{-\theta t}\theta^x}{\frac{1}{t^{x+1}}\Gamma(x+1,bt)}$$

$$= \frac{t^{x+1}e^{-\theta t}\theta^x}{\Gamma(x+1,bt)}, \tag{B.215}$$

where $\Gamma(x+1,\theta t)$ is an "incomplete" gamma function. (The complete gamma function is

$$\Gamma(\alpha,\infty) = \Gamma(\alpha) = \int_0^\infty e^{-y}y^{\alpha-1}dy, \text{ for } \alpha > 0.)$$

It is clear from (B.215) that a uniform prior does not yield a uniform posterior. A more widely used prior for θ is the gamma density, $G(\alpha,\beta)$, introduced in Section B.6.2. If $g(\theta)$ is a gamma density function, then we will see that the posterior density of θ given X is also a gamma density function. For this reason, the gamma distribution is considered the "natural" choice for a prior density of θ.

If X is a Poisson random variable with parameter $\nu = \theta t$, then the *natural conjugate* prior density function of θ is a gamma density function with parameters $\alpha, \beta > 0$,

$$G(\alpha,\beta) = \frac{1}{\Gamma(\alpha)\beta^\alpha}\theta^{\alpha-1}e^{-\theta/\beta}.$$

Using (B.213), we find the posterior density $g(\theta|x)$ when $g(\theta) = G(\alpha,\beta)$:

$$g(\theta|x) = \frac{(e^{-\theta t}(\theta t)^x/x!)g(\theta)}{\int_0^\infty (e^{-\theta t}(\theta t)^x/x!)g(\theta)d\theta}$$

$$= \frac{(e^{-\theta t}(\theta)^x)\frac{e^{-\theta/\beta}\theta^{\alpha-1}}{\Gamma(\alpha)\beta^\alpha}}{\frac{1}{\Gamma(\alpha)\beta^\alpha}\int_0^\infty e^{-\theta t}\theta^x e^{-\theta/\beta}\theta^{\alpha-1}d\theta}$$

$$= \frac{e^{-\theta(t+1/\beta)}\theta^{x+\alpha-1}}{\int_0^\infty e^{-\theta(t+1/\beta)}\theta^{x+\alpha-1}d\theta}.$$

Making the change of variable $y = \theta(t + \frac{1}{\beta})$ in the denominator, we get,

$$\int_0^\infty e^{-\theta(t+1/\beta)}\theta^{x+\alpha-1}d\theta$$

$$= \int_0^\infty e^{-y}(\frac{y}{t+\frac{1}{\beta}})^{x+\alpha-1}\frac{dy}{(t+\frac{1}{\beta})}$$

$$= \int_0^\infty \frac{1}{(t+\frac{1}{\beta})^{x+\alpha}}e^{-y}y^{x+\alpha-1}dy$$

$$= \frac{\Gamma(x+\alpha)}{(t+\frac{1}{\beta})^{x+\alpha}}.$$

(Be careful to note that this is not quite the marginal distribution of X, since we cancelled common factors before evaluating the denominator.) Now, the posterior density is

$$g(\theta|x) = \frac{e^{-\theta(t+\frac{1}{\beta})}\theta^{x+\alpha-1}}{\Gamma(x+\alpha)(t+\frac{1}{\beta})^{-(x+\alpha)}},$$

which is a gamma density function with parameters $\alpha_{new} = \alpha + x$ and $\beta_{new} = (t + \frac{1}{\beta})^{-1}$. In this case, we can see how the data truly update the prior density in the sense that it does not change its functional form.

Recall that if θ is $G(\alpha, \beta)$, then $E[\theta] = \alpha\beta$, and $V[\theta] = \alpha\beta^2$. If we have some estimates of the prior mean and prior variance, then the parameters can be chosen by solving the system

$$\alpha\beta = \text{(mean estimate)}$$
$$\alpha\beta^2 = \text{(variance estimate)}.$$

Since it may be difficult to pin down a best guess of the variance, two other methods might be used with better results. These are described fully in Martz and Waller [2], and were first presented in the papers, Martz and Waller [1] and Waller et al. [3].

The first method, from Martz and Waller [2], requires that an expert provide an upper limit (UL) and a lower limit (LL) for the failure rate, θ, such that

$$P[\theta < LL] = P[\theta > UL] = \frac{1 - p_o}{2},$$

where p_o is one of the values, .80, .90, or .95. (These are usual choices, since $p_o = .80$ gives 10th and 90th percentiles for LL and UL respectively, $p_o = .90$ gives the 5th and 95th percentiles, and $p_o = .95$ gives the 2.5th and 97.5th percentiles.) In general, LL is the $50(1-p_o)$th percentile, and UL is the $50(1 + p_o)$th percentile.

Then

$$
\begin{aligned}
P[LL < \theta < UL] &= \int_{LL}^{UL} g(\theta; \alpha, \beta) d\theta \\
&= \int_{LL}^{UL} \frac{1}{\Gamma(\alpha)\beta^\alpha} \theta^{\alpha-1} e^{-\theta/\beta} d\theta = p_o, \quad \text{(B.216)}
\end{aligned}
$$

or

$$
\begin{aligned}
\int_0^{LL} g(\theta; \alpha, \beta) d\theta &= \int_{UL}^{\infty} g(\theta; \alpha, \beta) d\theta \\
&= \frac{1 - p_o}{2}.
\end{aligned}
$$

By setting $LL/\beta = 1$, equation (B.216) can be rewritten as

$$\int_{LL}^{UL} \frac{1}{\Gamma(\alpha)(LL)^\alpha} \theta^{\alpha-1} e^{-\theta/(LL)} d\theta = p_o ,$$

and making the change of variable, $y = \theta/LL$, we get the integral

$$\int_1^{UL/LL} \frac{1}{\Gamma(\alpha)(LL)^\alpha} (LLy)^{\alpha-1} e^{-y} LL\, dy = p_o,$$

which is equal to

$$\int_1^{UL/LL} \frac{1}{\Gamma(\alpha)} y^{\alpha-1} e^{-y} dy = p_o . \quad \text{(B.217)}$$

The equation (B.217) is numerically solved for α. Once a value, α_o, is determined, a temporary lower limit, LL_o, is set, and equation (B.216) is solved for a temporary value of β, b_o. Then since β is a scale parameter, it can be found for any other LL from

$$\frac{\beta}{LL} = \frac{b_o}{LL_o} \Rightarrow \beta = \frac{LL}{LL_o}b_o.$$

A graph of α_o versus $\log(UL/LL)$ and a table of b_o versus α_o and p_o are given in ([2], pp. 700-705).

The second method, presented in [3], requires that any two percentiles, θ_1 and θ_2, are specified, so that

$$P[\theta < \theta_1] = p_1, \text{ and } P[\theta < \theta_2] = p_2.$$

These give two equations in two unknowns (α_o, β_o):

$$
\begin{aligned}
\int_0^{\theta_1} \frac{1}{\Gamma(\alpha)\beta^\alpha}e^{-x/\beta}x^{\alpha-1}dx &= p_1 \\
\int_0^{\theta_2} \frac{1}{\Gamma(\alpha)\beta^\alpha}e^{-x/\beta}x^{\alpha-1}dx &= p_2.
\end{aligned}
\tag{B.218}
$$

These are solved simultaneously for (α, β). A pair which simultaneously satisfies (B.218) is found by overlaying graphs of α_i and β_i for specified values θ_i and p_i, $(i = 1, 2)$. Such graphs are found in [3], and ([2], 707-712).

References

[1] Martz, Harry F. and Waller, R. A.(1979). "A Bayesian Zero-Failures (BAZE) Reliability Analysis," *Journal of Quality Technology*, v.11., pp. 128-138.

[2] Martz, H. F. and Waller, R. A. (1982). *Bayesian Reliability Analysis*, New York: John Wiley & Sons.

[3] Waller, R. A., Johnson, M.M., Waterman, M.S. and Martz, H.F. (1977). "Gamma Prior Distribution Selection for Bayesian Analysis of Failure Rate and Reliability," in *Nuclear Systems Reliability Engineering and Risk Assessment*, Philadelphia: SIAM, pp. 584-606.

Appendix C
Statistical Tables

APPENDIX C

C.1. Table of the Normal Distribution

	Values of $\frac{1}{\sqrt{2\pi}} \int_{-\infty}^{z} e^{-\frac{t^2}{2}} dt$									
z	.0	.01	.02	.03	.04	.05	.06	.07	.08	.09
.0	.50000	.50399	.50798	.51197	.51595	.51994	.52392	.52790	.53188	.53586
.1	.53983	.54380	.54776	.55172	.55567	.55962	.56356	.56749	.57142	.57535
.2	.57926	.58317	.58706	.59095	.59483	.59871	.60257	.60642	.61026	.61409
.3	.61791	.62172	.62552	.62930	.63307	.63683	.64058	.64431	.64803	.65173
.4	.65542	.65910	.66276	.66640	.67003	.67364	.67724	.68082	.68439	.68793
.5	.69146	.69497	.69847	.70194	.70540	.70884	.71226	.71566	.71904	.72240
.6	.72575	.72907	.73237	.73565	.73891	.74215	.74537	.74857	.75175	.75490
.7	.75804	.76115	.76424	.76730	.77035	.77337	.77637	.77935	.78230	.78524
.8	.78814	.79103	.79389	.79673	.79955	.80234	.80511	.80785	.81057	.81327
.9	.81594	.81859	.82121	.82381	.82639	.82894	.83147	.83398	.83646	.83891
1.0	.84134	.84375	.84614	.84849	.85083	.85314	.85543	.85769	.85993	.86214
1.1	.86433	.86650	.86864	.87076	.87286	.87493	.87698	.87900	.88100	.88298
1.2	.88493	.88686	.88877	.89065	.89251	.89435	.89617	.89796	.89973	.90147
1.3	.90320	.90490	.90658	.90824	.90988	.91149	.91309	.91466	.91621	.91774
1.4	.91924	.92073	.92220	.92364	.92507	.92647	.92785	.92922	.93056	.93189
1.5	.93319	.93448	.93574	.93699	.93822	.93943	.94062	.94179	.94295	.94408
1.6	.94520	.94630	.94738	.94845	.94950	.95053	.95154	.95254	.95352	.95449
1.7	.95543	.95637	.95728	.95818	.95907	.95994	.96080	.96164	.96246	.96327
1.8	.96407	.96485	.96562	.96638	.96712	.96784	.96856	.96926	.96995	.97062
1.9	.97128	.97193	.97257	.97320	.97381	.97441	.97500	.97558	.97615	.97670
2.0	.97725	.97778	.97831	.97882	.97932	.97982	.98030	.98077	.98124	.98169
2.1	.98214	.98257	.98300	.98341	.98382	.98422	.98461	.98500	.98537	.98574
2.2	.98610	.98645	.98679	.98713	.98745	.98778	.98809	.98840	.98870	.98899
2.3	.98928	.98956	.98983	.99010	.99036	.99061	.99086	.99111	.99134	.99158
2.4	.99180	.99202	.99224	.99245	.99266	.99286	.99305	.99324	.99343	.99361
2.5	.99379	.99396	.99413	.99430	.99446	.99461	.99477	.99492	.99506	.99520
2.6	.99534	.99547	.99560	.99573	.99585	.99598	.99609	.99621	.99632	.99643
2.7	.99653	.99664	.99674	.99683	.99693	.99702	.99711	.99720	.99728	.99736
2.8	.99744	.99752	.99760	.99767	.99774	.99781	.99788	.99795	.99801	.99807
2.9	.99813	.99819	.99825	.99831	.99836	.99841	.99846	.99851	.99856	.99861
3.0	.99865	.99869	.99874	.99878	.99882	.99886	.99889	.99893	.99896	.99900
3.1	.99903	.99906	.99910	.99913	.99916	.99918	.99921	.99924	.99926	.99929
3.2	.99931	.99934	.99936	.99938	.99940	.99942	.99944	.99946	.99948	.99950
3.3	.99952	.99953	.99955	.99957	.99958	.99960	.99961	.99962	.99964	.99965
3.4	.99966	.99968	.99969	.99970	.99971	.99972	.99973	.99974	.99975	.99976
3.5	.99977	.99978	.99978	.99979	.99980	.99981	.99981	.99982	.99983	.99983
3.6	.99984	.99985	.99985	.99986	.99986	.99987	.99987	.99988	.99988	.99989
3.7	.99989	.99990	.99990	.99990	.99991	.99991	.99992	.99992	.99992	.99992
3.8	.99993	.99993	.99993	.99994	.99994	.99994	.99994	.99995	.99995	.99995
3.9	.99995	.99995	.99996	.99996	.99996	.99996	.99996	.99996	.99997	.99997

C.2. Table of the Chi-Square Distribution

ν	Critical Values of $P = \frac{1}{2^{\nu/2}\Gamma(\nu/2)} \int_0^{\chi^2} x^{\nu/2-1} e^{-x/2} dx$										
	0.100	0.250	0.500	0.750	0.900	0.950	0.975	0.990	0.995	0.998	0.999
1	0.016	0.102	0.455	1.323	2.706	3.841	5.024	6.635	7.879	9.550	10.828
2	0.211	0.575	1.386	2.773	4.605	5.991	7.378	9.210	10.597	12.429	13.816
3	0.584	1.213	2.366	4.108	6.251	7.815	9.348	11.345	12.838	14.796	16.266
4	1.064	1.923	3.357	5.385	7.779	9.488	11.143	13.277	14.860	16.924	18.467
5	1.610	2.675	4.351	6.626	9.236	11.070	12.833	15.086	16.750	18.907	20.515
6	2.204	3.455	5.348	7.841	10.645	12.592	14.449	16.812	18.548	20.791	22.458
7	2.833	4.255	6.346	9.037	12.017	14.067	16.013	18.475	20.278	22.601	24.322
8	3.490	5.071	7.344	10.219	13.362	15.507	17.535	20.090	21.955	24.352	26.124
9	4.168	5.899	8.343	11.389	14.684	16.919	19.023	21.666	23.589	26.056	27.877
10	4.865	6.737	9.342	12.549	15.987	18.307	20.483	23.209	25.188	27.722	29.588
11	5.578	7.584	10.341	13.701	17.275	19.675	21.920	24.725	26.757	29.354	31.264
12	6.304	8.438	11.340	14.845	18.549	21.026	23.337	26.217	28.300	30.957	32.909
13	7.042	9.299	12.340	15.984	19.812	22.362	24.736	27.688	29.819	32.535	34.528
14	7.790	10.165	13.339	17.117	21.064	23.685	26.119	29.141	31.319	34.091	36.123
15	8.547	11.037	14.339	18.245	22.307	24.996	27.488	30.578	32.801	35.628	37.697
16	9.312	11.912	15.338	19.369	23.542	26.296	28.845	32	34.267	37.146	39.252
17	10.085	12.792	16.338	20.489	24.769	27.587	30.191	33.409	35.718	38.648	40.790
18	10.865	13.675	17.338	21.605	25.989	28.869	31.526	34.805	37.156	40.136	42.312
19	11.651	14.562	18.338	22.718	27.204	30.144	32.852	36.191	38.582	41.610	43.820
20	12.443	15.452	19.337	23.828	28.412	31.410	34.170	37.566	39.997	43.072	45.315
21	13.240	16.344	20.337	24.935	29.615	32.671	35.479	38.932	41.401	44.522	46.797
22	14.041	17.240	21.337	26.039	30.813	33.924	36.781	40.289	42.796	45.962	48.268
23	14.848	18.137	22.337	27.141	32.007	35.172	38.076	41.638	44.181	47.391	49.728
24	15.659	19.037	23.337	28.241	33.196	36.415	39.364	42.980	45.559	48.812	51.179
25	16.473	19.939	24.337	29.339	34.382	37.652	40.646	44.314	46.928	50.223	52.620
26	17.292	20.843	25.336	30.435	35.563	38.885	41.923	45.642	48.290	51.627	54.052
27	18.114	21.749	26.336	31.528	36.741	40.113	43.195	46.963	49.645	53.023	55.476
28	18.939	22.657	27.336	32.620	37.916	41.337	44.461	48.278	50.993	54.411	56.892
29	19.768	23.567	28.336	33.711	39.087	42.557	45.722	49.588	52.336	55.792	58.301
30	20.599	24.478	29.336	34.800	40.256	43.773	46.979	50.892	53.672	57.167	59.703

C.3. Table of Student's t Distribution

Critical Values of $P = \frac{\Gamma[(\nu+1)/2]}{\Gamma(\nu/2)\sqrt{\pi\nu}} \int_{-\infty}^{t} (1 + \frac{t^2}{\nu})^{-(\nu+1)/2} dt$								
ν	0.600	0.750	0.900	0.950	0.975	0.990	0.995	0.998	0.999
1	0.325	1	3.078	6.314	12.706	31.821	63.657	159.153	318.309
2	0.289	0.816	1.886	2.920	4.303	6.965	9.925	15.764	22.327
3	0.277	0.765	1.638	2.353	3.182	4.541	5.841	8.053	10.215
4	0.271	0.741	1.533	2.132	2.776	3.747	4.604	5.951	7.173
5	0.267	0.727	1.476	2.015	2.571	3.365	4.032	5.030	5.893
6	0.265	0.718	1.440	1.943	2.447	3.143	3.707	4.524	5.208
7	0.263	0.711	1.415	1.895	2.365	2.998	3.499	4.207	4.785
8	0.262	0.706	1.397	1.860	2.306	2.896	3.355	3.991	4.501
9	0.261	0.703	1.383	1.833	2.262	2.821	3.250	3.835	4.297
10	0.260	0.700	1.372	1.812	2.228	2.764	3.169	3.716	4.144
11	0.260	0.697	1.363	1.796	2.201	2.718	3.106	3.624	4.025
12	0.259	0.695	1.356	1.782	2.179	2.681	3.055	3.550	3.930
13	0.259	0.694	1.350	1.771	2.160	2.650	3.012	3.489	3.852
14	0.258	0.692	1.345	1.761	2.145	2.624	2.977	3.438	3.787
15	0.258	0.691	1.341	1.753	2.131	2.602	2.947	3.395	3.733
16	0.258	0.690	1.337	1.746	2.120	2.583	2.921	3.358	3.686
17	0.257	0.689	1.333	1.740	2.110	2.567	2.898	3.326	3.646
18	0.257	0.688	1.330	1.734	2.101	2.552	2.878	3.298	3.610
19	0.257	0.688	1.328	1.729	2.093	2.539	2.861	3.273	3.579
20	0.257	0.687	1.325	1.725	2.086	2.528	2.845	3.251	3.552
21	0.257	0.686	1.323	1.721	2.080	2.518	2.831	3.231	3.527
22	0.256	0.686	1.321	1.717	2.074	2.508	2.819	3.214	3.505
23	0.256	0.685	1.319	1.714	2.069	2.500	2.807	3.198	3.485
24	0.256	0.685	1.318	1.711	2.064	2.492	2.797	3.183	3.467
25	0.256	0.684	1.316	1.708	2.060	2.485	2.787	3.170	3.450
26	0.256	0.684	1.315	1.706	2.056	2.479	2.779	3.158	3.435
27	0.256	0.684	1.314	1.703	2.052	2.473	2.771	3.147	3.421
28	0.256	0.683	1.313	1.701	2.048	2.467	2.763	3.136	3.408
29	0.256	0.683	1.311	1.699	2.045	2.462	2.756	3.127	3.396
30	0.256	0.683	1.310	1.697	2.042	2.457	2.750	3.118	3.385
40	0.255	0.681	1.303	1.684	2.021	2.423	2.704	3.055	3.307
60	0.254	0.679	1.296	1.671	2.000	2.390	2.660	2.994	3.232
120	0.254	0.677	1.289	1.658	1.980	2.358	2.617	2.935	3.160
∞	0.253	0.675	1.282	1.645	1.960	2.327	2.576	2.879	3.091

C.4. Table of the $\mathcal{F}$ Distribution with $\alpha = .05$

$$1-\alpha = \int_0^F \frac{\Gamma((\nu_1+\nu_2)/2)(\nu_1/\nu_2)^{\nu_1/2}}{\Gamma(\nu_1/2)\Gamma(\nu_2/2)} \frac{x^{\nu_1/2-1}}{(1+\nu_1 x/\nu_2)^{(\nu_1+\nu_2)/2}}dx$$

Critical Values of the $\mathcal{F}$ Distribution when $\alpha = .05$

$\nu_2\backslash\nu_1$	1	2	3	4	5	6	8	10	20	30	60	∞
1	161.4476	199.5000	215.7073	224.5832	230.1619	233.9860	238.8827	241.8817	248.0131	250.0951	252.1957	254.3017
2	18.5128	19	19.1643	19.2468	19.2964	19.3295	19.3710	19.3959	19.4458	19.4624	19.4791	19.4956
3	10.1280	9.5521	9.2766	9.1172	9.0135	8.9406	8.8452	8.7855	8.6602	8.6166	8.5720	8.5267
4	7.7086	6.9443	6.5914	6.3882	6.2561	6.1631	6.0410	5.9644	5.8025	5.7459	5.6877	5.6284
5	6.6079	5.7861	5.4095	5.1922	5.0503	4.9503	4.8183	4.7351	4.5581	4.4957	4.4314	4.3654
6	5.9874	5.1433	4.7571	4.5337	4.3874	4.2839	4.1468	4.0600	3.8742	3.8082	3.7398	3.6693
7	5.5914	4.7374	4.3468	4.1203	3.9715	3.8660	3.7257	3.6365	3.4445	3.3758	3.3043	3.2302
8	5.3177	4.4590	4.0662	3.8379	3.6875	3.5806	3.4381	3.3472	3.1503	3.0794	3.0053	2.9281
9	5.1174	4.2565	3.8625	3.6331	3.4817	3.3738	3.2296	3.1373	2.9365	2.8637	2.7872	2.7072
10	4.9646	4.1028	3.7083	3.4780	3.3258	3.2172	3.0717	2.9782	2.7740	2.6996	2.6211	2.5384
11	4.8443	3.9823	3.5874	3.3567	3.2039	3.0946	2.9480	2.8536	2.6464	2.5705	2.4901	2.4050
12	4.7472	3.8853	3.4903	3.2592	3.1059	2.9961	2.8486	2.7534	2.5436	2.4663	2.3842	2.2967
13	4.6672	3.8056	3.4105	3.1791	3.0254	2.9153	2.7669	2.6710	2.4589	2.3803	2.2966	2.2070
14	4.6001	3.7389	3.3439	3.1122	2.9582	2.8477	2.6987	2.6022	2.3879	2.3082	2.2229	2.1313
15	4.5431	3.6823	3.2874	3.0556	2.9013	2.7905	2.6408	2.5437	2.3275	2.2468	2.1601	2.0664
16	4.4940	3.6337	3.2389	3.0069	2.8524	2.7413	2.5911	2.4935	2.2756	2.1938	2.1058	2.0102
17	4.4513	3.5915	3.1968	2.9647	2.8100	2.6987	2.5480	2.4499	2.2304	2.1477	2.0584	1.9610
18	4.4139	3.5546	3.1599	2.9277	2.7729	2.6613	2.5102	2.4117	2.1906	2.1071	2.0166	1.9175
19	4.3807	3.5219	3.1274	2.8951	2.7401	2.6283	2.4768	2.3779	2.1555	2.0712	1.9795	1.8787
20	4.3512	3.4928	3.0984	2.8661	2.7109	2.5990	2.4471	2.3479	2.1242	2.0391	1.9464	1.8438
30	4.1709	3.3158	2.9223	2.6896	2.5336	2.4205	2.2662	2.1646	1.9317	1.8409	1.7396	1.6230
40	4.0847	3.2317	2.8387	2.6060	2.4495	2.3359	2.1802	2.0772	1.8389	1.7444	1.6373	1.5098
60	4.0012	3.1504	2.7581	2.5252	2.3683	2.2541	2.0970	1.9926	1.7480	1.6491	1.5343	1.3903
120	3.9201	3.0718	2.6802	2.4472	2.2899	2.1750	2.0164	1.9105	1.6587	1.5543	1.4290	1.2553
∞	3.8424	2.9966	2.6058	2.3728	2.2150	2.0995	1.9393	1.8316	1.5716	1.4602	1.3194	1.0000

C.5. Table of the F Distribution with $\alpha = .01$

$$1 - \alpha = \int_0^F \frac{\Gamma((\nu_1+\nu_2)/2)(\nu_1/\nu_2)^{\nu_1/2}}{\Gamma(\nu_1/2)\Gamma(\nu_2/2)} \frac{x^{\nu_1/2-1}}{(1+\nu_1 x/\nu_2)^{(\nu_1+\nu_2)/2}}\,dx$$

Critical Values of the F Distribution when $\alpha = .01$

$\nu_2\backslash\nu_1$	1	2	3	4	5	6	8	10	20	30	60	∞
1	4052	4999	5403	5624	5763	5858	5981	6055	6208	6260	6313	6365
2	98.5025	99	99.1662	99.2494	99.2993	99.3326	99.3742	99.3992	99.4492	99.4658	99.4825	99.4991
3	34.1162	30.8165	29.4567	28.7099	28.2371	27.9107	27.4892	27.2287	26.6898	26.5045	26.3164	26.1263
4	21.1977	18.0000	16.6944	15.9770	15.5219	15.2069	14.7989	14.5459	14.0196	13.8377	13.6522	13.4642
5	16.2582	13.2739	12.0600	11.3919	10.9670	10.6723	10.2893	10.0510	9.5526	9.3793	9.2020	9.0215
6	13.7450	10.9248	9.7795	9.1483	8.7459	8.4661	8.1017	7.8741	7.3958	7.2285	7.0567	6.8811
7	12.2464	9.5466	8.4513	7.8466	7.4604	7.1914	6.8400	6.6201	6.1554	5.9920	5.8236	5.6506
8	11.2586	8.6491	7.5910	7.0061	6.6318	6.3707	6.0289	5.8143	5.3591	5.1981	5.0316	4.8599
9	10.5614	8.0215	6.9919	6.4221	6.0569	5.8018	5.4671	5.2565	4.8080	4.6486	4.4831	4.3118
10	10.0443	7.5594	6.5523	5.9943	5.6363	5.3858	5.0567	4.8491	4.4054	4.2469	4.0819	3.9111
11	9.6460	7.2057	6.2167	5.6683	5.3160	5.0692	4.7445	4.5393	4.0990	3.9411	3.7761	3.6062
12	9.3302	6.9266	5.9525	5.4120	5.0643	4.8206	4.4994	4.2961	3.8584	3.7008	3.5355	3.3648
13	9.0738	6.7010	5.7394	5.2053	4.8616	4.6204	4.3021	4.1003	3.6646	3.5070	3.3413	3.1695
14	8.8616	6.5149	5.5639	5.0354	4.6950	4.4558	4.1399	3.9394	3.5052	3.3476	3.1813	3.0080
15	8.6831	6.3589	5.4170	4.8932	4.5556	4.3183	4.0045	3.8049	3.3719	3.2141	3.0471	2.8723
16	8.5310	6.2262	5.2922	4.7726	4.4374	4.2016	3.8896	3.6909	3.2587	3.1007	2.9330	2.7565
17	8.3997	6.1121	5.1850	4.6690	4.3359	4.1015	3.7910	3.5931	3.1615	3.0032	2.8348	2.6565
18	8.2854	6.0129	5.0919	4.5790	4.2479	4.0146	3.7054	3.5082	3.0771	2.9185	2.7493	2.5692
19	8.1849	5.9259	5.0103	4.5003	4.1708	3.9386	3.6305	3.4338	3.0031	2.8442	2.6742	2.4923
20	8.0960	5.8489	4.9382	4.4307	4.1027	3.8714	3.5644	3.3682	2.9377	2.7785	2.6077	2.4240
30	7.5625	5.3903	4.5097	4.0179	3.6990	3.4735	3.1726	2.9791	2.5487	2.3860	2.2079	2.0079
40	7.3141	5.1785	4.3126	3.8283	3.5138	3.2910	2.9930	2.8005	2.3689	2.2034	2.0194	1.8062
60	7.0771	4.9774	4.1259	3.6490	3.3389	3.1187	2.8233	2.6318	2.1978	2.0285	1.8363	1.6023
120	6.8509	4.7865	3.9491	3.4795	3.1735	2.9559	2.6629	2.4721	2.0346	1.8600	1.6557	1.3827
∞	6.6374	4.6073	3.7836	3.3210	3.0191	2.8038	2.5130	2.3227	1.8801	1.6983	1.4752	1.0476

An environmentally friendly book printed and bound in England by www.printondemand-worldwide.com

PEFC Certified

This product is
from sustainably
managed forests
and controlled
sources

www.pefc.org

PEFC/16-33-415

This book is made of chain-of-custody materials; FSC materials for the cover and PEFC materials for the text pages.

#0001 - 100216 - C0 - 234/156/24 [26] - CB - 9781584882428